Concepts in Viral Pathogenesis II

Concepts in
Viral Pathogenesis II

Edited by
Abner Louis Notkins
Michael B.A. Oldstone

With 31 Figures

Springer-Verlag
New York Berlin Heidelberg
London Paris Tokyo

ABNER LOUIS NOTKINS, M.D.
Laboratory of Oral Medicine, National Institute of Dental Research,
NIH, Bethesda, Maryland 20892, U.S.A.

MICHAEL B.A. OLDSTONE, M.D.
Department of Immunopathology, Scripps Clinic and Research
Foundation, La Jolla, California 92037, U.S.A.

Cover photograph: Computer graphic representation of the recently solved 3-D structure
of poliomyelitis virus and its antibody binding sites. White dot surfaces highlight the
antigenic sites of poliovirus. A portion of the viral icosahedral structure (straight lines)
containing a pentamer of the major coat proteins (viral protein: VP) is shown. VP1 is
blue, VP2 is yellow, and VP3 is red. Figure kindly supplied by Dr. James M. Hogle,
Research Institute of Scripps Clinic, La Jolla, California.

Library of Congress Cataloging in Publication Data
Concepts in viral pathogenesis II.
   Includes bibliographies and index.
   1. Host-virus relationships. 2. Virus diseases.
I. Notkins, Abner Louis. II. Oldstone, Michael B.A.
III. Title: Concepts in viral pathogenesis 2.
[DNLM: 1. Virus Diseases. 2. Viruses—pathogenicity.
QW 160 C744]
QR482.C654 1986   616'.0194   86-13082

Typeset by David Seham Associates, Inc., Metuchen, New Jersey.
Printed and bound by Arcata Graphics/Halliday, West Hanover, Massachusetts.
Printed in the United States of America.

9 8 7 6 5 4 3 2 1

ISBN 0-387-96322-7 Springer-Verlag New York Berlin Heidelberg
ISBN 3-540-96322-7 Springer-Verlag Berlin Heidelberg New York

Preface

The favorable reception of the format and contents of Volume I of *Concepts in Viral Pathogenesis* encouraged us to prepare Volume II. As indicated in the preface to the first volume, we felt that the current proliferation of scientific information made it difficult for even the most diligent reader to keep up with the latest developments in his/her own field, let alone other areas of interest. Review articles are one solution, but they, too, have become so voluminous and detailed that they often defeat the very purpose for which they were intended. We attempted to remedy this problem by using a different format. In Volume I, we assembled over fifty mini-reviews/editorials, 1,000–2,000 words in length, on research that was at the cutting edge of virology. This format allowed the authors to provide interpretive, up-to-date information in a brief and easily readable form. Emphasis was on current thinking and unifying concepts rather than a compendium of the literature or presentation of detailed data. Volume II of *Concepts of Viral Pathogenesis* employs the same format as Volume I. It consists of 49 mini-reviews/editorials, all on entirely different topics than covered in Volume I. As in Volume I, the articles are organized systematically so as to provide a conceptual core of up-to-date information on mechanisms by which viruses cause disease.

The book is divided into ten sections. It begins with chapters on viral structure and function, viral constructs, oncogenes, transfection, differentiation, and the molecular basis of viral tropism. The next section focuses on immune recognition of viruses and new principles in viral immunology and immunopathology. As in the previous volume, there are sections on evolving concepts in viral pathogenesis as illustrated by selected plant and animal models and selected human viral diseases. Principles and important viral-cell interactions are emphasized rather than a detailed description of individual viruses. The book concludes with five chapters on new trends in viral diagnosis and epidemiology and ten chapters on vaccines and antriviral therapy.

Abner Louis Notkins, M.D.
Bethesda, Maryland

Michael B.A. Oldstone, M.D.
La Jolla, California

Contents

Oncogenes, Transfection, and Differentiation

Viral Tropism and Entry into Cells

Immune Recognition of Viruses

New Principles in Viral Immunology and Immunopathology

Evolving Concepts in Viral Pathogenesis
Illustrated by Selected Plant and Animal Models

Evolving Concepts in Viral Pathogenesis Illustrated by Selected Diseases in Humans

New Trends in Diagnosis and Epidemiology

Vaccines and Antiviral Therapy

Contributors

STUART A. AARONSON, Laboratory of Cellular and Molecular Biology, National Cancer Institute, NIH, Building 37, Room 1E-24, Bethesda, MD 20892, U.S.A.

M. J. ANDERSON, Department of Medical Microbiology, University College, London, University Street, London WCIE 6JJ, England

WARREN A. ANDIMAN, Department of Pediatrics, Yale University School of Medicine, 333 Cedar Street, New Haven, CT 06510, U.S.A.

BART G. BARRELL, Medical Research Council, Laboratory of Molecular Biology, Hills Road, Cambridge, CB2 2QH, England

RONALD A. BARRY, Department of Biochemistry and Biophysics, University of California, San Francisco, CA 94143, U.S.A.

WILLIAM E. BIDDISON, Neuroimmunology Branch, National Institute of Neurological and Communicative Disorders and Stroke, NIH, Building 10, Room 5B-16, Bethesda, MD 20892, U.S.A.

DAVID H.L. BISHOP, NERC Institute of Virology, Mansfield Road, Oxford OX1 3SR, England

J. MICHAEL BISHOP, Department of Microbiology and Immunology, The George Williams Hooper Research Foundation, University of California Medical Center, San Francisco, CA 94143, U.S.A.

DANI P. BOLOGNESI, Department of Surgery, Duke University Medical Center, Box 2926, Durham, NC 27710, U.S.A.

THOMAS J. BRACIALE, Department of Pathology, Washington University School of Medicine, 660 South Euclid Avenue, St. Louis, MO 63110, U.S.A.

VIVIAN L. BRACIALE, Department of Pathology, Washington University School of Medicine, 660 South Euclid Avenue, St. Louis, MO 63110, U.S.A.

JOHN BRADY, Laboratory of Molecular Virology, Division of Cancer Etiology, National Cancer Institute, NIH, Building 41, Room A109, Bethesda, MD 20892, U.S.A.

FRED BROWN, Wellcome Biotechnology Ltd., Langley Court, Beckenham, Kent BR3 3BS, England

M. CHOW, Department of Applied Biological Sciences, Massachusetts Institute of Technology and The Whitehead Institute for Biomedical Research, Cambridge, MA 02139, U.S.A.

MAN SUNG CO, Department of Microbiology and Molecular Genetics, Harvard Medical School, 25 Shattuck Street, Boston, MA 02115, U.S.A.

PETER L. COLLINS, Laboratory of Infectious Diseases, National Institute of Allergy and Infectious Diseases, NIH, Room 100, Building 7, Bethesda, MD 20892, U.S.A.

CARLO M. CROCE, The Wistar Institute of Anatomy and Biology, 36th and Spruce Streets, Philadelphia, PA 19104, U.S.A.

RICHARD L. CROWELL, Department of Microbiology and Immunology, Hahnemann University School of Medicine, Broad and Vine Street, Philadelphia, PA 19102, U.S.A.

DAVID C. DIAMOND, Department of Microbiology, State University of New York at Stony Brook, Stony Brook, NY 11794, U.S.A.

T.O. DIENER, Plant Virology Laboratory, Plant Protection Institute, U.S. Department of Argriculture, Agricultural Research Center, Beltsville, MD 20705, U.S.A.

EMILIO A. EMINI, Division of Virus and Cell Biology Research, Merck, Sharp and Dohme Research Laboratories, West Point, PA 19486, U.S.A.

PAUL J. FARRELL, The Ludwig Institute for Cancer Research, Medical Research Council Centre, Hills Road, Cambridge, CB2 2QH, England

LIONEL FEIGENBAUM, Laboratory of Molecular Virology, Division of Cancer Etiology, National Cancer Institute, NIH, Building 41, Room A109, Bethesda, MD 20892, U.S.A.

MARK A. FEITELSON, Fox Chase Cancer Center, Institute for Cancer Research, 7701 Burholme Avenue, Philadelphia, PA 19111, U.S.A.

FRANK FENNER, John Curtin School of Medical Research, The Australian National University, Canberra, ACT 2601, Australia

BERNARD N. FIELDS, Department of Microbiology and Molecular Genetics, Harvard Medical School, 25 Shattuck Street, Boston, MA 02115, U.S.A.

DAVID J. FILMAN, Department of Molecular Biology, Research Institute of Scripps Clinic, 10666 N. Torrey Pines Road, La Jolla, CA 92037, U.S.A.

PETER J. FISCHINGER, Laboratory of Tumor Cell Biology, National Cancer Institute, NIH, Building 37, Room 6A09, Bethesda, MD 20892, U.S.A.

DAVID FITZGERALD, Laboratory of Molecular Biology, Division of Cancer Biology and Diagnosis, National Cancer Institute, NIH, Bethesda, MD 20892, U.S.A.

DAVID A. FUCCILLO, Microbiological Associates, Inc., 5221 River Road, Bethesda, MD 20816, U.S.A.

ROBERT S. FUJINAMI, Department of Pathology, University of California, San Diego, La Jolla, CA 92093, U.S.A.

ROBERT C. GALLO, Laboratory of Tumor Cell Biology, National Cancer Institute, NIH, Building 37, Room 6A-09, Bethesda, MD 20892, U.S.A.

MARK I. GREENE, Department of Pathology, Harvard Medical School, 25 Shattuck Street, Boston, MA 02115, U.S.A.

ION GRESSER, Laboratory of Viral Oncology, Institut de Recherches Scientifiques sur le Cancer, BP 8, 94802 Villejuif, France

DIANE E. GRIFFIN, The Johns Hopkins University School of Medicine, 500 N. Wolfe Street, Baltimore, MD 21205, U.S.A.

ASHLEY T. HAASE, Department of Microbiology, University of Minnesota, Room 1460, Mayo Memorial Building, 420 Delaware Street, S.E., Minneapolis, Minnesota 55455, U.S.A.

DOROTHEE HERLYN, The Wistar Institute of Anatomy and Biology, 36th and Spruce Streets, Philadelphia, PA 19104, U.S.A.

JAMES M. HOGLE, Department of Molecular Biology, Scripps Clinic and Research Foundation, 10666 N. Torrey Pines Road, La Jolla, CA 92037, U.S.A.

MORIO HOMMA, Department of Microbiology, Kobe University School of Medicine, 5-1 Kusunoki-cho 7-chome, Chuo-ku, Kobe 650, Japan

PETER M. HOWLEY, Laboratory of Tumor Virus Biology, National Cancer Institute, NIH, Bethesda, MD 20892, U.S.A.

K-H. LEE HSU, Department of Microbiology and Immunology, Hahnemann University School of Medicine, Broad and Vine Street, Philadelphia, PA 19102, U.S.A.

STEVEN JACOBSON, Neuroimmunology Branch, National Institute of Neurological and Communicative Disorders and Stroke, NIH, Building 10, Room 5B-16, Bethesda, MD 20892, U.S.A.

RICHARD T. JOHNSON, The Johns Hopkins University School of Medicine, 500 N. Wolfe Street, Baltimore, MD 21205, U.S.A.

BEN Z. KATZ, Yale University School of Medicine, 333 Cedar Street, New Haven, CT 06510, U.S.A.

OLEN KEW, Division of Viral Diseases, Centers for Disease Control, 1600 Clifton Road, N.E., Atlanta, GA 30333, U.S.A.

GEORGE KHOURY, Laboratory of Molecular Virology, Division of Cancer Etiology, National Cancer Institute, NIH, Building 41, Room A109, Bethesda, MD 20892, U.S.A.

EDWIN D. KILBOURNE, Department of Microbiology, Mount Sinai School of Medicine, City University of New York, One Gustave L. Levy Place, New York, NY 10029, U.S.A.

GEORGE KLEIN, Department of Tumor Biology, Karolinska Institute, S-104 01 Stockholm, Sweden

HILARY KOPROWSKI, The Wistar Institute of Anatomy and Biology, 36th and Spruce Streets, Philadelphia, PA 19104, U.S.A.

RICHARD A. LERNER, Department of Molecular Biology, Scripps Clinic & Research Foundation, 10666 N. Torrey Pines Road, La Jolla, CA 92037, U.S.A.

HOWARD LIPTON, Department of Neurology, Northwestern University Medical School, 303 East Chicago Avenue, Chicago, IL 60611, U.S.A.

DOUGLAS R. LOWY, Laboratory of Cellular Oncology, National Cancer Institute, NIH, Building 37, Room 1B-26, Bethesda, MD 20892, U.S.A.

JOSEPH B. McCORMICK, Special Pathogens Branch, Division of Viral Diseases, Centers for Disease Control, 1600 Clifton Road, N.E., Atlanta, GA 30333, U.S.A.

MICHAEL P. McKINLEY, Department of Neurology, University of California, San Francisco, CA 94143, U.S.A.

ROGER MELVOLD, Department of Microbiology–Immunology, Northwestern University Medical School, 303 East Chicago Avenue, Chicago, IL 60611, U.S.A.

GEORGE MILLER, Yale University School of Medicine, 333 Cedar Street, New Haven, CT 06510, U.S.A.

STEPHEN MILLER, Department of Microbiology–Immunology, Northwestern University Medical School, 303 East Chicago Avenue, Chicago, IL 60611, U.S.A.

BERNARD MOSS, Laboratory of Viral Diseases, National Institute of Allergy and Infectious Diseases, Bethesda, MD 20892, U.S.A.

ERLING NORRBY, Department of Virology, Karolinska Institute, School of Medicine, S-105 21 Stockholm, Sweden

ABNER LOUIS NOTKINS, Laboratory of Oral Medicine, National Institute of Dental Research, NIH, Bldg. 30, Room 121, Bethesda, MD 20892, U.S.A.

BALDEV NOTTAY, Division of Viral Diseases, Centers for Disease Control, 1600 Clifton Road, N.E., Atlanta, GA 30333, U.S.A.

PETER C. NOWELL, Department of Pathology and Laboratory Medicine, School of Medicine, University of Pennsylvania, Philadelphia, PA 19104, U.S.A.

MICHAEL B.A. OLDSTONE, Department of Immunology, Scripps Clinic and Research Foundation, 10666 N. Torrey Pines Road, La Jolla, CA 92037, U.S.A.

IRA PASTAN, Laboratory of Molecular Biology, Division of Cancer Biology and Diagnosis, National Cancer Institute, NIH, Bethesda, MD 20892, U.S.A.

J.R. PATTISON, Department of Medical Microbiology, University College London, University Street, London WCIE 6JJ, England

STANLEY B. PRUSINER, Department of Neurology, HSE-781, University of California, San Francisco, CA 94143, U.S.A.

L. RATNER, Laboratory of Tumor Cell Biology, National Cancer Institute, NIH, Building 37, Room 6A-09, Bethesda, MD 20892, U.S.A.

DOUGLAS D. RICHMAN, Departments of Pathology and Medicine, Veterans Administration Medical Center, University of California, San Diego, La Jolla, CA 92093, U.S.A.

JAMES F. ROONEY, Laboratory of Oral Medicine, National Institute of Dental Research, NIH, Building 30, Room 121, Bethesda, MD 20892, U.S.A.

NAVA SARVER, Revlon Health Care, Meloy Laboratories, Springfield, VA 22151, U.S.A.

PREM SETH, Laboratory of Molecular Biology, Division of Cancer Biology and Diagnosis, National Cancer Institute, NIH, Bethesda, MD 20892, U.S.A.

JOHN L. SEVER, National Institute of Neurological and Communicative Disorders and Stroke, NIH, Building 36, Room 5D06, Bethesda, MD 20892, U.S.A.

THOMAS M. SHINNICK, Dept. of Molecular Biology, Scripps Clinic and Research Foundation, 10666 N. Torrey Pines Road, La Jolla, CA 92037, U.S.A.

PETER J. SOUTHERN, Department of Immunology, Scripps Clinic and Research Foundation, 10666 N. Torrey Pines Road, La Jolla, CA 92037, U.S.A.

MICHIAKI TAKAHASHI, Department of Virology, Research Institute for Microbial Diseases, Osaka University, 3-1, Yamadaoka, Suita, Osaka 565, Japan

SATVIR S. TEVETHIA, Department of Microbiology, The Pennsylvania State University, College of Medicine, Hershey, PA 17033, U.S.A.

STEVEN R. TRONICK, Laboratory of Cellular and Molecular Biology, National Cancer Institute, NIH, Building 37, Room 1E-24, Bethesda, MD 20892, U.S.A.

ELADIO VIÑUELA, Centro de Biología Molecular (CSIC-UAM), Facultad de Ciencias, Universidad Autonoma, Canto Blanco, 28049 Madrid, Spain

GEOFFREY M. WAHL, The Salk Institute, 10010 N. Torrey Pines Road, La Jolla, CA 92037, U.S.A.

GAIL W. WERTZ, Department of Microbiology and Immunology, University of North Carolina, School of Medicine, Chapel Hill, NC 27514, U.S.A.

HEINER WESTPHAL, Section on Animal Viruses, Laboratory of Molecular Genetics, National Institute of Child Health and Human Development, NIH, Building 6, Room 338, Bethesda, MD 20892, U.S.A.

MARK WILLINGHAM, Laboratory of Molecular Biology, Division of Cancer Biology and Diagnosis, National Cancer Institute, NIH, Bethesda, MD 20892, U.S.A.

IAN A. WILSON, Department of Molecular Biology, Scripps Clinic and Research Foundation, 10666 N. Torrey Pines Road, La Jolla, CA 92037, U.S.A.

ECKARD WIMMER, Department of Microbiology, State University of New York at Stony Brook, Stony Brook, NY 11794, U.S.A.

NEAL S. YOUNG, Cell Biology Section, National Heart, Lung and Blood Institute, NIH, Building 10, Room 7C-108, Bethesda, MD 20892, U.S.A.

HARALD ZUR HAUSEN, Deutsches Krebsforschungszentrum, Im Neuenheimer Feld 280, 6900 Heidelberg, Federal Republic of Germany

Viral Structure and Function

CHAPTER 1
The Three-dimensional Structure of Poliovirus: Implications for Virus Evolution, Assembly, and Immune Recognition

JAMES M. HOGLE, M. CHOW, AND DAVID J. FILMAN

In the past year, the structures of two animal viruses have been determined at near atomic resolution [1,2]. The viruses, poliomyelitis virus type 1 (Mahoney strain) and human rhinovirus 14, are both members of the Picornaviridae family, a large family of small spherical RNA viruses that also includes hepatitis A virus, the coxsackieviruses, and foot-and-mouth disease virus. Although poliovirus and rhinovirus are the first animal viruses to be mapped at high resolution, these studies have their roots in the earlier determinations of several plant virus structures [3–6]. The application of x-ray crystallographic techniques to animal viruses is particularly exciting because the animal viruses in general, and poliovirus in particular, are far better characterized biologically. These structures thus provide for the first time the opportunity to study the structural basis for immune recognition (and neutralization), host and tissue specificity, and viral pathogenesis. In the future, this work will also facilitate structural studies of other picornaviruses (several of which have already been crystallized) and the extension of the techniques to other families of animal viruses.

Poliovirus

First described in 1908 as the causative agent of poliomyelitis [7], poliovirus has been the subject of intensive study since the demonstration that the virus could be grown in cultured cells [8]. The poliovirion is approximately 310 Å in diameter, and 8.4 million daltons in molecular mass. The capsid is composed entirely of sixty copies each of four capsid proteins VP1, VP2, VP3, and VP4 (with molecular weights 33K, 30K, 26K, and 7.5K, respectively) arranged on a T = 1 icosahedral surface. The capsid encloses one molecule of single-stranded plus sense RNA of 2.5 million daltons (approximately 7500 nucleotides)[9]. The RNA contains a single, large, open reading frame from which a 220-kilodalton polyprotein is translated. All known viral proteins are derived from this polyprotein through processing by viral proteases. In an early step in the processing, the polyprotein is cleaved into three large fragments. The fragment derived from the amino terminus (P1-1a) contains the capsid protein sequences in the order VP4–VP2–VP3–VP1. The processing of P1-1a to VP0, VP3, VP1 is linked with the early stages of virus assembly. The final cleavage of VP0 to VP4 and VP2 occurs late in assembly after the encapsidation of RNA [10].

Structure of Capsid Proteins

The main chains of the four capsid proteins are shown in Figure 1.1. VP1, VP2, and VP3 show a striking structural homology. Each is composed of a conserved "core" with variable "elaborations." The cores consist of an eight-stranded antiparallel β barrel with two flanking helices (Figure 1.1A). Four of the β strands (B, I, D, and G in Figure 1.1A) make up the large twisted β sheet that forms the front and bottom surfaces of the barrel. The remaining four strands (C, H, E, and F in Figure 1.1A) make up the shorter flatter β sheet that forms the back surface of the barrel. The strands of the front and back surfaces are connected at one end by short loops, giving the barrel the shape of a triangular wedge. Although the cores of the three large capsid proteins are very similar, the connecting loops and amino- and carboxy-terminal strands are dissimilar. In particular, VP1 contains a large insertion (residues 207–237) in the loop connecting strands G and H of the β barrel, as well as an unusually large loop (residues 96–104) connecting the top two strands (B and C) of the β barrel. VP2 has a large insertion (residues 127–185) in the loop connecting the E strand of the barrel to the "back" helix. The largest insertion in VP3 (residues 53–69) occurs in the middle of the B strand of the β barrel.

The component cores, connecting loops, and terminal extensions of the three large proteins each appears to play a different role in the structure of the intact virus. The cores make up the continuous shell of the virus; the connecting loops and carboxy-terminal extensions make up many of the major features on the external surface of the virion; and the amino-terminal extensions cover much of the internal surface of the capsid. The small protein VP4 (shown with VP2 in Figure 1.1C) is similar in many respects to the amino-terminal extensions of VP1 and VP3. Like them, it is rather extended in structure, and occupies a totally internal position, forming a significant portion of the internal surface of the capsid. Because VP4 is covalently linked to VP2 until very late in virion assembly, it can be regarded as the detached amino-terminal extension of VP2.

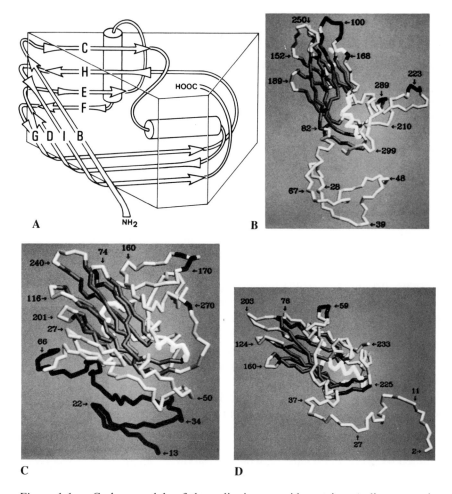

Figure 1.1. α-Carbon models of the poliovirus capsid proteins. **A** diagrammatic representation of the conserved folding pattern of the major capsid proteins; **B** VP1; **C** VP2 and VP4; **D** VP3. The structurally conserved "cores" of the proteins each consists of a radial (back) helix, a tangential (front) helix, and an eight-stranded antiparallel β barrel. The helices are shown in white in **B, C,** and **D,** and are indicated by cylinders in **A.** Strands of the β barrels are shaded dark gray in **B, C,** and **D,** and are indicated by arrows in **A.** Single-letter designations for the β strands in **A** are consistent with previous descriptions of the icosahedral plant viruses [23]. In **B, C,** and **D,** the variable loops and amino- and carboxy-terminal extensions are labeled by residue numbers, and are shaded medium-gray, except for the sites of monoclonal release mutations, which are black. VP4 is indicated in black at the bottom of **C.**

The internal position of VP4 is interesting because this protein is lost during several of the structural transitions that the poliovirion undergoes (including those induced by heating to 55°C for 10 minutes and by attachment to susceptible cells). The extrusion of an internal portion of a viral capsid protein has also been seen in the expansion of the plant viruses tomato bushy stunt virus (TBSV) and turnip crinkle virus (TCV) (S. C. Harrison, personal communication).

Virion Structure

The arrangement of the capsid proteins in the virion shell is depicted in Figures 1.2 and 1.3. VP1 is located adjacent to the particle fivefold axis, with the narrow end of the β barrel closest to the axis. A pronounced tilt of the β barrel outward along the fivefold results in the exposure of the top three connecting loops at the narrow end of the barrel and the formation of a prominent surface protrusion extending to a radius of 165 Å. VP2 and VP3 alternate around the particle threefold axis, also with the narrows end of the barrels closest to the axis, but with a less pronounced tilt. As a result, only the top two loops at the narrow ends of the barrels of VP2 and VP3 are exposed, and the surface protrusion at the threefold extends to a lesser radius (150 Å). This protrusion is ringed by two sets of outward projections (or promontories). The larger projection is formed primarily by the large insertion in VP2 (residues 127–185) with contributions from the carboxy terminus of VP2, the large insertion in VP1 (residues 207–237), and residues 271–295 near the carboxy terminus of VP1. This large projection extends to a radius of 165Å. The smaller projection is formed by the insertion in VP3 (residues 53–70) and extends to 155 Å radius. The peaks at the fivefold axes are surrounded by broad deep valleys, whereas the threefold plateaus are separated by saddle surfaces at the twofold axes.

Significance

Virus Evolution

As predicted from earlier sequence comparison, the structures of the Mahoney strain of type 1 poliovirus and of rhinovirus 14 are strikingly similar. The "cores" of the capsid proteins are nearly identical, and the major structural differences occur in loops that are exposed on the surface of the virions (particularly the top loop of VP1 and the large insertions in VP1 and VP2). The sequence and structural similarities clearly indicate a close evolutionary relationship between the two viruses, and point to the danger of classifying viruses within families on the basis of pathology, or physical characteristics such as acid stability or permeability to ions. One particularly interesting hypothesis is that the (functionally defined) rhinoviruses might have been derived from different enteroviruses upon loss of acid stability, thus explaining the large number of rhinovirus serotypes.

A much more surprising observation is that the structure of the poliovirus (and hence the rhinovirus) capsid proteins are similar in structure to the capsid proteins of the icosahedral plant viruses. Indeed, the packing of capsid proteins in poliovirions is similar to the packing of proteins in the "S domains" of the T = 3 plant viruses. Obviously, this structural similarity indicates an evolutionary relationship, although the exact nature of the relationship is unclear.

Assembly

The poliovirus structure also has significant implications for the assembly of picornaviruses. The amino-terminal extensions of VP1, VP2 (considering VP4), and VP3 form an extensive network on the inner surface of the capsid shell. The network is particularly strong in fivefold related protomers, and less strong between adjacent pentamers. This finding is consistent with the proposal that pentameric associations of subunits are an intermediate in capsid assembly, as originally proposed from physical characterization of assembly intermediates isolated from infected cells. It also suggests that the amino-terminal network may play a role in controlling the assembly process. One particularly interesting portion of the network is a structure formed by the interaction of five copies of the amino terminus of VP3. Five chains wrap around the fivefold axis, forming a highly twisted tube stabilized by parallel β interactions. This structure, which can form only upon pentamer formation, may serve to direct and/or stabilize the formation of the pentameric intermediate. An analogous structure (called a β *annulus*) that has been observed in the T = 3 plant viruses is also formed from the amino-terminal extensions of the capsid proteins, and also appears to play a crucial role in directing capsid assembly. Although the cores and amino-terminal extensions of the plant and animal viruses have analogous functions, it is significant that the β annuli of the T = 3 plant viruses occur at threefold axes rather than at the fivefold axes, underscoring the differences in the assembly of T = 3 as opposed to T = 1 virions.

The disposition of the termini generated by the processing of the P1-1a protomer also has interesting implications for assembly. The carboxy terminus of VP4 is very close to the the first ordered residue at the amino terminus of VP2 (residue 5). The four disordered residues could comfortably fill the gap. However, both termini are located in the interior of the virion, inaccessible to exogenous proteases. Because this cleavage is thought to occur late in assembly (subsequent to encapsidation of RNA), either there must be a substantial conformational rearrangement subsequent to the cleavage, or the cleavage must be "autocatalytic." In contrast, the amino

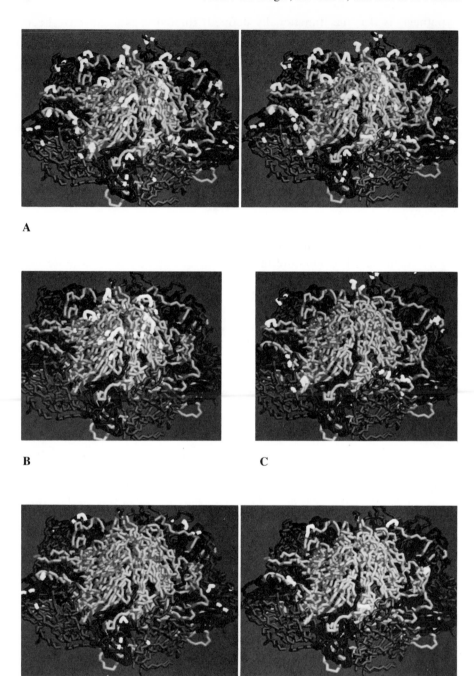

terminus of VP2 is spatially distant from the carboxy terminus of VP3, and the amino terminus of VP3 is distant from the carboxy terminus of VP1; indeed the carboxy termini are located on the exterior, whereas the amino termini are in the interior of the virion. The termini must, therefore, undergo substantial rearrangement subsequent to the cleavages. The apparent role of the amino-terminal extensions in directing and stabilizing the formation of pentamers confirms that proteolytic processing and subsequent rearrangement play an important role in the control of virion assembly.

Antigenic Sites

The antigenic sites of poliovirus have been identified by two techniques: mapping mutants resistant to monoclonal antibodies [11–16] and identification of synthetic peptides capable of eliciting or priming for neutralizing antibodies [17–22]. The antigenic sites identified by the monoclonal release mutations are indicated in Figures 1.1, 1.2, and 1.3. The sites can be clustered by proximity (where *proximity* is defined as being sufficiently close to be within the "footprint" of an antigen binding site of an Fab') into four general areas (Figures 1.2b–e, and 1.3b–e). Cluster 1 (Figures 1.2b, 1.3b) contains residues in the top two loops in the VP1 barrel (residues 89–103, and residue 254) and a residue from the loop connecting strand E with the back helix (residue 168). Cluster 2 (Figures 1.2c, 1.3c) contains residues from the large insertion in VP1 (residues 222–224), from the large insertion of VP2 (residues 166, 169, and 170), and from the carboxy terminus of VP2 (residue 270). Cluster 3 (Figures 1.2d, 1.3d) contains residues from the top loop of the VP2 barrel (residue 72), from the top loop of the VP3 barrel (residues 71–73), and residues from the insertion in VP3 (residues 58–60). Cluster 4 (Figures 1.2e, 1.3e) contains residues near the carboxy terminus of VP1 (residues 284–287). Cluster 4 is sufficiently close to both cluster 2 and cluster 3 to be part of either site, or to bridge the two sites. (Recently characterized monoclonal release mutants from type 3 poliovirus indicate that clusters 3 and 4 can form parts of the same antigenic site [16].) A distribution of antigenic sites similar

Figure 1.2. The organization of the major capsid proteins of poliovirus around the fivefold axis. Five copies each of VP1 (light-gray), VP2 (medium-gray), and VP3 (dark-gray) are shown arranged in a pentamer. VP4 is located on the inner surface of the capsid, and is not visible. Sites of individual monoclonal release mutations (shown in white) are located on radial projections on the outer surface. A Stereo pair, with all mutation sites shown. As described in the text, these mutations are grouped by proximity into four clusters: **B** cluster 1, **C** cluster 2, **D** cluster 3, **E** cluster 4. (Figures 1.1. and 1.2. were produced with the assistance of D.L. Bloch, using the molecular surface program RAMS2.)

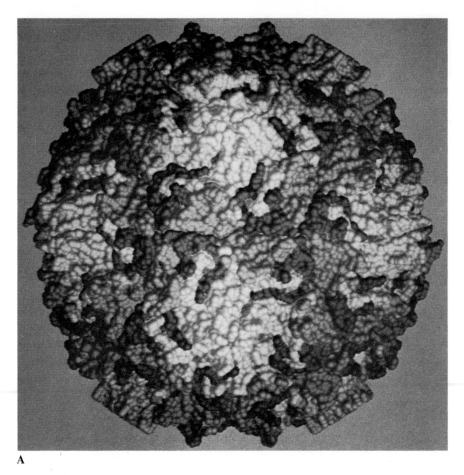

A

Figure 1.3. The organization of the major capsid proteins on the outer surface of poliovirus. VP1 is light-gray; VP2 is medium-gray; VP3 is dark-gray. VP4 is located on the inner surface of the capsid, and is not visible. In **A,** the major features of the outer surface are apparent. A large-ribbed peak is formed at each fivefold axis by the cores of VP1, due to the pronounced tilt of the VP1 β barrels along the axis. At each threefold axis, the β barrels of VP2 and VP3 alternate around the axis, forming a smaller peak that is ringed by promontories. The promontories are formed by residues from the major insertions and carboxy-terminal extensions of the three large capsid proteins. Peaks at the fivefold axes are surrounded by broad valleys, whereas the peaks at the threshold axes are separated by saddle surfaces. In **B, C, D,** and **E,** the sites of individual monoclonal release mutations are shown in white. As described in the text, these mutations are grouped by proximity into four clusters: **B** cluster 1, **C** cluster 2, **D** cluster 3, **E** cluster 4. (Molecular surface representations were calculated by A.J. Olson, using the programs AMS [25] and RAMS [24].)

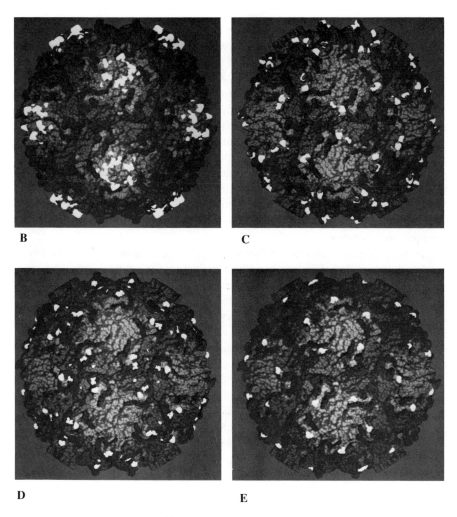

B

C

D

E

Figure 1.3 (continued).

to that in poliovirus has been observed in rhinovirus 14 [2]. An intriguing but as yet unexplained finding is that whereas most of the escape mutations from type 1 poliovirus map to sites 2 and 3, nearly all of the escape mutations in type 3 map to site 1 (specifically to residues 89–104 in the top loop of VP1).

Note that all of the regions except for region 4 are comprised of amino acids from several different portions of a polypeptide, and that two of the clusters contain amino acids from more than one polypeptide chain. In several instances, the clustering of such sequentially distant residues into individual sites has been confirmed by experimental results. Thus, there are neutralizing monoclonal antibodies that cannot neutralize those viruses that have mutated in residues 89–103, 254, or 168 of VP1 (cluster 1). Similarly, there are antibodies that do not neutralize mutants having changes either at residue 222 of VP1 or at residue 170 of VP2 (cluster 2).

All of the mutation sites are located in highly exposed loops on the virus surface, where they are readily accessible to antibody binding. These sites can be expected to accommodate point mutations locally, without disrupting structure elsewhere in the virion. Because the sites of monoclonal escape mutation in poliovirus appear to correspond to antibody binding sites, there is no need to propose that any of the mutations act by disrupting a spatially distant antibody binding site. The mapping of antibody binding sites to exposed loops on the surface of an antigen has also been observed in the influenza virus hemagglutinin and neuraminidase (see I. A. Wilson, Chapter 2, this volume). The decoration of a virus surface with exposed (and mutable) loops may thus be a common mechanism by which animal viruses escape immune surveillance.

Synthetic peptides capable of eliciting or priming for a neutralizing response have been identified that overlap all four clusters. In several instances, two peptides from different portions of the linear sequence that were each capable of eliciting a neutralizing response both mapped into the same cluster. This observation suggests that the distinction between a conformational (discontinuous) and a sequential (continuous) antigenic site may be somewhat artificial, because an immunogenic peptide (usually thought to indicate a sequential determinant) may form a portion of a conformational site.

There are several peptides capable of eliciting a neutralizing response that do not map in or near the clusters identified by monoclonal release mutations. Surprisingly, these peptides (three from the amino-terminal extension of VP1 and one from the bottom loop of the closed end of the barrel of VP1) are in the interior of the mature virion. The ability of these buried peptides to induce a neutralizing response may prove to be important in understanding conformational changes involved in neutralization.

Future Prospects

Knowledge of the structure of poliovirus has begun to provide insights into the evolution and assembly of the virus and into the relationship of virus structure to immune recognition. We believe that the potential of the structure is just beginning to be realized. In particular, the structure will be useful for the design and interpretation of experiments in several areas, for example, those designed to elucidate the mechanism of neutralization, and to identify the structural factors responsible for host and tissue tropism and the attenuation of neurovirulence. We also anticipate that the structure will be crucial for the design and interpretation of site-directed mutagenesis experiments using infectious DNA copies of the genome to construct mutants in the capsid region.

Acknowledgments

This work was supported by NIH grant AI-20566 to J. M. Hogle, and in part by a grant from the Rockefeller Foundation and by NIH grant AI-22346 to D. Baltimore (Whitehead Institute for Biomedical Research). D. J. Filman was supported in part by training grants NS-07078 and NS-12428. Preliminary crystallographic studies that led to this work were supported by NIH grant CA-13202 to S. C. Harrison. We would like to thank Richard Lerner, David Baltimore, and Stephen Harrison for their interest and support throughout various phases of this work; and Joseph Icenogle, Carl Fricks, Michael Oldstone, and Ian Wilson for helpful comments and discussions. This is publication no. MB4203 of the Research Institute of Scripps Clinic.

References

1. Hogle JM, Chow M, Filman DJ (1985) Three-dimensional structure of poliovirus at 2.9 Å resolution. Science 229:1358–1365
2. Rossmann MG, Arnold E, Erickson JW, Fankenberger EA, Griffith JP, Hecht H-J, Johnson JE, Kamer G, Luo M, Mosser AG, Rueckert RR, Sherry B, Vriend G (1985) Structure of a human common cold virus and functional relationship to other picornaviruses. Nature 317:145–153
3. Harrison SC, Olson AJ, Schutt CE, Winkler FK, Bricogne G (1978) Tomato bushy stunt virus at 2.9 Å resolution. Nature 276:368–373
4. Abad-Zapatero C, Abdel-Mequid SS, Johnson JE, Leslie AGW, Rayment I, Rossmann MG, Suck D, Tsukihara T (1980) Structure of southern bean mosaic virus at 2.8 Å resolution. Nature 286:33–39
5. Liljas L, Unge T, Jones TA, Fridborg K, Lovgren S, Skoglund U, Strandberg B (1982) Structure of satellite tobacco necrosis virus at 3.0 Å resolution. J Mol Biol 159:93–108
6. Hogle JM, Maeda A, Harrison SC (1986) Structure and assembly of turnip crinkle virus, I: X-ray crystallographic structure analysis at 3.2 Å resolution. J Mol Biol (in press)
7. Landsteiner K, Popper E (1908) Mikroskopische Praparate von einem menschlichen und zwei Affenruckenmarken. Wien Klin Wochenshr 21:1830
8. Enders JF, Weller TH, Robbins FC (1949) Cultivation of the Lansing strain of poliomyelitis virus in cultures of various human embryonic tissues. Science 109:85–87
9. Rueckert RR (1976) On the structure of morphogenesis of picornaviruses. In Smith IE, Jones BC (eds) Comprehensive Virology, vol 6. Plenum Publishing, New York, pp 131–213
10. Pallansch MA, Kew OM, Semler BL, Omilianowski DR, Anderson CW, Wimmer E, Rueckert RR (1984) Protein processing map of poliovirus. J Virol 49:873–880
11. Emini EA, Jameson BA, Lewis AJ, Larsen GR, Wimmer E (1982) Poliovirus neutralization epitopes: Analysis and localization with neutralizing monoclonal antibodies. J Virol 43:997–1005

12. Crainic R, Couillin P, Blondel B, Cabau N, Boue A, Horodniceanu F (1983) Natural variation of poliovirus neutralization epitopes. Infect Immun 41:1217–1225

13. Minor PD, Schild GC, Bootman J, Evans DMA, Ferguson M, Reeve P, Spitz M, Stanway G, Cann AJ, Hauptmann R, Clarke LD, Mountford RC, Almond JW (1983) Location and primary structure of a major antigenic site for poliovirus neutralization. Nature 301:674–679

14. Minor PD, Evans DMA, Ferguson M, Schild GC, Westorp G, Almond JW (1985) Principal and subsidiary antigenic site of VP1 involved in the neutralization of poliovirus type 3. J Gen Virol 65:1159–1165

15. Diamond DC, Jameson BA, Bonin J, Kohara M, Abe S, Itoh H, Komatsu T, Arita M, Kuge S, Nomoto A, Osterhaus ADME, Crainic R, Wimmer E (1985) Antigenic variation and resistance to neutralization in poliovirus type 1. Science 229:1090–1093

16. Minor PD, Ferguson M, Evans DMA, Icenogle JP (1986) Antigenic structure of polioviruses of serotypes 1, 2, and 3. (In preparation)

17. Chow M, Yabrov R, Bittle J, Hogle J, Baltimore D (1985) Synthetic peptides from four separate regions of the poliovirus type 1 capsid protein VP1 induce neutralizing antibodies. Proc Natl Acad Sci USA 82:910–914

18. Emini EA, Jameson BA, Wimmer E (1983) Priming for and induction of anti-poliovirus neutralizing antibodies by synthetic peptides. Nature 304:699–703

19. Emini EA, Jameson BA, Wimmer E (1984) Identification of a new neutralization antigenic site on poliovirus coat protein VP2. J Virol 52:719–721

20. Emini EA, Jameson BA, Wimmer E (1984) Identification of multiple neutralization antigenic sites on poliovirus type 1 and the priming of the immune response with synthetic peptides. In Lerner, RA, Chanock, RM (eds) Modern Approaches to Vaccines. Cold Spring Harbor Laboratory, Cold Spring Harbor, New York, pp 65–75

21. Ferguson M, Evans DMA, Magrath DI, Minor PD, Almond JW, Schild GC (1985) Induction by synthetic peptides of broadly reactive, type-specific neutralizing antibody to poliovirus type 3. Virology 143:505–515

22. Jameson BA, Bonin J, Murray MG, Wimmer E, Kew O (1985) Peptide-induced neutralizing antibodies to poliovirus. In Lerner, RA, Chanock, RM, Brown, F (eds) Vaccines 85. Cold Spring Harbor Laboratory, Cold Spring Harbor, New York, pp 191–198

23. Rossmann MG, Abad-Zapatero C, Murthy MRN, Liljas L, Jones TA, Strandberg B (1983) Structural comparisons of some small spherical plant viruses. J Mol Biol 165:711–736

24. Connolly ML (1985) Depth–buffer algorithms for molecular modelling. J Mol Graphics 3:19–24

25. Connolly ML (1983) Analytical molecular surface calculation. J Appl Crystallogr 16:548–558

CHAPTER 2
The Three-dimensional Structure of Surface Antigens from Animal Viruses

IAN A. WILSON

Significant advances have been made recently in our understanding of the structural basis of immune recognition of viral antigens and intact animal viruses. X-ray crystallographic analyses have provided the three-dimensional structures to two viral antigens: i.e., the hemagglutinin (HA) [1–4] and neuraminidase (NA) [5–7] of influenza virus and of two animal viruses, the poliomyelitis virus [8] and rhinovirus 14 [9] of the picornavirus family. The structure of the poliomyelitis virus is detailed by Hogle, Chow, and Filman in Chapter 1 of this book. The main goal in determining these structures was to correlate their three-dimensional structure and biological function with their ability to evade neutralization by the immune system. The main achievements to date have been primarily concerned with understanding viral recognition by the humoral system. This has been facilitated by the ability to target anti-protein or anti-peptide mouse monoclonal antibodies against specific antigens and viruses. In addition, the ability to follow the evolution of a virus from a defined pandemic strain has enabled evaluation of natural antigenic variation and selection in the human population. By comparison, the analysis of recognition of viral antigens by the cellular system has been more technologically difficult; however, rapid advances in this direction are being made at present, with the ability both to clone T cells and to understand antigen presentation and processing. In this chapter, I review the structures of the influenza virus surface antigens, which are undoubtedly the best characterized in terms of structure, function, and antigenic variation. It is clear from such structural studies that x-ray crystallography can contribute greatly to our understanding of the molecular basis of immune recognition. In addition, it will be obvious that such advances are possible only through successful collaborations of researchers in virology, immunology, molecular biology, and crystallography.

X-ray Crystallography of Viral Antigens

Viral antigen crystal structural determinations have, in general, been difficult to accomplish, as can be seen from the relative paucity of solved structures. The major difficulty confronted by researchers attempting to solve x-ray structures of biological macromolecules is the crystallization of the protein or nucleic acid material. The crystallization of viral antigens that are integral membrane proteins has been possible thus far, only via their proteolytic cleavage from the viral membrane, which leaves behind their small hydrophobic membrane anchor. The hemagglutinin and neuraminidase of influenza virus have both been crystallized by such techniques. The second major problem has been the difficulty of obtaining sufficient amounts of material (10–100 mg) for crystallization studies. It is to be hoped that cloning and expression of viral proteins will soon allow large amounts of pure, soluble viral proteins to become available for such x-ray studies. The final major factor restricting rapid progress in this field is the lack of purity of the protein material itself, due to heterogeneity of the viral proteins and, in particular, of their carbohydrate, which leads to problems in protein crystallization.

The Three-dimensional Structure and Function of the Hemagglutinin and the Neuraminidase

In the past five years, the structures of the two influenza virus surface antigens, the hemagglutinin (HA) [1,2] and the neuraminidase (NA) [5,6], have been reported to high resolution. Both cleaved proteins have been shown to be antigenically and functionally intact so that the resulting antigen crystal structures can reasonably be expected to provide structural information on their viral function, antigenic variation, and immune recognition.

The two influenza surface antigens have completely distinct viral activities and have very different three-dimensional structures. However, they do have similar features, which can be compared in terms of their structure, function, and antigenic variation. The HA structure determination was unique in its use of only one heavy atom derivative with threefold noncrystallographic symmetry averaging and solvent flattening to solve the structure [1]. The structure of the HA spike projects 135 Å from the membrane surface and is composed of a fascinating structural motif of a globular head and fibrous tail [1] (Figures 2.1, 2.2). A receptor binding site to attach the virus to sialated host-cell receptors was proposed in the globular head at the distal end of the antigen [1] (Figures 2.1, 2.2) and confirmed later by receptor binding variants of the HA [10]. The membrane fusion activity of influenza occurs at around pH 5.5 [11] and is associated with the fibrous tail of the HA. A conformational change has been identified in the HA at this pH, giving rise to the appearance of the previously buried fusion peptide (HA2 1–20) at the trimer surface [12] (Figure 2.2). This fusion-triggered activation is thought to be important in infection by enabling membrane fusion in the endosome after initial cell

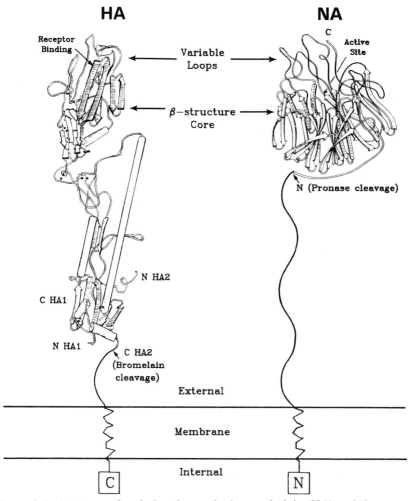

HA   **NA**

Receptor Binding

Variable Loops

β–structure Core

Active Site

N (Pronase cleavage)

N HA2

C HA1

N HA1

C HA2 (Bromelain cleavage)

External

Membrane

Internal

C   N

Figure 2.1. Anatomy of a viral antigen—the hemagglutinin (HA) and the neuraminidase (NA) of influenza virus. The schematic drawings are modified with permission from the published structures of the HA [1] and NA [7] antigens solved by x-ray crystallography. The two structures are shown on approximately the same scale, with arrows representing β strands and cylinders representing α helices. The locations of the proteolytic cleavage sites used to obtain soluble antigens for crystallization are shown. The HA is cleaved close to the viral membrane (HA2 175), whereas the NA has a 100-Å-long stalk removed (NA 1–74/77). The HA is anchored in the membrane by its carboxyl end, whereas the NA is anchored in the membrane by its amino end. The conformations of the remaining membrane-bound structures have not been determined, and their representation here is only to illustrate some general features. N and C indicate the amino and carboxy termini for the two polypeptide chains of HA, HA1 (328 residues), and HA2 (225 residues) and the NA (469 residues). The viral antigens extend approximately 135–150 Å from the viral membrane. The locations of the receptor binding site and active site, and fusion peptide, are also indicated. Both binding sites are located in clefts at the ends of the β-structure cores and are surrounded by protruding loops, which are the major sites of antigenic variation.

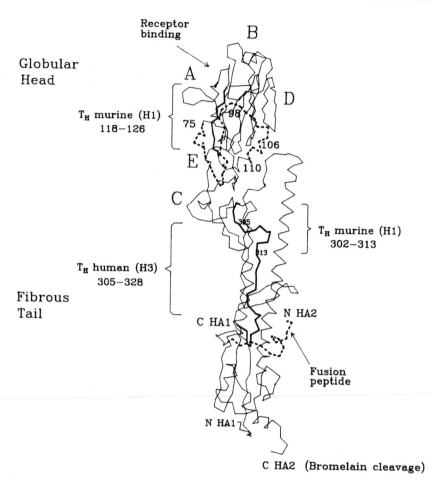

Figure 2.2. Antigenic sites of influenza virus hemagglutinin (an α-carbon drawing of the A/Hong Kong/68 HA is shown, as in the coordinates of Wilson, Skehel, and Wiley [1]). The proposed antibody-combining sites (A–E) [2, 4] are on the globular head and both T_H cell sites [18, 19] are shown on the carboxy terminal of HA1 and HA2. The location in the HA monomer of a chemically synthesized peptide (75–110) against which 20 monoclonal antibodies were raised is shown by a dotted line [21]. The locations of the receptor binding site and fusion peptide (dotted) are indicated.

binding through the HA receptor binding site and viral uptake by endocytosis [13]. Additional structural information on fusion has shown pH-sensitive HA variants to cluster not only around the fusion peptide but also around residues in the trimer interface [14]. However, the actual mechanism of fusion remains unclear and awaits new approaches or probes of conformational changes such as anti-peptide or anti-protein monoclonal antibodies, which can be used to follow structural changes during the time course of the fusion process.

The function of the neuraminidase is more debatable. The NA is an enzyme, specific for terminal sialic acid removal from carbohydrate-containing molecules. The NA is thought to remove the sialic acid from the influenza virus surface to prevent virus aggregation and in addition to help virus release from infected cells. The proteolytically cleaved tetramerical head is 100 × 100 × 60 Å and differs from the HA in that approximately 100 Å of the spindly stalk that is attached to NA and that connects the amino end to the membrane has been proteolytically removed (Figure 2.1). The structural determination was also novel in that an average of two N2 subtype structures, the A/RI/5$^+$/57 and A/Tokyo/3/67, were used in the phase calculation employing both multiple isomorphous replacement and noncrystallographic symmetry averaging [5]. The structure in itself is truly remarkable, being a new class of structure, a β-sheet propeller with large loops extending from the central shaft (Figure 2.2). The active site has been identified in a cleft in the center of each globular head of the tetramer and contains many conserved charged residues [6,7].

Antigenic Variation of the Hemagglutinin and Neuraminidase

It is important to consider that the ability of a viral antigen to vary its external morphologic and immunologic appearance is inherently correlated with its ability to conserve viral function. Some structures seem better designed to match these conflicting requirements, and both influenza viral antigens seemed well suited for such variation. Both are composed of a globular head with a conserved β-structure core; they have many exposed loops extending to the surface furthest from the viral membrane where amino acid insertions, deletions, and substitutions or carbohydration additions can occur (Figure 2.1). These antigenic loops are particularly apparent in the schematic of the structure in Figure 2.2 (for example HA, sites A, B, and C) and in Figure 2.1, the many loops of NA. Both sialic acid binding sites are at the center of these protruding loops. These completely conserved pockets of residues are either on the surface (HA) or in a deep pocket (NA) and are not accessible to neutralization by antibodies [1] (Figure 2.1). On the other hand, the HA differs fron the NA in that it has a second functional domain, the fibrous tail, which is required for fusion and trimer assembly. The tail is very conserved, presumably as a consequence of no strong selection pressure for antigenic variation, as compared to the more exposed globular head.

Humoral Recognition of the Hemagglutinin and Neuraminidase

From 1968 to 1979, analysis of natural variants (H3 strains) and of laboratory variants [2,4] identified four to five areas on the surface of the HA as antigenic determinants (Figure 2.2). At least one mutation in each of these sites seemed to be required for the virus to escape neutralization by the immune system [2]. These antibody-combining sites are generally composed of looplike structures and discontinuous determinants of the complex, assembled, top-ographical type (Figure 2.2, sites A–E). The loops in the neuraminidase have similarly been identified as the major sites of antigenic variation for both natural and laboratory selected variants [6]. However, the rate of natural variation of the NA is significantly slower than that of the HA and correlates with the serologic dominance of the anti-HA response in influenza virus in-fection. Structural analysis of an HA variant (Gly-Asp 146) has shown that a single amino acid substitution causes, in fact, no major conformational changes on the HA except for the generation of a single new salt bridge that is sufficient to enable the virus to escape neutralization [15]. The role of the carbohydrate in antigenic variation has been the subject of much speculation. Evidence from the HA structure has shown that not only the protein but also the carbohydrate is important in mediating antigenicity. Analysis of car-bohydrate locations in natural variants [1,3,4] and in a laboratory variant [16] has shown carbohydrate to have an important role in masking and un-masking protein determinants.

Antigenic variation of a different subtype (H1), the A/PR/8/34 virus, has shown the location of amino acid changes in the HA of 35 mutant viruses selected by mouse monoclonal antibodies to be strikingly similar to those areas earlier mapped on the A/Hong Kong/68 structure [17]. We have used one of these mutants to solve a crystallization problem (Arg-Ile 224 HA1) and to obtain crystals of the PR/8 HA (in collaboration with Dr. Walter Ger-hard, Wistar Institute), even though the parent strain with only one amino acid difference did not crystallize. It seems that the HA structure in different subtypes must also contain the basic, central, conserved β-structure core and fibrous tail for viral function but must vary significantly in the protruding loops, as seen in Figures 2.1 and 2.2.

Cellular Recognition of the Hemagglutinin

T cell recognition of the surface antigens of influenza virus has been analyzed most extensively in two strains of the H1 and H3 subtypes. T-helper cell determinants of the H1 subtype have been mapped to two to three peptide sequences on HA1 (Figure 2.2) [18]. T cell clones induced by the H3 subtype A/Texas/1/77 have been shown to recognize two chemically synthesized peptides and also to induce tolerance in T cell clones [19] (Figure 2.2). At the present the T cell determinants seem to be more limited in both location and extent than the B cell determinants, and possibly map to similar locations in different subtypes.

Anti-peptide HA antibodies

When synthetic peptides of an H3 HA (A/Vic/3/75) have been used as im-munogens, essentially the entire surface of the HA can be mapped as antigenic by anti-peptide antibodies [20]. However, these antibodies have not been shown to neutralize the virus and are markedly different from the anti-protein antibodies. The question as to whether neutralizing antibodies cause conformational changes in the antigen, perturb biological function such as receptor binding and fusion, or cause general changes in the virion has still to be determined. Indeed, anti-peptide antibodies against residues 75–110 of HA1 ([21], see also Figure 2.2) have good affinity constants with the immunizing peptide (Ka~10^8) (Bergmann, Klingler, and Wilson, unpublished) and monomeric HA1 (Ka~10^6) but may not bind quickly enough to the intact trimeric virus to prevent viral infection, especially if conformational changes in the antigen are required before antibody binding occurs. It is at present still difficult to understand how a free peptide can mimic its cognate structure when it is embedded in the intact protein structure. It is clear that this dual recognition cannot necessarily be artifactual and a consequence of binding assays such as enzyme-linked immunosorbent assay (ELISA), as successes have been reported recently for neutralization of viruses by anti-peptide antibodies. Neutralizing and/or protective anti-peptide antibodies have been reported for foot-and-mouth disease [22], hepatitis B [23], influenza [24], and poliovirus [25,26]. A discussion of results with synthetic peptides for one of these structures, poliovirus, is found in Chapter 1.

Influenza Virus Antigens as General Viral Antigen Models

Defense against microorganisms depends on the ability of the immune system to both recognize and neutralize foreign antigens. Viruses, such as influenza virus, have the ability either to alter significantly their surface glycoproteins by genetic recombination or reassortment [27,28,29] or to successively vary an existing antigen by accumulating point mutations in their RNA, which are expressed as amino acid changes on the glycoprotein surface. The structures of the influenza virus antigens have enabled us to understand the molecular basis of this recognition and to advance our knowledge of antigenic variation. However, we do not yet fully understand why some regions are more immunogenic or antigenic and how some regions (e.g., in the NA), although immunogenic, do not induce neutralizing antibodies. The poliovirus and rhinovirus structures support the suggestion in influenza that protruding, probably flexible, loops that can accommodate extensive mutations play a major role in enabling a virus to escape immune recognition. By the same argument, it is these loops that become the major target of such immune attack. Other organisms, however, have evolved different methods for such escape. African trypanosomes, for example, completely change their surface antigens at periodic intervals by DNA recombinational mechanisms [30]. Malaria, on the other hand, has evolved different host stages with different tandemly repeating amino acid sequences as the immunodominant regions of its surface antigens [31,32]. It is to be expected that other viral antigens may also differ in their immune defense from influenza either in a subtle or less subtle way. In particular, animal viruses such as polio, which have no viral membrane, have additional limitations on antigenic variation, because they must also assemble an outer protein shell with its surface antigens. However, the general principles involved in variation of the antigenic structure of these and other membrane-bound antigens are likely to be similar to those found for influenza.

The future of x-ray crystallography of viral antigens at present looks encouraging but depends heavily upon obtaining large quantities of pure antigen and also upon continuing advances in crystallization methods such as those seen recently in membrane protein crystallography [33]. Nevertheless, x-ray crystallography has already shown in a spectacular way how we can significantly advance our understanding of the structure of viral pathogens and their recognition by the immune system.

Acknowledgments

The use of the unpublished coordinates of Wilson, Skehel (National Institute for Medical Research, Mill Hill, London), and Wiley (Harvard University), is acknowledged. I thank Peggy Graber for manuscript preparation, Dan Bloch for preparation of the figures, and Dr. Peter Colman for permission to use the neuraminidase schematic.

This is publication number MB 4153 from the Research Institute of Scripps Clinic.

References

1. Wilson IA, Skehel JJ, Wiley DC (1981) Structure of the haemagglutinin membrane glycoprotein of influenza at 3A resoltuion. Nature 289:73–78
2. Wiley DC, Wilson IA, Skehel JJ (1981) Structural identification of the antibody-binding sites of Hong Kong influenza haemagglutinin and their involvement in antigenic variation. Nature 289:373–378
3. Wilson IA, Ladner RC, Skehel JJ, Wiley DC (1983) The structure and role of the carbohydrate moieties of influenza virus haemagglutinin. Biochem Soc Trans 11:145–147
4. Wiley DC, Wilson IA, Skehel JJ (1984) The haemagglutinin membrane glycoprotein of influenza virus. In McPherson A (ed) Biological Macromolecules and Assemblies, vol 1. John Wiley & Sons, New York, pp. 299–336
5. Varghese JN, Laver WG, Colman PM (1983) Structure of the influenza virus glycoprotein antigen neuraminidase at 2.9 Å resolution. Nature 303:35–40
6. Colman PM, Varghese JN, Laver WG (1983) Structure of the catalytic and antigenic sites in influenza virus neuraminidase. Nature 303:41–44
7. Colman PM, Ward CW (1985) Structure and diversity of influenza virus neuraminidase. Curr Top Microbiol Immunol 114 (in press)
8. Hogle JM, Chow M, Filman DJ (1985) Three dimensional structure of poliovirus at 2.9 Å resolution. Science 229 (4720):1358–1365
9. Rossmann MG, Arnold E, Erickson JW, Frankenberger EA, Griffith JP, Hecht H-J, Johnson JE, Kamer G, Luo M, Mosser AG, Rueckert R, Sherry B, Vriend

G (1985) Structure of a human common cold virus and functional relationship to other picornaviruses. Nature 317:145–153

10. Rogers GN, Paulson JC, Daniels RS, Skehel J J, Wilson IA, Wiley DC (1983) Single amino acid substitutions in influenza hemagglutinin change receptor binding specificity. Nature 304:76–78

11. White J, Matlin, K, Helenius, A (1981) Fusion of Semliki Forest virus with the plasma membrane can be induced by low pH. J Cell Biol 87:264–272

12. Skehel JJ, Bayley PM, Brown EB, Martin SR, Waterfield MD, White JM, Wilson IA, Wiley DC (1982) Changes in the conformation of influenza virus haemagglutinin at the pH optimum of virus-mediated membrane fusion. Proc Natl Acad Sci USA 79:968–972

13. White J, Killian M, Helenius A (1983) Membrane fusion proteins of enveloped animal viruses. Q Rev Biophys 16:151–195

14. Daniels RS, Downie JC, Hay AJ, Knossow M, Skehel JJ, Wang ML, Wiley DC (1985) Fusion mutants of the infectious virus haemagglutinin glycoprotein. Cell 40(2):431–439

15. Knossow M, Daniels RS, Douglas AR, Skehel JJ, Wiley DC (1984) Three-dimensional structure of an antigenic mutant of the influenza virus hemagglutinin. Nature 311:678–680

16. Skehel JJ, Stevens DJ, Daniels RS, Douglas AR, Knossow M, Wilson IA, Wiley DC (1984) A carbohydrate side-chain in the haemagglutinin of Hong Kong influenza virus inhibits recognition by a monoclonal antibody. Proc Natl Acad Sci USA 81:1771–1783

17. Caton AJ, Brownlee GG, Yewell JW, Gerhard W (1982) The antigenic structure of the influenza virus A/PR/8/34 haemagglutinin (H1 subtype). Cell 31:417–427

18. Hurwitz JL, Heber-Katz E, Hackett CJ, Gerhard W (1984) Characterization of the murine T_H response to influenza virus hemagglutinin: Evidence for three major specificities. J Immunol 133:3371–3377

19. Lamb JR, Skidmore BJ, Green N, Chiller J, Feldmann M (1983) Induction of tolerance in influenza virus-immune T lymphocyte clones with synthetic peptides of influenza hemagglutinin. J Exp Med 157:1434–1447

20. Green N, Alexander H, Olson A, Alexander S, Shinnick TM, Sutcliffe JG, Lerner RA (1982) Immunogenic structure of the influenza virus haemagglutinin. Cell 28:477–487

21. Wilson IA, Niman HL, Houghten RA, Cherenson AR, Connolly ML, Lerner RA (1984) The structure of an antigenic determinant in a protein. Cell 37:767–778

22. Rowlands DJ, Clarke BE, Carroll AR, Brown F, Nicholson BH, Bittle JL, Houghten RA, Lerner RA (1983) Chemical basis of antigenic variation in foot-and-mouth disease virus. Nature 306:694–697

23. Gerin JL, Alexander H, Wai-Kuo Smith J, Purcell RH, Dappolito E, Engle R, Green N, Sutcliffe JG, Shinnick TM, Lerner RA (1983) Chemically synthesized peptides of hepatitis B surface antigen duplicate the d/y specificities and induce subtype-specific antibodies in chimpanzees. Proc Natl Acad Sci USA 80:2365–2369

24. Shapira M, Jibson M, Muller G, Arnon R (1984) Immunity and protection against influenza virus by synthetic peptide corresponding to antigenic sites of haemagglutinin. Proc Natl Acad Sci USA 81:2461–2465

25. Emini EA, Bradford AJ, Wimmer E (1983) Priming for an induction of anti-poliovirus neutralizing antibodies by synthetic peptides. Nature 304:699–703

26. Chow M, Bittle JL, Hogle J, Baltimore D (1984) Antigenic determinants in polio-virus capsid protein VP1. *In* Chanock RM, Lerner RA (eds) Modern Approaches to Vaccines. Cold Spring Harbor Laboratory, Cold Spring Harbor, New York pp. 257–261
27. Burnett FM (1955) Principles of Animal Virology. Academic Press, New York, p 282
28. Kilbourne ED (1969) Bull WHO 41:643–645
29. Pays, E, Delauw M-F, Van Assel S, Laurent M, Vervoort T, Van Meirvenne N, Steinert M (1983) Modifications of a *Trypanosoma* b. *brucei* antigen gene repertoire by different DNA recombinational mechanisms. Cell 35:721–731
30. Webster RG, Laver WG (1975) *In* Kilbourne ED (ed) The Influenza Viruses and Influenza. Academic Press, New York, pp 269–314
31. Godsen GN, Ellis J, Svec P, Schlesinger DH, Nussenzweig V (1983) Identification and chemical synthesis of a tandemly repeated immunogenic region of *Plasmodium knowles* circumsporozoite protein. Nature 305:29–33
32. Coppel RL, Cowman AF, Lingelbach KR, Brown GV, Saint RB, Kemp DJ, Anders RF (1983) Isolate-specific S-antigen of *Plasmodium falciparum* contains a repeated sequence of eleven amino acids. Nature 306:751–756
33. Hartmut M (1983) Crystallization of membrane proteins. TIBS 8(1):56–59

CHAPTER 3

Using Nucleotide Sequence Determination to Understand Viruses

BART G. BARRELL AND PAUL J. FARRELL

Our understanding of the molecular biology of viruses has always advanced through a combination of genetics, biochemistry, and structural analysis. For many viruses, genetic and biochemical approaches have been hampered by inadequate tissue culture systems for growth of the virus or hazard to the investigator of working with a particularly pathogenic virus. The advent of molecular cloning has permitted nucleotide sequence analysis of almost any virus. Even if only very small amounts of viral nucleic acid can be obtained, this can be cloned and produced in unlimited amounts. As a result of recent increases in the speed of nucleic acid sequencing, particularly the shotgun strategy using the M13 cloning/dideoxynucleotide chain-termination method [1], it is now quite common for the first detailed information on a particular viral genome to come from DNA sequencing. This technology is being applied to the larger, less well understood viruses such as the herpesviruses and has resulted in the sequences of large regions of herpes simplex virus [2], varicella zoster virus [3], human cytomegalovirus, and the complete sequence of Epstein–Barr virus (EBV) [4]. The complete sequences of all these human herpesviruses (genome sizes of the order of 100–240 kilobase pairs) will be known within the next few years. Because the DNA sequencing is breaking new ground with little or no previous genetics, it is useful and important to assess what information can be interpreted from a DNA sequence and how this knowledge can be used to design further experiments to understand these sequences. An important point is the quality of the DNA sequence. A sequence should be determined at least once on both strands and should not rely on restriction map data to assemble the sequence. Strand-specific artifactual problems on sequencing gels occur as a result of sequence and secondary structure, particularly with guanosine–cytosine (GC) -rich sequences, and there are many traps for the unwary.

Prediction of Gene Structure

Once the nucleotide sequence has been determined, computer-assisted methods are used to search it for features known to be relevant to gene structure. Our knowledge of these signals is far from perfect, and we are not able to fully predict gene structure solely from nucleotide sequence. Nevertheless, logical combinations of transcription promoter sequences, open reading frames, splicing signals, and polyadenylation signals are powerful and useful predictions, and we shall examine these.

Open Reading Frames

These are defined as any region of any of the six possible translations of a nucleotide sequence uninterrupted by stop codons (UGA, UAG, UAA). Especially in the larger viral genomes (e.g., herpesviruses), both strands of the DNA molecule carry genes. Sometimes genes overlap on the same DNA strand, but the coding regions of genes almost never overlap on opposite strands. Small overlaps at the beginning and ends of genes are common and presumably reflect the pressure on a viral genome to remain compact, though it is possible that these fulfill some role in expression. Examples of overlapping genes are the large and small T antigens in the early region of SV40. Here the overlap is in the same translation phase. In the Bam HI-L region of EBV, two genes overlap in different translation phases.

The search for open reading frames is best done with a computer, and a useful first step is to highlight the largest reading frame in any one region. If there is little or no splicing, then this is likely to be close to the actual arrangement of coding sequences. Next the sequences surrounding the open reading frame are examined for potential gene expression signals, promoters, and polyadenylation sites.

Polyadenylation Sites

The great majority of eukaryotic mRNAs are polyadenylated 10–30 bases downstream of the sequence AAUAAA. AUUAAA is occasionally found, and some other sequences have been noted [5]. PolyA addition is quite often at the A of the sequence PyA, and sequences rich in TGT and TTT are often found in the 30 bases or so downstream of the addition site. It seems that AAUAAA is probably the signal for cleavage of the pre-mRNA to generate the 3' end to which polyA is added. The further sequence requirements for these reactions are being defined. The polyA sequence is not always found immediately downstream of the coding region; it may be one or more reading frames downstream because of the presence of overlapping 3' coterminal mRNAs which use a common polyadenylation site.

Promoters

Eukaryotic virus promoters have a complex structure at the DNA sequence level [6]. The herpesviruses, adenoviruses, papovaviruses, and retroviruses are transcribed mainly by the cellular RNA polymerase II (pol II), giving rise to the mRNAs of the virus. Pol II promoters are made up of several sequence elements, including (a) the upstream elements, which are most important for promoter function; and (b) the TATA box, which may be important for aligning polymerase at the correct start point. There is as yet no general consensus sequence for upstream promoter elements, but most pol II promoters have a recognizable TATA box about 30 bases upstream of the transcription start point. The canonical TATA box sequence is TATA^T_AA^T_A, but there is much variation and a few well studied promoters completely lack a TATA box sequence. Searching for promoters is thus reduced to screening for good homologies to the TATA consensus and then picking out those that seem to be logically placed with respect to open reading frames. Some eukaryotic promoters have an upstream sequence called the CAT box. This is generally 60—90 bp upstream of the RNA start, with a consensus sequence GGCCAATCT.

Initiation Codons

Once candidates for a promoter and reading frame have been found, it is possible (in the absence of splicing) to predict the initiation codon. This will usually be [7] the first ATG downstream of the transcription start which has either a purine at position -3 to the ATG and/or a G at position 4, i.e., ^A_GXXATGG.

Splicing

Predicting a splicing pattern is normally impossible when DNA sequence data are used alone, because there is too much redundancy in the consensus donor and acceptor sequences (particularly in the latter), and there is no way of predicting which donor will splice to which acceptor (particularly when splicing can occur over large distances).
The consensus sequences are:

Donor: ^C_A A G/G T Pu A G T
Acceptor: (Py)$_n$G PyA G/G T

The underlined sequences are obligatory, and there should be at least 12 bases between the obligatory AG of the acceptor and any upstream AG. In many cases, the acceptor also has the lariat sequence PyXPyTPuAPy 15–40 bases upstream of the junction [8].

Statistical Analysis of Coding Sequences

It is possible to apply other criteria to substantiate the prediction that an open reading frame is indeed coding for a protein. The simplest test is on grounds of length alone. In a random sequence, termination codons should occur on average once in approximately every 20 codons, so that long open reading frames are unlikely to have arisen by chance. Protein sequences do not, on average, contain an equal distribution of amino acids [9]; thus the coding sequence will have a biased distribution of codons which can be detected by computer [10]. Many viral genomes have a biased base composition, for example, the high GC content of some herpesviruses. Where this occurs in the coding sequence of a protein of normal composition, the excess bases cannot be accommodated in the first two positions of the codon, which mostly specifies the amino acid, and so they tend to accumulate in the third (degenerative) position of the codon. This asymetric distribution can be detected by computer analysis and is characteristic of open reading frames that are actually coding, rather than ones found simply by chance.

Prediction of Function by Homology with Other Proteins

The number of protein sequences known is growing rapidly as a result of gene sequencing; these are available as protein or nucleic acid sequences in computer data banks (Genbank and the EMBL library). Protein data banks are more useful for looking for homologies with unknown sequences, the main data bases being the Protein Information Resource and the Doolittle database. Several computer programs are available to find homologies with unknown sequences [11,12]. These are very useful when closely related sequences from, say, retroviruses are being compared, but similarities between structural proteins of unrelated viruses are unlikely. Virally encoded enzymes are more likely to be conserved. For example, thymidine kinases, DNA polymerases, ribonucleotide reductases, and protein kinases contain recognizable structural motifs because enzyme structures have certain common features. Homology may occur throughout the sequences or may be confined to a single region such as a nucleotide binding site.

Predicting Function from Sequence Data

Predicting the function of a gene product from sequence data alone is impossible except in rare cases. However, one class of viral proteins, i.e., membrane glycoproteins, is easily recognized from primary sequence data. These have distinct hydrophobic signal sequences, Asn-xxx-Thr or Ser (NXT/S) which is the substrate for N-linked glycosylation [13], and have a hydrophobic membrane anchor region near the C terminus of the amino acid sequence. A simple computer program that plots the hydropathicity will give an easily recognized profile for this type of molecule. As these are targets for the host's immune system, they are obviously important in constructing diagnostic and vaccination agents. Indeed, with a capacity to sequence viral genomes quickly, DNA sequencing alone can be an efficient way of locating and determining the sequence of membrane antigens.

Mapping Transcripts Predicted from the DNA Sequence

The M13 clones produced during the sequence analysis offer an extremely convenient way to make probes to study the transcription. If the universal primer is used to prime synthesis across the insert, highly radioactive single-stranded probes ("prime-cut probes") of known sequence can be produced [14]. These can be used in three ways:

1. As probes for mRNA that has been fractionated on agarose gels and transferred to nitrocellulose filters ("Northern blots"). This method gives information on the size and number of transcripts in a particular region of the genome. By probing mRNA made at different times in the viral life cycle, it is possible to ascertain the time of appearance of a transcript.
2. As probes to hybridize to mRNA in S1 mapping experiments [15]. This permits accurate mapping of boundaries of exons in mRNAs onto the DNA sequence.
3. For primer extension analysis, to determine 5' ends of mRNAs [16].

Synthetic Peptides and Expression Vectors

One of the most powerful approaches for identifying gene products encoded by a particular region of DNA involves making antibodies against the predicted protein sequence[17]. A knowledge of the DNA sequence and location of open reading frames permits insertion of an appropriate viral restriction fragment into one of the many bacterial or yeast expression vectors. These vectors permit production of a large amount of a modified protein containing part or all of the viral protein under study. Antibodies can be raised against this hybrid protein, which will often recognize the native viral protein to permit its further study. An alternative is to synthesize chemically the peptides (usually about ten amino acids in length) that correspond to parts of the protein sequence deduced from the relevant DNA sequence. Antibodies can then be raised from these peptides after they are coupled to appropriate carriers [18]. The fact that such hybrid proteins and synthetic peptides are immunogenic and that the antibodies frequently react with the native viral proteins permits their use as virus-free vaccines and diagnostic reagents.

Conclusion

DNA sequence analysis and transcription mapping by means of the M13 clones generated during the sequencing provides a powerful approach to the study of viruses. Because of their small size and the ease of purification of their nucleic acid, viruses are ideal subjects to study in this way, and the DNA sequencing techniques available now can readily cope with the genomes of even the largest viruses. It is now quite possible for one person to sequence over 50 kb in a year, so that sequencing of a herpesvirus genome will soon become a Ph.D. project in a well-equipped DNA sequencing laboratory. We have reviewed the ways a DNA sequence can be analyzed and how these predictions can be used to design further experiments to study the transcription. Limited predictions about gene structure can be made purely from nucleotide sequence; however, because of considerable uncertainties in those predictions, it is still necessary to determine gene structure directly wherever possible. The determination of the nucleotide sequence provides the basic framework for building an understanding of the virus.

References

1. Bankier AT, Barrell BG (1983) Shotgun DNA sequencing. *In* Techniques in the Life Sciences. Elsevier, Ireland, vol B5, pp 1–35
2. McGeoch DJ, Dolan A, Donald S, Rixon FJ (1985) Sequence determination and genetic content of the short unique region in the genome of herpes simplex virus type 1. J Mol Biol 181:1–13
3. Davison AJ (1983) DNA sequence of the U_s component of the varicella zoster virus genome. EMBO J 2:2203–2209
4. Baer R, Bankier AT, Biggin MD, Deininger PL, Farrell PJ, Gibson TJ, Hatfull GS, Satchwell SC, Seguin C, Tuffnell PS, Barrell BG (1984) DNA sequence and expression of the B95-8 Epstein–Barr virus genome. Nature 310:207–211
5. Birnstiel ML, Busslinger M, Strub K (1985) Transcription termination and 3' processing: The end is in site. Cell 41:349–359
6. Eukaryotic transcription: The role of cis- and trans-acting elements in initiation (1985) *In* Gluzman Y (ed). Cold Spring Harbor Laboratory, Cold Spring Harbor, New York (in press)
7. Kozak M (1984) Point mutations close to the AUG initiator codon affect the efficiency of translation of rat preproinsulin in vivo. Nature 308:241–246
8. Keller W (1984) The RNA lariat: A new ring to the splicing of mRNA precursors. Cell 39:423–425
9. Dayhoff MO (1969) Atlas of protein sequence and structure. National Biomedical Foundation, Silver Springs, Maryland
10. Staden R (1984) Measurements of the effects that coding for a protein has on a DNA sequence and their use for finding genes. Nucleic Acids Res 12:551–567
11. Wilbur WJ, Lipman DJ (1983) Rapid similarity searches of nucleic acid and protein data banks. Proc Natl Acad Sci USA 80:726–730
12. Staden R (1982) An interactive graphics program for comparing and aligning nucleic acid and amino acid sequences. Nucleic Acids Res 10:2951–2961

13. Neuberger A, Gottschalk A, Marshall RD, Spiro RG (1972) *In* Gottschalk A (ed) The glycoproteins: Their composition, Structure and Function. Elsevier, Amsterdam, pp 450–490
14. Biggin M, Farrell PJ, Barrell BG (1984) Transcription and DNA sequence of the Bam HI-L fragment of B95-8 Epstein–Barr virus. EMBO J 3:1083–1090
15. Berk AJ, Sharp PA (1978) Spliced early mRNAs of SV40. Proc Natl Acad Sci USA 75:1274–1278
16. Bina-Stein M, Thoren M, Salzman N, Thompson JA (1979) Rapid sequence determination of late SV40 16S mRNA leader by using inhibitors of reverse transcriptase. Proc Natl Acad Sci USA 76:731–736
17. Shinnick TM, Sutcliffe JG, Lerner RA (1984) Antibodies to synthetic peptide immunogens as probes for virus protein expression and function. *In* Notkins AL, Oldstone MBA (eds) Concepts in Viral Pathogenesis. Springer-Verlag, New York, 1984, pp 361–365
18. Sutcliffe JG, Shinnick TM, Green N, Lerner RA (1983) Antibodies that react with predetermined sites on proteins. Science 219:660–666

CHAPTER 4
Viruses with Ambisense RNA Genomes

DAVID H.L. BISHOP

Viruses are grouped on the basis of the type of their genetic information
(RNA or DNA), its form (single-stranded, or double-stranded), and the procedures employed for virus replication [1]. A variety of strategies of replication are used by viruses to reproduce in cells. For the single-stranded
RNA genome viruses (other than members of the Retroviridae which have
a DNA intermediate in the replication cycle), the strategy involves either
(a) the synthesis of proteins by translation of the viral RNA (and in some
cases derivatives of that sequence that are made during the replication process), i.e., the viral-sense RNAs function as mRNA species (positive-stranded
viruses); or (b) the transcription of viral-complementary mRNA species from
the genome, i.e., viral RNAs do not function as mRNA species (negative-
stranded viruses). For all the negative-stranded viruses, the synthesis of
mRNA species at the onset of infection is achieved by viral-coded enzymes
that are present in the infecting virus particles. Double-stranded RNA viruses
(e.g., members of the Reoviridae) also use a virion polymerase to copy into
mRNA one strand of each duplex of genomic RNA.

Examples of positive-stranded virus families are the Picornaviridae, Cal-
iciviridae, Coronaviridae, Togaviridae (alphaviruses and flaviviruses), and
most of the plant virus groups that have a single or segmented RNA genome
(exceptions include the negative-stranded plant rhabdoviruses). Positive-
stranded viruses include certain enveloped viruses (togaviruses, coronaviruses) and nonenveloped viruses (picornaviruses, caliciviruses, and most of
the plant virus groups). The negative-stranded RNA viruses are all enveloped
(Rhabdoviridae, Orthomyxoviridae, Paramyxoviridae, Bunyaviridae, and
Arenaviridae). Among these negative-stranded viruses are viruses with a
single species of RNA (rhabdoviruses, paramyxoviruses) and viruses with

7–8 species of RNA (orthomyxoviruses), 3 species (bunyaviruses), or 2 species (arenaviruses).

In this article, the evidence and implications of an ambisense coding strategy for the negative-stranded arenaviruses and at least one group of bunyaviruses are discussed. The term *ambisense* is used to indicate that gene products are translated from viral-complementary as well as viral-sense mRNA sequences. The differences in organization of genetic information that have been observed between members of different genera of viruses *(Phlebovirus, Bunyavirus)* belonging to the Bunyaviridae family are discussed in the context of the perceived hierachy of virus relationships within that group.

Arenaviruses

Members of the family Arenaviridae include lymphocytic choriomeningitis (LCM), Lassa (the etiologic agent of Lassa fever), Mopeia, Mobala, Junin (the agent of Argentine hemorrhagic fever), Machupo (the agent of Bolivian hemorrhagic fever), Pichinde, Tacaribe, Tamiami, Latino, Amapari, Flexal, and Parana viruses [2,3]. All are transmitted in nature by rodents (*Mus, Calomys, Mastomys, Oryzomys, Sigmodon* species), or, for Tacaribe virus, fruit-eating bats (*Artibeus* species) [1]. In the normal nonfatal transuterine, transovarian, or neonatally acquired infection of the host species, persistent life-long virus infections are common [2]. Viremia and, or viruria contribute to the syndrome and to the transmission cycle. The acquisition of virus by an adult species usually results in a temporary acute phase of infection with virus removal in the survivors. The virulence of the virus in the adult rodent species depends on the virus, the host species, and the route of inoculation. Infection within the rodent nest may involve milk, saliva, and/or urine transmission of virus to the offspring [1]. Although virus spread to other species is uncommon, human infections in nature (LCM, Lassa, Junin, Machupo, Flexal), and in the laboratory (e.g., LCM, Lassa, Pichinde, Tacaribe), have been documented [4]. The symptoms and severities of the ensuing human diseases vary with the virus.

Interest in arenaviruses has centered on the persistent nature of the normal rodent infection, from both immunologic and molecular viewpoints [2—5]. The effect of virus replication on the functions of infected cells in different organs of the host species is also under study [6]. Persistent infections are readily established in vitro, and results are reviewed elsewhere [5].

Molecular studies [2,3] have shown that arenaviruses are enveloped and pleomorphic (50–300 nm in diameter), with glycoprotein projections on the surface and two internal nucleocapsid structures. These nucleocapsids consist of viral RNA (large, L, ca. 2×10^6 daltons; small, S, ca. 1×10^6 daltons), nucleocapsid protein (N), and small amounts of a $180–200 \times 10^3$-dalton polypeptide (L protein) that may be a transcriptase/replicase component. Frequently ribosomes are incorporated within the virus envelope. The existence of ribosomes within virus particles is not surprising in view of (a) the pleomorphic character of virions, (b) the failure of arenaviruses to inhibit effectively host-cell macromolecular syntheses, and (c) the observation of ribosomes at the sites of virus maturation (the cell plasma membrane). It is, however, a characteristic feature that sets arenaviruses apart from members of other virus families [1]. Experiments involving the growth of Pichinde virus in cells in which the ribosomes are temperature sensitive (ts) have shown that the virion ribosomes are not essential for virus infectivity because other cells can be infected at nonpermissive temperatures with viruses possessing ts ribosomes [7].

Genetic Attributes of Arenaviruses

Genetic experiments have shown that reassortant (recombinant) viruses can be generated between distinguishable strains of LCM virus, or Pichinde virus [8,9]. Two recombination groups of mutants have been identified commensurate with the existence of two species of viral RNA. In addition, virus particles that are diploid (at least with respect to the S RNA) have been detected among virus populations [9]. Again this finding is not suprising in view of the pleomorphic character of the virions. Diploidy (polyploidy) may contribute to the biological stability of arenaviruses in nature. Certainly the difficulties encountered in recovering ts mutants of arenaviruses could be explained by this character.

The observation that there are three virus proteins and only two viral RNA species raises questions concerning the RNA coding arrangement. Studies of reassortant Pichinde viruses have shown that the L RNA codes for the L protein and that the S RNA codes for both the viral glycoproteins and the N protein [10,11]. Sequence analyses of DNA copies of Pichinde and LCM viral S RNA species and studies of the induced RNA species recovered from infected cells have confirmed the latter result and demonstrated a novel coding arrangement in the S RNA [12,13]. The N protein is coded in a subgenomic viral-complementary sequence corresponding to the 3' half of the S RNA. The glycoprotein precursor (GPC) is coded in a subgenomic viral-sense sequence corresponding to the 5' half of the RNA. To characterize this coding arrangement, the term *ambisense* RNA has been coined [12]. The central intergenic regions of the S RNA species of LCM and Pichinde viruses have unique sequences that can be arranged in hairpin configurations [13]. As determined by appropriate oligonucleotide annealing experiments, these intergenic sequences appear to function in transcription termination for both N and GPC mRNA species (M. Galinski and D. H. L. Bishop, unpublished data), although exactly how this is achieved, and how N, L, and GPC mRNA syntheses are initiated are not known.

Ambisense Coding Strategy of Arenaviruses

The presence of RNA polymerase activites in arenavirus preparations [2] suggests that after attachment, penetration, and uncoating in permissive cells, arenaviruses synthesize N and L mRNA species (primary transcription). Presumably these are translated to give the N and L proteins. It is assumed that with the availability of these proteins, RNA replication takes place (Figure 4.1). The ambisense strategy of the arenavirus S RNA species precludes the synthesis of the subgenomic S-coded GPC mRNA species until after viral RNA replication has been initiated. This is in stark contrast to the strategy employed by the negative-stranded rhabdovirus, paramyxovirus, and orthomyxovirus groups. For those viruses with a single species of genomic RNA (i.e., rhabdoviruses and paramyxoviruses), all the mRNA species are made in a coordinated manner by transcription of viral RNA both during primary transcription and (in larger quantities) after RNA replication has been initiated (secondary transcription). For orthomyxoviruses, although the genome is segmented so that the selected synthesis of certain mRNA species might be possible, no evidence for such discrimination has been obtained.

The data obtained by analyses of the species and amounts of intracellular S-coded RNA induced by arenaviruses indicate that N mRNA is made before GPC mRNA. At later stages of an arenavirus infection when both are present, the subgenomic N mRNA species predominate ([12], and unpublished data). These observations support the hypothesis that N mRNA synthesis is not coordinate with GPC mRNA synthesis. The time course of synthesis of L mRNA species has not been investigated, but presumably it is like that of the N mRNA. If so, then one advantage of this strategy is that N and L proteins are made at the beginning of the infection when they are needed.

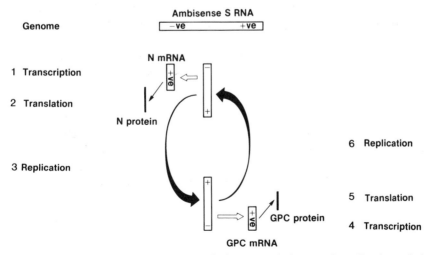

Figure 4.1. A schematic of the transcription, translation, and replication of the arenavirus S RNA species.

GPC mRNA and protein are not made until they are needed, i.e., after the initiation of RNA replication. Just as the continued synthesis of N mRNA from the template viral RNA may compete with viral-complementary (replicative) RNA synthesis, so too the synthesis of GPC mRNA may compete with the viral RNA synthesis. Conceivably for both, mRNA synthesis and RNA replication may occur concomitantly, with other viral gene products (e.g., N protein) involved in their regulation.

Bunyaviridae

The family Bunyaviridae includes some 250 virus serotypes, varieties, and variants that have been categorized into five genera *(Bunyavirus, Phlebovirus, Uukuvirus, Nairovirus,* and *Hantavirus)* [14]. Little is known about the strategy of infection of nairoviruses and hantaviruses. The available molecular information indicates that phleboviruses and uukuviruses are similar to each other in many properties; however, the coding arrangement of the three RNA species of uukuviruses has not been reported.

Analyses of the three viral RNA species of bunyaviruses indicate that proteins are coded in viral-complementary RNA sequences. For the small (S) RNA, two proteins (nucleoprotein, N, and a nonstructural protein NS_S) are coded in overlapping reading frames in a single mRNA species [15]. For the middle (M)-sized RNA, the glycoprotein precursor is also coded in a single mRNA species [16]. Incomplete data indicate that the large (L) RNA gene product is likewise coded in viral-complementary sequences [17]. Primary transcription of at least the S mRNA species has been demonstrated with protein synthesis inhibitors and the appropriate single-stranded DNA probes [18]—an observation commensurate with the existence of a virion transcriptase [19]. Bunyaviruses are therefore considered to be negative-stranded viruses.

Phleboviruses have an ambisense S RNA coding strategy [20]. A nucleo-protein (N) is coded in a subgenomic, viral-complementary, mRNA that is synthesized (presumably by a virion transcriptase) in the presence of protein synthesis inhibitors (primary transcription) [21]. The N mRNA species corresponds in sequence to the 3' half of the S RNA. A nonstructural protein (NS_S) is coded in a viral-sense, subgenomic, S RNA sequence [20] that is made only after the initiation of translation and (presumably) RNA replication [21]. The NS_S mRNA species corresponds to the 5' half of the viral RNA. The phlebovirus M RNA species codes for the glycoprotein precursor in a viral-complementary mRNA species [22]. The coding arrangement in the phlebovirus L RNA is not known, but presumably it codes for the virion transcriptase/replicase.

Appropriate hybridization analyses using single-stranded DNA probes have shown that the phlebovirus NS_S mRNA species is produced in much lower yields than the N mRNA (or viral S RNA). These data are interpreted to indicate that the two S mRNA syntheses are not coordinate and that the molar requirements for NS_S protein are low.

The ambisense strategy of the phlebovirus S RNA is in marked contrast to that of the bunyavirus S RNA. For the latter, the NS_S protein can be made at any time during the course of infection. For the former, viral replication must occur before the phlebovirus NS_S protein can be made. The functions of the two NS_S proteins are not known. Whether they have similar roles, and whether the bunyavirus NS_S protein is in fact required before RNA replication occurs, are questions that have not been answered.

The Perceived Hierarchy of Evolutionary Relationships in the Bunyaviridae

In view of the differences of strategy of coding in the S RNA of bunyaviruses and phleboviruses, should the viruses be considered members of the same virus family? All members of the Bunyaviridae have similar structural characteristics and mature in the Golgi apparatus of infected cells [1]. The reason and molecular basis for this site of maturation are not known. The viruses have three species of RNA: The L RNA has been deduced to code for the transcriptase/replicase; the M RNA, for the glycoprotein precursor; and the S RNA, for N protein plus a nonstructural protein [14]. For bunyaviruses and phleboviruses, evidence has been obtained for similar mRNA transcription mechanisms involving the acquisition of primers from (presumably) cellular mRNA species [18,21]. Taken together, these observations support the assignment of bunyaviruses and phleboviruses to the same family. It can be argued that none of these characteristics is sufficient to distinguish virus families. In other virus families [1], there are different morphological forms even for the same virus species (e.g., filamentous and spherical virus particles of myxoviruses). Different morphological structures and morphogenetic processes have been described for members of different genera of reoviruses (e.g., rotaviruses and reoviruses). Also the number of RNA segments and structural proteins may vary for different members of a virus family (e.g., orthomyxoviruses, reoviruses). If one were to choose one character of viruses that is likely to be more conserved than another, then it would surely be the processes of RNA transcription and replication that are employed by viruses (e.g., the use of primers, VPg, de novo mRNA transcription initiation, etc.).

In order to indicate the evolutionary significance of the ambisense strategy of phleboviruses (and uukuviruses?), it has been suggested [14] that they be categorized to a separate subfamily (i.e., Phlebovirinae) by comparison to the bunyaviruses (subfamily: Bunyavirinae). If uukuviruses have a similar S RNA ambisense coding strategy to that of phleboviruses, then the Phlebovirinae subfamily can have two component genera *(Phlebovirus, Uukuvirus)*. Whether hantaviruses and nairoviruses will resemble bunyaviruses, or phleboviruses, or exhibit other coding strategies, remains to be determined.

The Origins of Viruses with Ambisense RNA

Finally, a word on the question of how viruses with an ambisense coding strategy may have arisen. Unless one invokes an origin from a DNA source in which the arrangement of proteins coded on opposite strands of nucleic acid has been maintained, the simplest explanation is that a chimeric RNA was derived at some stage in virus evolution. Such a chimeric RNA could have been formed during RNA replication and could represent a consolidation of genetic information (i.e., a virus with four RNA species producing a phlebovirus with three RNAs, or a virus with three RNAs giving rise to an arenavirus with only two RNAs). If this occurred, then one might have to argue that in bunyaviruses the gene was lost. Such consolidation of genetic information could occur if a viral replicase copied the coding strand of one RNA species and then, instead of terminating, continued RNA synthesis on the noncoding strand of another. This would result in a chimeric RNA composed of two genes coded on opposite strands of the RNA. A variation of this idea is the incorporation of new genes from other sources (RNA species derived from cells or other viruses or other parasites). Alternatively, the replicase could, after copying the coding species of a viral RNA, backcopy the RNA it had made (or another similar species). This would result in gene duplication that through evolution would allow one of the genes to evolve into a new gene product (possibly involving a pseudogene intermediate), without loss of the original gene function. Whatever the origins, the ambisense coding arrangements that have been observed for arenaviruses and phleboviruses open yet another dimension to the way in which viruses replicate in cells.

References

1. Matthews REF (1982) Classification and nomenclature of viruses. Intervirol 17:1–200
2. Rawls WE, Leung W-C (1979) Arenaviruses. *In* Fraenkel-Conrat H, Wagner RR (eds) Comprehensive Virology vol 14. Plenum Press, New York, p 157
3. Bishop DHL, Compans RW (1985) Biochemistry of arenaviruses. Curr Top Microbiol Immunol 114:153–175
4. Casals J (1975) Arenaviruses. Yale J Biol Med 48:115–140
5. Oldstone MBA, Ahmed R, Buchmeier MJ, Blount P, Tishon A (1985) Perturbation of differentiated functions during viral infection in vivo 1. Relationship of lymphocytic choriomeningitis virus and host strains to growth hormone deficiency. Virology 142:158–174
6. Rawls WE, Chan MA, Gee SR (1981) Mechanisms of persistence in arenaviruses infections: A brief review. Can J Microbiol 27:568–574
7. Leung W-C, Rawls WE (1977) Virion-associated ribosomes are not required for the replication of Pichinde virus. Virology 81:174–176
8. Vezza AC, Bishop DHL (1977) Recombination between temperature-sensitive mutants of the arenavirus Pichinde. J Virol 24:712–715

9. Romanowski V, Bishop DHL (1983) The formation of arenaviruses that are genetically diploid. Virology 126:87–95
10. Vezza AC, Cash P, Jahrling P, Eddy G, Bishop DHL (1980) Arenavirus recombination: The formation of recombinants between prototype Pichinde and Pichinde Munchique viruses and evidence that arenavirus S RNA codes for N polypeptide. Virology 106:250–260
11. Harnish DG, Dimock K, Bishop DHL, Rawls WE (1983) Gene mapping in Pichinde virus: Assignment of viral polypeptides to genomic L and S RNAs. J Virol 46:638–641
12. Auperin DD, Romanowski V, Galinski M, Bishop DHL (1984) Sequencing studies of Pichinde arenavirus S RNA indicate a novel coding strategy, an ambisense viral S RNA. J Virol 52:897–904
13. Romanowski V, Matsuura Y, Bishop DHL (1985) The complete sequence of the S RNA of lymphocytic choriomeningitis virus (WE strain) compared to that of Pichinde virus. Virus Res 3 (in press)
14. Bishop DHL (1985) The genetic basis for describing viruses as species. Intervirology (in press)
15. Bishop DHL, Gould KG, Akashi H, Clerx-van Haaster CM (1982) The complete sequence and coding content of snowshoe hare bunyavirus small (S) viral RNA species. Nucleic Acids Res 10:3703–3713
16. Eshita Y, Bishop DHL (1984) The complete sequence of the M RNA of snowshoe hare bunyavirus reveals the presence of internal hydrophobic domains in the viral glycoproteins Virology 137:227–240
17. Clerx-van Haaster CM, Akashi H, Auperin DD, Bishop DHL (1982) Nucleotide sequence analyses and predicted coding of bunyavirus genome RNA species. J Virol 41:119–128
18. Eshita Y, Ericson B, Romanowski V. Bishop DHL (1985) Analyses of the mRNA transcription processes of snowshoe hare bunyavirus S and M RNA species. J Virol (in press)
19. Bouloy M, Colbere F, Krams-Ozden S, Vialat P, Garapin AC, Hannoun C (1975) Activité RNA polymerasique associée à un Bunyavirus (Lumbo). C R Séances Acad Sci 280D:213–215
20. Ihara T, Akashi H, Bishop DHL (1984) Novel coding strategy (ambisense genomic RNA) reveled by sequence analyses of Punta Toro phlebovirus S RNA. Virology 136:293–306
21. Ihara T, Matsuura Y, Bishop DHL (1985) Analyses of the mRNA transcription processes of Punta Toro phlebovirus (Bunyaviridae). Virology (submitted)
22. Ihara T, Smith J, Dalrymple JM, Bishop DHL (1985) Complete sequences of the glycoprotein and M RNA of Punta Toro phlebovirus compared to those of Rift valley fever virus. Virology 144:246–259

CHAPTER 5
Human Respiratory Syncytial Virus Genome and Gene Products

PETER L. COLLINS AND GAIL W. WERTZ

Human respiratory syncytial (RS) virus is recognized as the major cause of serious lower respiratory tract disease in young children [1,2]. In 1978, it was classified in the genus *Pneumovirus* of the Paramyxoviridae family with the proviso that because so little was known about the virus at that time, future information might alter this classification [3].

In past years, the poor growth in cell culture and virion instability of respiratory syncytial virus severely hindered efforts to obtain a definitive description of the genome structure and to identify all of the gene products of the virus. Recently, however, as a result of cDNA cloning of the mRNAs of RS virus, a complete catalog of the genes and gene products and a transcriptional map of the viral genome have been generated [4,5]. This work has shown that RS virus differs from other paramyxoviruses in (a) its number of genes (it has 10, as compared to 6 or 7 for the other paramyxoviruses); (b) its gene order; and (c) its possession of several proteins that have no direct counterparts among the other paramyxoviruses. In particular, the major glycoprotein G of RS virus which has neither hemagglutinin nor neuraminidase activities, is distinct from any previously described negative-strand virus glycoprotein [6].

Respiratory Syncytial Virus RNAs

The genome of RS virus is a single negative-sense strand of RNA, having a molecular weight of 5.6×10^6 [7]. cDNA clones of the mRNAs from RS virus-infected cells were prepared and used as probes in Northern blot analysis of infected cell mRNAs to show that there are 10 unique RS virus mRNAs in infected cells [4,8]. Hybridization-selection and in vitro translation of in-

Table 5.1. Human respiratory syncytial virus mRNAs and proteins.

mRNA[a]	Protein encoded[b]	Amino acids[c]	Mature MW $\times 10^3$ daltons[d]	Modification[e]	Location[f]
7	L	(ND)	~200	—	nucleocapsid
5	F_0	574	68–70	glycosylated F_0 cleaved to F_1 and F_2	membrane
4	N	391	42.6	—	nucleocapsid
3b	22K	194	22	—	envelope associated
3a	M	256	28.7	—	envelope associated
2b	G	298	84–90	glycosylated	membrane
2a	P	241	27.1	phosphorylated	nucleocapsid
1c	1c	139	15.5	—	nonstructural
1b	1b	124	14.6	—	nonstructural
1a	1a	64	7.5	—	virion (?)

[a]Refs. 4, 8; [b]Ref. 8; [c]Based on sequence analysis, 6, 9–16; [d]By gel electrophoresis (see Ref. 17) or calculated from sequence data in (c) above; [e]Ref. 17; [f]Ref. 17.

dividual mRNAs determined the polypeptide coding assignments for each mRNA and showed that the RS virus genome coded for 10 unique polypeptides [8] (Table 5.1). In addition, cDNAs of 9 of the 10 RS virus genes have now been sequenced, providing detailed information about the encoded proteins and indicating that each gene codes for a single polypeptide [6,9–16].

Conserved Sequences

Comparison of the sequences at the 5' and 3' ends of the RS virus mRNAs has identified conserved sequences [9]. The sequence 5' GGGGCAAAU . . . 3' is present at the 5' termini of all mRNAs, and the sequence 5' AGU$\overset{A}{U}$A(N)$_{0-2}$ $\overset{A}{U}$ polyA is present at the 3' end of the RSV mRNAs. Conserved sequences have also been noted at the 5' and 3' termini of the Sendai and vesicular stomatitis virus (VSV) mRNAs.

Proteins of Respiratory Syncytial Virus

Eight RS virus-specific proteins are found associated with purified RS virions; the remaining two proteins, 1B and 1C, specified by the RS virus genome are nonstructural proteins and are found only in infected cells [17]. In the virus particle, the negative-stranded RNA is found in the form of a nucleocapsid structure, having associated with it 3 of the 10 viral proteins: the nucleocapsid protein, N (391 amino acids); the phosphoprotein, P (241 amino acids); and the presumed RNA-dependent RNA polymerase, L [14,15,17].

The viral nucleocapsid structure is surrounded by a membrane which has four viral proteins associated with it [17]: these are the matrix protein, M (256 amino acids); a 22-kilodalton (K) protein (194 amino acids); the fusion protein, F (574 amino acids); and the major glycoprotein, G (298 amino acids) [6,9,12,13,16]. The M and 22K proteins have been shown by detergent and salt-stripping studies to be removed from the envelope in a similar fashion; these proteins are thought to underlie the lipid bilayer as extrinsic membrane matrix proteins [17]. The characterization of the 22K protein of RS virus as a fourth membrane-associated protein identified it as a novel type of membrane protein without counterpart among the other paramyxoviruses. Finally, the 1A protein (64 amino acids) has been found associated, in low quantities, with purified virions; its location is unknown [10,17].

The F and G proteins are the major surface glycoproteins of RS virions. The complete nucleotide sequences for these two important viral antigens have recently been determined [6,9,16]. The F mRNA is 1899 nucleotides long and has a single open reading frame which codes for a protein of 574 amino acids with a calculated molecular weight of 63,453. The major structural features of the F protein include an amino terminal signal sequence (residues 1–22), a hydrophobic carboxy-terminal anchor sequence (residues 525–550), five potential acceptor sites for N-linked carbohydrate chains, and an internal site (residues 131–136) for the proteolytic cleavage that generates the disulfide-linked F_1 and F_2 subunits [9,16]. Thus, the RS virus fusion protein has overall structural similarity to other paramyxovirus fusion proteins [18].

In contrast to F, the G protein of RS virus has little or no similarity to the second glycoprotein, the HN protein, of other paramyxoviruses [6,19]. Indeed, the RS virus G protein appears to be a protein that is distinct from any viral glycoprotein described to date. The mature G protein has a molecular weight of 84,000 to 90,000; it lacks both hemagglutinin and neuraminidase activities [6]. The sequence of the G mRNA is 918 nucleotides long and contains a single open reading frame that encodes a polypeptide of 298 amino acids with a calculated molecular weight of 32,500, a finding that agrees with the molecular weight of 36,000 estimated for the in vitro translation product of the G mRNA [6]. This suggests that approximately 60% of the molecular weight of the mature glycoprotein is contributed by carbohydrate. The majority of the carbohydrate of G is resistant to the action of tunicamycin, suggesting that the majority of the carbohydrate residues are attached via O-glycosidic bonds [6,20]. Consistent with this observation is the fact that 30.6% of the amino acid sequence of G is composed of the hydroxyamino acids serine and threonine, which are known to be the sites of O-glycosylation [6]. Because O-linked carbohydrate chains are frequently small, it is estimated that in order to achieve 60% of the molecular weight of the mature G, 40–80 O-linked carbohydrate side chains might be required to be attached to the G protein. Therefore, the high proportion of serine and threonine residues is consistent with the extensive glycosylation in G. In addition, there are four potential sites for the attachment of N-linked oligosaccharides, and available evidence using inhibitors of glycosylation indicate that at least two of these sites may be used [6,20].

Other remarkable features of the G glycoprotein were revealed by hydrophobicity analyses of the G protein amino acid sequence. These studies revealed no N-terminal signal sequence or C-terminal anchor sequence. Instead, the G protein has a single hydrophobic region between amino acid residues 38 and 66. This region is postulated to serve as the transmembrane anchor region of G; its location predicts that the C terminus of G is located on the external side of the membrane [6]. Consistent with this, 77 of the 90 potential carbohydrate attachment sites are located in this C-terminal portion of the molecule. This orientation would resemble that of the SV5 HN protein [19]. However, unlike that protein, the mature RS virus G protein contains greater than 60% of its molecular weight as carbohydrate, the majority of which is O-linked. Taken together, the combination of features described above suggests that the G protein is unique among viral glycoproteins described to date. This unusual structure and the extensive glycosylation of the G protein may play a major role in the unusual pathogenesis of RS virus disease.

Genome Organization

A transcriptional map for RS virus has been obtained by two independent methods: (a) size and sequence analyses of polycistronic read-through RNAs [4], and (b) UV transcriptional mapping [5]. Identical results have been obtained by both methods.

The UV mapping results also have provided evidence that a single promoter exists for RS virus transcription and that the order of transcription proceeding from the 3' end of the genome is 3' 1C-1B-N-P-M-1A-G-F-22K-L 5' [5]. These results show that in contrast to other negative-stranded RNA viruses, the nucleocapsid gene, N, is not first in the genome order of RS virus, but is preceded by two small nonstructural protein genes, 1C and 1B. In addition, the gene order is interrupted by the insertion of the 1A gene between the M and G genes and the 22K gene between the F and L genes.

With the availability of complete sequences for nine mRNAs, specific oligonucleotides have been synthesized, hybridized to genomic RNA, and used to direct dideoxynucleotide sequencing across intergenic regions (Collins et al.; Dickens and Wertz, unpublished data). These sequencing studies provided evidence for the physical order of genes and showed that the nine viral genes were nonoverlapping and arranged in the order 3' 1C-1B-N-P-M-1A-G-F-22K 5', the same as that observed for the transcriptional map [5]. In addition, these data also showed that the intergenic regions of RS virus vary in length (from 1 to 52 nucleotides) and do not have detectable sequence conservation. Thus, this information indicates that the transcriptional order of genes is the same as the physical order of genes. A diagram of the gene order of RS virus is presented in Figure 5.1.

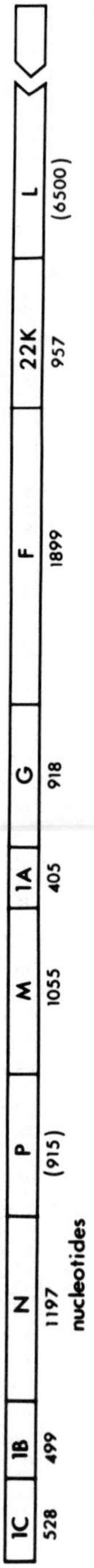

Figure 5.1. Gene order for human respiratory syncytial virus.

Summary and Discussion

RS virus contains 10 genes encoding 10 unique proteins. Although the virus is similar to other paramyxoviruses in overall nucleocapsid structure, mode of transcription, requirement for protein synthesis in replication, and structure and function of its fusion protein, this virus shows marked differences in (a) the number of genes; (b) the organization of these genes on the genome; (c) sequences at intergenic regions; (d) structure of the major glycoprotein, G; and (e) the identification of two independently coded nonstructural polypeptides. The recent advances made in identifying the gene products of RS virus and availability now of cDNA clones and complete sequence information for 9 of the 10 RS virus genes afford numerous new avenues to investigate the roles of each of the RS virus gene products in infection and pathogenesis of this unusual virus.

For example, antibodies raised against synthetic peptides prepared from the predicted protein sequences can be used to identify intracellular locations, membrane orientations, and possible intracellular interactions among the viral proteins. It may be possible to identify lethal or conditional lethal viral mutants by complementation with cDNAs engineered for expression. cDNAs expressed by vectors, such as live vaccinia virus [21], can be used to introduce viral antigens singly and in combination into cell cultures or live animal model systems in order to assess the contributions of viral antigens to the immune response to the virus. And finally, using available sequences to prepare oligonucleotide primers will make possible rapid sequencing of the glycoprotein genes of naturally occurring and antigenic variant strains in order to identify regions important to antibody binding and to determine active sites.

References

1. Chanock RM, Kim HW, Brandt CD, Parrott RH (1982) *In* Evans (ed) Viral Infections of Humans: Epidemiology and Control. Plenum Publishing, New York, pp 471–489
2. Stott EJ, Taylor G (1985) Respiratory syncytial virus: Brief review. Arch Virol 84: 1–59
3. Kingsbury DW, Bratt MA, Choppin PW, Hanson RP, Hosaka Y, ter Meulen V, Norrby E, Plowright W, Rott R, Wunner WH (1978) Intervirology 10:137–152
4. Collins PL, Wertz GW (1983) cDNA cloning and transcriptional mapping of the nine mRNAs encoded by the genome of human respiratory syncytial virus. Proc Natl Acad Sci USA 80:3208–3212
5. Dickens LE, Collins PL, Wertz GW (1984) Transcriptional mapping of human respiratory syncytial virus. J Virol 52:364–369
6. Wertz GW, Collins PL, Huang YT, Gruber G, Levine S. Ball LA (1985) Nucleotide sequence of the G protein of human respiratory syncytial virus reveals an unusual type of viral membrane protein. Proc Natl Acad Sci USA 82:4075–4079

7. Huang Y, Wertz GW (1982] The genome of respiratory syncytial virus is a negative stranded RNA that codes for at least seven messenger RNA species. J Virol 43:150–157

8. Collins PL, Huang YT, Wertz GW (1984) Identification of a tenth mRNA of respiratory syncytial virus and assignment of polypeptides to the 10 viral genes. J Virol 49:572–578

9. Collins PL, Huang YT, Wertz GW (1984) Nucleotide sequence of the gene encoding the fusion (F) glycoprotein of human respiratory syncytial virus. Proc Natl Acad Sci USA 81:7683–7687

10. Collins PL, Wertz GW (1985) The 1A protein gene of human respiratory syncytial virus: Nucleotide sequence of the mRNA and a related polycistronic transcript. Virology 141:283–291

11. Collins PL, Wertz GW (1985) Nucleotide sequences of the 1B and 1C nonstructural protein mRNAs of human respiratory syncytial virus. Virology 143:442–451

12. Collins PL, Wertz GW (1985) The envelope-associated 22K protein of human respiratory syncytial (RS) virus: Nucleotide sequence of the mRNA and a related polytranscript. J Virol 54:65–71

13. Satake M, Venkatesan S (1984) Nucleotide sequence of the gene encoding the respiratory syncytial virus matrix protein. J Virol 50:92–99

14. Satake M, Elango N, Venkatesan S (1984) Sequence analysis of the respiratory syncytial virus phosphoprotein gene. J Virol 52:991–994

15. Collins PL, Anderson K, Langer SJ, Wertz GW (1985) Correct sequence for the major nucleocapsid (N) protein mRNA for respiratory syncytial virus. Virology 146:69–77

16. Elango N, Satake M, Coligan J, Norrby E, Camargo E, Venkatesan S (1985) Nucleic Acids Res 13:1559–1574

17. Huang YT, Collins PL, Wertz GW (1985) Characterization of the ten proteins of human respiratory syncytial virus: Identification of a fourth envelope associated protein. Virus Res 2:157–173

18. Choppin PW, Scheid A (1980) The role of viral glycoproteins in adsorption, penetration and pathogenicity of viruses. Rev Infect Dis 2:40–61

19. Hiebert S, Paterson R, Lamb RA (1985) Hemagglutinin–neuraminidase protein of the paramyxovirus simian virus 5: Nucleotide sequence of the mRNA predicts an N-terminal membrane anchor. J Virol 54:1–6

20. Gruber C, Levine S (1985) Respiratory syncytial virus polypeptides. IV. The oligosaccharides of the glycoproteins. J Gen Virol 66:417–432

21. Ball LA, Young KK, Anderson K, Collins P, Wertz G (1986) Expression of the major glycoprotein G of human respiratory syncytial virus from recombinant vaccinia virus vectors, Prod Natl Acad Sci USA 83:246–250

Viral Constructs

CHAPTER 6
Bovine Papillomavirus DNA Vectors

NAVA SARVER AND PETER M. HOWLEY

Viral vectors have proven to be important tools in recombinant DNA research for the introduction and expression of foreign genes in mammalian cells. Several animal viruses, (including SV40, adenovirus, retroviruses, parvoviruses, vaccinia virus, and baculoviruses) have now been successfully employed for this purpose. The use of the bovine papillomavirus (BPV) DNA as a viral vector is also well documented and is the subject of this review.

BPV is capable of transforming certain mammalian cells including rat, bovine, hamster, and mouse cell lines in culture [1,2]. The viral genome is maintained in such cells as a stable, multicopy extrachromosomal plasmid [3]. The following features describe several properties of the BPV vector system:

1. The plasmid DNA can be readily separated from the host chromosomal DNA by using physical means. If the plasmid contains a bacterial origin of replication and a selective marker, it can be rescued directly in bacterial cells.
2. The extrachromosomal plasmid state of the vector eliminates potential problems associated with the integration of the vector into inactive regions of the host chromosome or with insertion of the DNA into critical regions of the cellular genome.
3. The uniform extrachromosomal sequence environment of the BPV minichromosomes should decrease the cell-to-cell variation in expression levels of exogenous genes cloned in these plasmids.
4. The multicopy mode of plasmid replication provides for natural amplification of cloned genes and, consequently, the possibility of higher levels of expression.

5. The ability of BPV DNA to transform susceptible rodent cells provides a selective phenotype for detecting cells harboring the viral vector and potentially provides a permanent genetically engineered cell line containing, and possibly expressing, the exogenous gene of interest.

To date, no tissue culture system has been identified that can support the productive growth of the virus in vitro. Consequently, there is a paucity of information bearing on the genetic structure and biology of this group of viruses. Nonetheless, molecular cloning [4], DNA sequence analysis [5], and transcription analyses [6,7] have provided the necessary information for a functional mapping of the viral genome.

Genomic Organization

The viral genome (7945 base pairs, bp) is functionally divided into two regions. The transforming region consists of 69% of the genome (BPV_{69T}) and contains all the sequence information required for transformation[8] and stable plasmid replication [3]. This is the only region actively transcribed in transformed cells. At the 5' end of BPV_{69T} is a 1000-bp region, referred to as the noncoding region (NCR), which contains an origin for DNA replication [9], a DNase-hypersensitive region [10], and transcription regulatory elements [11–13]. The 3' end contains the polyadenylation site for the early transcripts and a transcriptional enhancer which, like other viral and cellular enhancers, can activate heterologous promoters in a position- and orientation-independent manner [14]. As will be discussed below, this distal enhancer is not required for the expression of the viral genes involved in cellular transformation or plasmid maintenance. Recently, a second viral enhancer has been mapped to the 5' promoter region of BPV_{69T}. Unlike the 3' distal enhancer, the 5' enhancer is trans-activated by the E2 viral gene product [15].

Only one strand of the BPV1 genome is transcribed in transformed cells and in the productively infected cells of a bovine fibropapilloma [12, 16, 17]. Eight open reading frames (ORFs) are present in BPV_{69T}, and transcripts mapping to this region are generated by differential splicing [6,7]. There are two independent transforming genes in BPV [7,13,18]. Each of these genes, when inserted behind a strong heterologous promoter, can independently induce transformation of mouse cells [7,18]. These transforming functions have now been mapped to the 5' end of the E6 ORF [7,18] and to the 3' end of the E5 ORF ([19,20], Groff and Lancaster, unpublished results; Schiller and Lowy, unpublished results). Mutations in the E1 ORF result in the integration of the viral genome into the host chromosome, indicating the requirement for the E1 gene product in extrachromosomal replication [13,21]. Mutations in the E7 ORF result in a lower copy number of the viral plasmids, indicating that it encodes for a function that controls plasmid copy number [21].

The remaining 31% of the genome (BPV_{31NT}) contains the two ORFs (L2 and L1) that are expressed only in differentiating epithelial cells of productive fibropapillomas [2,16]. BPV_{31NT} is, therefore, not required for either transformation or plasmid replication and is not transcribed in transformed cells. However, its presence stabilizes the plasmid in mammalian cells and enhances its ability to transform susceptible cells [22].

Inhibition of BPV DNA Replication by Prokaryotic Sequences

Initial studies using BPV DNA cloned in pBR322 demonstrated a cis inhibition by the prokaryotic DNA on BPV DNA-mediated transformation [8]. This necessitated the removal of pBR322 sequences prior to transformation, which in turn, precluded the shuttling of the DNA between mammalian and bacterial cells. The plasmid pML2 [23] is a derivative of pBR322 that has lost the sequences inhibitory for complete BPV DNA-mediated transformation. Indeed, the complete BPV genome cloned in pML2 can replicate as a stable plasmid in either bacterial or mouse cells [22]. It is noteworthy that pML2 is not absolutely inert when linked to BPV DNA sequences. This point is exemplified by the fact that pML2 sequences are still cis-inhibitory to DNA-mediated transformation by the 69% subgenomic region of BPV, a fragment that is by itself sufficient for transformation [22]. Another example of cis inhibition by pML2 sequences has recently been reported, in which the complete BPV genome deleted only of the distal 3′ enhancer and linked to pML2 was impaired in its ability to transform mouse cells [24]. The same enhancer-deleted viral DNA fragment that was physically separated from pML2 was as efficient in inducing cellular transformation as the intact viral genome.

Several other BPV shuttle vectors have also been described [25–27] each consisting of PBV_{69T}, a prokaryotic DNA origin of replication, and an antibiotic-selective marker. In addition, each of these vectors contains a DNA segment such as the human β-globin gene or BPV_{31NT} which, by a yet undefined mechanism, stabilizes the plasmid and facilitates its ability to transform and to be expressed in mammalian cells.

Foreign Genes: Expression and Regulation

In the first report demonstrating the efficacy of BPV DNA as a mammalian vector [28], the complete rat preproinsulin gene I was linked to BPV_{69T} and introduced into mouse C127 cells via calcium phosphate coprecipitation. The hybrid DNA in mouse cells selected by virtue of their transformed phenotype persisted as a plasmid and was identical in its physical structure to that of the input DNA. Rat preproinsulin-specific transcripts were properly spliced and had authentic 5' and 3' termini. Posttranslational cleavage of the pre-prohormone to the proinsulin had also occurred faithfully in the engineered cells, and the proinsulin was secreted into the cultured medium.

A variety of eukaryotic and prokaryotic gene products destined for either an extracellular or an intracellular location have now been successfully expressed in rodent cells using BPV vectors. Additionally, several inducible genes have been assessed for their potential to be properly regulated. The mouse metallothionein genes when maintained in a BPV plasmid, can be induced by the heavy metals cadmium and zinc [26,29]. It should be noted that although these genes are also induced in vivo with glucocorticoids, such induction is not observed in the BPV system. In contrast, the mouse mammary tumor virus (MMTV), which contains a glucocorticoid-responsive element in its long terminal repeat (LTR), is properly regulated with steroids when that element is maintained as part of a BPV vector [30]. The regulation of human β-interferon (IFNβ) expression by viral infection and polyribinosomic acid treatment has also been demonstrated in mouse cells transformed by recombinant BPV DNA [31,32].

Adaptation of the BPV System to a Large-scale Protein Production

Recently, the utility of a BPV vector for the continuous expression of human IFNγ was demonstrated [33]. Human IFNγ-engineered C127 cells secreted more than 10^5 U/ml in 24 hr per 10^5 cells, a rate that was sustained by a single culture for at least 85 days. Large-scale purification of the protein resulted in 77% recovery of highly purified IFNγ ($>10^8$ U/mg protein). Greater than 80% of the recombinant protein was found extracellularly, indicating efficient transport across the cell membrane. The recombinant protein remained biologically stable in the culture medium during five days of continuous incubation, an observation of particular importance in the economical, large-scale production of recombinant proteins.

The recombinant protein produced was biologically indistinguishable from natural IFNγ, in that it conferred antiviral protection on appropriate host cells and induced the expression of class II major histocompatibility (MHC) antigens in both HeLa and endothelial cells. The recombinant product was neutralized by anti-human IFNγ serum. Its identity with authentic human IFNγ was further illustrated by the fact that a monoclonal antibody that recognizes natural IFNγ isolated from induced peripheral blood leukocytes, but not bacterially produced recombinant human IFNγ, completely neutralized the IFNγ obtained from the engineered C127 cells.

Interestingly, isolated foci of transformed cells harboring the recombinant DNA manifested considerable heterogeneity in the level of IFNγ expression. This contrasts sharply with the more homogeneous levels of expression ob-

served with single-cell clones that were derived from a single focus of transformed cells. Because neither the copy number nor the extent of molecular
rearrangements appeared to be a contributing factor, the observed heterogeneity may depend on a yet unidentified variation in the host cells.

Gene Amplification

Another practical innovation to allow increased levels of foreign gene
expression that is presently being explored is the use of gene amplification
(G. Pavlakis, personal communication). A BPV–human-growth-hormone recombinant DNA containing the metallothionein gene was introduced into
mouse cells, and the cells were treated with a stepwise-increasing concentration of zinc (20–80 μM). Cells resistant to high levels of the heavy metal
were shown to contain as many as 1000 copies of the recombinant plasmid.
Many of these cells also secreted elevated levels of the growth hormone,
suggesting that one way of optimizing foreign gene expression may be at the
level of gene copy number.

Positional Dependence of Gene Expression

We have recently addressed the question of whether the position occupied
by the foreign gene inserted in a BPV vector affects the level of gene expression [34]. Various BPV shuttle vectors were constructed, all consisting of
the complete viral genome cloned in pML2d, and the rat preproinsulin gene
I (rI_1) inserted at either of the BPV/pML2d junctions in the two possible
transcriptional orientations. Thus, the gene was positioned either adjacent
to or away from the viral enhancer located at the 3' end of the transforming
region. Transformants containing the rI_1 gene adjacent to the viral enhancer
expressed the gene, whereas cells with the insert away from the element did
not. There were no consistent differences in the plasmid copy number or in
the extent of genomic rearrangements that could account for the difference
in insulin expression. We concluded that the expression of the rI_1 (which is
normally expressed in a tissue-specific manner in pancreatic islet cells) was
due to transcriptional activation by the BPV-1 viral enhancer element and
that the intervening BPV or pML2d sequences can block this enhancer-mediated activation. In agreement with enhancer-dependent activation, a rat
preproinsulin gene located in a blocked position (i.e., not adjacent to the
BPV 3' distal enhancer) could be activated by a DNA fragment containing
the SV40, murine sarcoma virus (MSV), or BPV enhancer element when
inserted adjacent to the rI_1 gene. Thus, a gene that is not normally expressed
in a particular cell may be expressed in a BPV vector when the gene is
placed adjacent to a viral enhancer.

Size Considerations

Because the BPV system does not result in a productive infection, a packaging constraint on the size of a DNA is not a problem. In general, recombinant plasmids ranging from 7.2 to 17 kb in size can be maintained in transformed cells. However, DiMaio et al. [35] reported that a 24-kb BPV–human–HLA recombinant DNA was integrated into the host chromosome. Whether this observation reflects a size constraint on extrachromosomal replication or the involvement of certain sequences in promoting integration is not known. To minimize the overall size of the hybrid molecules, we have designed several vectors truncated in sequences derived from the nontransforming BPV_{31NT} region. These plasmids are indistinguishable from full-length BPV vectors in their transforming and plasmid replication properties.

Another approach is to limit the BPV elements in the plasmid to those required in cis for plasmid maintenance. The chimeric DNA is then introduced into a cell line providing all the plasmid maintenance functions that are required in trans. Two independent plasmid maintenance sequences have now been identified in the BPV genome and are being tested for their potential use as replicons [36].

Dominant Selection

A significant development in the BPV system has been the incorporation into the hybrid DNA molecule of a dominant selective marker, most notably, that of the aminoglycoside phosphotransferase gene of TN5 [37,38]. This gene confers resistance to G418 (an aminoglycoside analogue of neomycin) and potentially expands the range of cells capable of supporting BPV plasmid replication beyond those capable of manifesting the transformed phenotype. Additionally, a dominant selectible marker should be instrumental in the development of a nontransforming, replication-competent BPV vector. To this end, a BPV replicon containing the neomycin-resistant gene was recently described [21]. Once fully developed, such vectors should be suitable for studies of gene expression in differentiated cells whose specialized functions or desired growth properties might otherwise be compromised when a transforming vector is used. The removal of the viral transforming function should also permit study of other transforming genes cloned into a BPV replicon.

References

1. Boiron J, Levy JP, Thomas M, Friedman JC, Bernard J (1964) Some properties of bovine papillomavirus. Nature 201:423–424
2. Dvoretzky I, Shober R, Lowy DR (1980) Focus assay in mouse cells for bovine papillomavirus. Virology 103:369–375
3. Law M-F, Lowy DR, Dvoretzky I, Howley PM (1981) Mouse cells transformed by bovine papillomavirus contain only extrachromosomal viral DNA sequences. Proc Natl Acad Sci USA 78:2727–2731

4. Howley PM, Law M-F, Heilman CA, Engel LW, Alonso MC, Lancaster WD, Israel MA, Lowy DR (1980) Molecular characterization of papillomavirus genomes. *In* Viruses in Naturally Occurring Cancers. Cold Spring Harbor Laboratory, Cold Spring Harbor, New York, pp 233–247

5. Chen EY, Howley PM, Levinson AD, Seeburg PH (1982) The primary structure and genetic organization of the bovine papillomavirus (BPV) type 1. Nature 299:529–534

6. Stenlund A, Zabielski J, Ahola H, Moreno-Lopez J, Petterson U (1985) The messenger RNAs from the transforming region of bovine papillomavirus type 1. J Mol Biol 182:541–554

7. Yang Y-C, Okayama H, Howley PM (1985) Bovine papillomavirus contains multiple transforming genes. Proc Natl Acad Sci USA 82:1030–1034

8. Lowy DR, Dvoretzky I, Shober R, Law M-F, Engel L, Howley PM (1980) In vitro tumorigenic transformation by a defined sub-genomic fragment of bovine papillomavirus DNA. Nature 287:72–74

9. Waldeck W, Rosel F, Zentgraf H (1984) Origin of replication in episomal bovine papillomavirus type 1 DNA isolated from transformed cells. Eur Mol Biol Org J 3:2173–2178

10. Rosl F, Waldeck W, Sauer G (1983) Isolation of episomal bovine papillomavirus chromatin and identification of a DNase I-hypersensitive region. J Virol 46:567–574

11. Ahola H, Stenlund J, Moreno-Lopez J, Petterson U (1983) Sequences of bovine papillomavirus type 1 DNA—functional and evolutionary implications. Nucelc Acids Res 11:2639–2650

12. Heilman CA, Engel L, Lowy DR, Howley PM (1982) Virus specific transcription in bovine papillomavirus transformed mouse cells. Virology 119:22–34

13. Sarver N, Rabson MS, Yang Y-C, Byrne JC, Howley PM (1984) Localization and analysis of bovine papillomavirus type 1 transforming functions. J Virol 52:377–388

14. Lusky M, Berg L, Weiher H, Botchan M (1983) The bovine papillomavirus contains an activator of gene expression at the distal end of the transcriptional unit. Mol Cell Biol 3:1108–1122

15. Spalholz BA, Yang Y-C, Howley PM (1985) Transactivation of a bovine papillomavirus transcriptional regulatory element by the E2 gene product. Cell 42:183–191

16. Antmann E, Sauer G (1982) Bovine papillomavirus transcription: Polyadenylated RNA species and assessment of the direction of transcription. J Virol 43:59–66

17. Engel LW, Heilman CA, Howley PM (1983) Transcriptional organization of the bovine papillomavirus type 1. J Virol 47:516–528

18. Schiller JT, Vass WC, Lowy DR (1984) Identification of a second transforming region in bovine papillomavirus DNA. Proc Natl Acad Sci USA 81:7880–7884

19. DiMaio D, Metherall J, Neary K, Guralski D (1985) Genetic analysis of cell transformation by bovine papillomavirus. *In* Howley, PM, Broker TR (eds) Papilloma Viruses: Molecular and Clinical Aspects. Alan R. Liss, New York pp.437–456

20. Yang Y-C, Spalholz BA, Rabson MS, Howley PM (1985) Dissociation of transforming and transactivation functions for bovine papillomavirus type 1. Nature 318:575–577

21. Lusky M, Botchan M (1985) Genetic analysis of the bovine papillomavirus type 1 transacting replication factors. J Virol 53:955–965

22. Sarver N, Byrne JC, Howley PM (1982) Transformation and replication in mouse cells of a bovine papillomavirus/pML2 plasmid vector that can be rescued in bacteria. Proc Natl Acad Sci USA 79:7147–7151

23. Lusky M, Botchan M (1981) Inhibitory effect of specific pBR322 DNA sequences upon SV40 replication in simian cells. Nature 293:79–81

24. Howley PM, Schenborn ET, Lund E, Byrne JC, Dahlberg JE (1985) The bovine papillomavirus distal "enhancer" is not *cis*-essential for transformation or for plasmid maintenance. Mol Cell Biol 5:3310–3315

25. DiMaio D, Treisman R, Maniatis T (1982) A bovine papillomavirus vector which propagates as an episome in both mouse and bacterial cells. Proc Natl Acad Sci USA 79:4030–4034

26. Karin M, Cathala G, Nguyen-Huu MC (1983) Expression and regulation of a mouse metallothionein gene carried on an autonomously replicating shuttle vector. Proc Natl Acad Sci USA 80:4040–4044

27. Kushner PJ, Levinson BB, Goodman HM (1982) A plasmid that replicates in both mouse and *E. coli* cells. J Mol Appl Genet 1:527–538

28. Sarver N, Gruss P, Law M-F, Khoury G, Howley PM (1981) Bovine papillomavirus DNA—a novel eukaryotic cloning vector. Mol Cell Biol 1:486–496

29. Pavlakis GN, Hamer DH (1983) Regulation of a metallothionein–growth hormone hybrid gene in bovine papilloma virus. Proc Natl Acad Sci USA 80:397–401

30. Ostrowski MC, Richard-Foy H, Wolford RG, Berard DS, Hager GL (1983) Glucocorticoid regulation of transcription at an amplified episomal promoter. Mol Cell Biol 3:2048–2057

31. Mitrani-Rosenbaum S, Maroteaux L, Mory Y, Revel M, Howley PM (1983) Inducible expression of the human interferon β_1 gene linked to a bovine papillomavirus DNA vector and maintained extrachromosomally in mouse cells. Mol Cell Biol 3:233–240

32. Zinn K, Mellon P, Maniatis T (1982) Regulated expression of an extrachromosomal human β-interferon gene in mouse cells. Proc Natl Acad Sci USA 79:4897–4901

33. Sarver N, Ricca GA, Hood M, Link J, Tarr GC, Drohan WN (1985) Sustained high-level expression of recombinant human gamma interferon using a bovine papillomavirus vector. *In* Howley PM, Broker TR (eds) Papillomaviruses: Molecular and Clinical Aspects. Alan R. Liss, New York pp. 515–527

34. Sarver N, Muschel R, Byrne JC, Khoury G, Howley PM (1985) Enhancer dependent expression of the rat preproinsulin gene in BPV-1 vectors. Mol Cell Biol 5:3507–3516

35. DiMaio D, Corbin V, Sibley E, Maniatis T (1984) High-level expression of a cloned HLA heavy chain gene introduced into mouse cells on a bovine papillomavirus vector. Mol Cell Biol 4:340–350

36. Lusky M, Botchan M (1984) Characterization of the bovine papillomavirus plasmid maintenance sequences. Cell 36:391–401

37. Law M-F, Byrne JC, Howley PM (1983) A stable bovine papillomavirus hybrid plasmid that expresses a dominant selective trait. Mol Cell Biol 3:2110–2115

38. Matthias DD, Bernard HU, Scott A, Hashimoto-Gotoh T, Schutz G (1983) A bovine papillomavirus vector with a dominant resistance marker replicates extrachromosomally in mouse and *E. coli* cells. Eur Mol Biol Org J 2: 1487–1492

CHAPTER 7
Genetic Engineering of Vaccinia Virus Vectors: Development of Live Recombinant Vaccines

BERNARD MOSS

Edward Jenner's historic demonstration, that inoculation with material from vaccinia vesicles prevented subsequent smallpox infection, led to the eventual eradication of a dreaded disease. Because the last natural case of smallpox occurred eight years ago, the need for vaccination is over. Recently, however, the heretical notion that vaccinia virus might be used for immunoprophylaxis of other diseases was suggested [1–4]. Simply stated, by isolating a gene (or cDNA) from one microorganism and inserting it into the genome of vaccinia virus, a live recombinant vaccine might be produced. The feasibility of constructing such recombinants and the demonstration of their ability to protectively immunize experimental animals against a variety of viruses were soon demonstrated [5–12]. In this article, I shall outline procedures for using vaccinia virus as a vector, summarize the properties of some recombinant viruses, and consider their potential uses.

Molecular Biology of Vaccinia Virus

Vaccinia, a member of the Poxviridae family, is a large complex virus that replicates in the cytoplasm of infected cells [13]. The genome consists of a 185,000-bp linear duplex DNA molecule with hairpin ends. Enzymes, including a DNA-dependent RNA polymerase, are packaged within the infectious virus particle. Upon its entry into the cell, early genes are transcribed by the viral polymerase, and translation products are soon detected. The latter include a DNA polymerase and other enzymes needed for replication of the genome. After the onset of DNA replication, expression of many early genes stops and expression of late genes begins. Regulation occurs primarily at the transcriptional level, although posttranscriptional mechanisms also may play a role.

Vaccinia Virus as a Vector

Some of the biological features of vaccinia virus provide obstacles to its development as a vector for expression of foreign genes. The large size of the genome makes it necessary to rely on homologous recombination for insertion of foreign DNA. This procedure depends on infrequent recombinational events occurring in infected cells between vaccinia DNA flanking the foreign gene and homologous DNA within the vaccinia genome. For efficient expression of foreign genes, vaccinia virus transcriptional regulatory sequences have to be used. Of course, these alterations of the vaccinia genome should not result in loss of infectivity. To facilitate these procedures, a series of plasmid vectors have been constructed, based on the scheme outlined by Mackett et al. [3]. Such plasmids contain a DNA segment (with a vaccinia virus transcriptional promoter region and one or more restriction endonuclease sites for insertion of a foreign gene) that is flanked by DNA from a nonessential region of the vaccinia genome. The choice of vaccinia promoter determines the time and level of expression, whereas the flanking DNA determines the site of insertion. Insertion into the thymidine kinase TK) locus provides a simple method of selecting recombinant virus that has a TK^- phenotype. In practice, cells are infected with vaccinia virus and transfected with the above plasmid. The virus progeny are then incubated with a monolayer of TK^- cells, and 5-bromodeoxyuridine is added to the agar overlay to prevent plaque formation of TK^+ parental virus. Other screening methods that do not rely on special cell lines also have been developed [14]. Alternatively, DNA hybridization [4] or specific antibody binding [6,8] may be used for primary identification of recombinant virus plaques.

Expression of Foreign Genes by Vaccinia Virus Vectors

Immunoprecipitation with specific antiserum followed by polyacrylamide gel electrophoresis has been used to characterize a variety of polypeptides expressed by vaccinia vectors. These include the hepatitis B virus surface antigen (HBsAg) [5], influenza hemagglutinin (Influ HA) [6], the vesicular stomatitis G protein (VSV G) [10], rabies virus G protein [12], and the herpes simplex virus type 1 glycoprotein D (HSV-1gD) [11]. In addition, all of the above polypeptides were shown to be glycosylated and inserted into the plasma membrane except for HBsAg, which was appropriately secreted as a 22-nm particle.

Immune Response of Vaccinated Animals

Animals vaccinated with live recombinants expressing HBsAg [5,8], Influ HA [6,7], HSV-1 gD [11], rabies virus G [12], VSV G [10], and plasmodial circumsporozoite [13] produce antibodies that react with or neutralize the corresponding virus or protozoa. Furthermore, mice inoculated with vaccinia virus recombinants that express Influ HA or nucleoprotein (N) genes were primed to produced cytotoxic T cells (CTL) [15,16]. In addition, cells infected with vaccinia virus recombinants that express Influ HA or N served as targets for CTLs produced in influenza virus-infected mice. Interestingly, the CTL response directed to HA was subtype specific, whereas the response to N showed cross-specificity.

Protection Against Challenge

Vaccination with live recombinant viruses has been shown to protect, completely or partially, (a) hamsters, against lower respiratory infection with influenza [6]; (b) mice, against lethal intraperitoneal challenge with HSV-1 [8,11] or HSV-2 [11], development of a latent trigeminal infection with HSV-1 [11], lethal intracranial infection with rabies virus [10], or lethal intravenous infection with VSV [10]; (c) cattle against lingual infection with VSV [10]; and (d) chimpanzees against hepatitis caused by intravenous infection with hepatitis B virus [9].

Uses of Vaccinia Virus Recombinants

Vaccinia virus is useful as an expression vector for studying the synthesis and processing of proteins and the immunological response of animals. In particular, these recombinants provide a novel way of determining the target antigens for cell-mediated immunity. In view of the ability to protectively immunize animals against a variety of infectious disease, the use of vaccinia virus recombinants for medical or veterinary purposes may be feasible. The capacity of the virus for at least 25,000 base pairs of additional DNA indicates that there should be no technical obstacles to making polyvalent vaccines [17]. Experience with smallpox immunization suggests that such recombinant vaccines would be economical to produce and simple to administer. The absence of a refrigeration requirement is a feature that would be important if the vaccine were used in undeveloped tropical areas of the world.

References

1. Weir JP, Bajszar G, Moss B (1982) Mapping of the vaccinia virus thymidine kinase gene by market rescue and by cell-free translation of selected mRNA. Proc Natl Acad Sci USA 79:1210–1214
2. Nakano E, Panicali D, Paoletti E (1982) Molecular genetics of vaccinia virus: Demonstration of marker rescue Proc Natl Acad Sci USA 79:1593–1596
3. Mackett M, Smith GL, Moss B (1982) Vaccinia virus: A selectable eukaryotic cloning and expression vector Proc Natl Acad Sci USA 79:7415–7419
4. Panicali D, Paoletti E (1982) Construction of poxviruses as cloning vectors; Insertion of the thymidine kinase gene of herpes simplex virus into the DNA of infectious vaccinia virus. Proc Natl Acad Sci USA 79:4927–4931

5. Smith GL, Mackett M, Moss B (1983) Infectious vaccinia virus recombinants that express hepatitis B virus surface antigen. Nature 302:490–495

6. Smith GL, Murphy BR, Moss B (1983) Construction and characterization of an infectious vaccinia virus recombinant that expresses the influenza hemagglutinin gene and induces resistance to influenza virus infection in hamsters. Proc Natl Acad Sci USA 80:7155–7159

7. Panicali D, Davis SW, Weinberg, RL, Paoletti E (1983) Construction of live vaccines using genetically engineered poxviruses: Biological activity of recombinant vaccinia virus expressing influenza virus hemagglutinin Proc Natl Acad Sci USA 80:5364–5368

8. Paoletti E, Lipinskas BR, Samsonoff C, Mercer S, Panicali D (1984) Construction of live vaccines using genetically engineered poxviruses: Biological activity of vaccinia virus recombinants expressing the hepatitis B virus surface antigen and the herpes simplex virus glycoprotein D. Proc Natl Acad Sci USA 81:193–197

9. Moss B, Smith GL, Gerin JL, Purcell R (1984) Live recombinant vaccinia virus protects chimpanzees against hepatitis B. Nature 311:67–69

10. Mackett M, Yilma TY, Rose JA, Moss B (1985) Vaccinia virus recombinants express vesicular stomatitis virus genes and protectively immunize mice and cattle. Science 227:433–435

11. Cremer MA, Mackett M, Wohlenberg C, Notkins AL, Moss B (1985) Vaccinia virus recombinant expressing herpes simplex type 1 glycoprotein D prevents latent herpes in mice. Science 228:737–739

12. Wiktor TJ, Mcfarlan RI, Reagan KJ, Dietzschold B, Curtis PJ, Wunner NH, Kieny MP, Lathe R, Lecocq J-P, Mackett M, Moss B, Koprowski H (1984) Protection from rabies by a vaccinia virus recombinant containing the rabies virus glycoprotein gene. Proc Natl Acad Sci USA 81:7194–7198

13. Moss B (1985) Replication of poxviruses. *In* Fields BN, Chanock RM, Roizman B (eds) Virology. Raven Press, New York, pp 685–703

14. Chakrabarti S, Brechling K, Moss B (1985) Vaccinia virus expression vector: Co-expression of β-galactosidase provides visual screening of recombinant virus plaques. Mol. Cell. Biol. 5:3403–3409

15. Bennink JR, Yewdell JW, Smith GL, Moller C, Moss B (1984) Recombinant vaccinia virus primes and stimulates influenza virus HA-specific CTL. Nature 311:578–579

16. Yewdell JW, Bennink JR, Smith GL, Moss B (1985) Influenza A virus nucleoprotein is a major target antigen for cross-reactive anti-influenza A virus cytotoxic T lymphocytes. Proc Natl Acad Sci USA 82:1785–1789

17. Smith GL, Moss B (1983) Infectious poxvirus vectors have capacity for at least 25,000 base pairs of foreign DNA. Gene 25:21–28

CHAPTER 8
Viral Enhancer Elements

JOHN BRADY, LIONEL FEIGENBAUM, AND GEORGE KHOURY

In the past five years since their discovery, enhancer elements have been studied intensively in a large number of laboratories (see refs. 1,2). The prototype enhancer element, the 72-bp repeat of SV40, was first shown to be an essential set of sequences required for the efficient transcription of SV40 early genes [3–5]. Although there is some variability in the sequences that we have come to know as enhancers, in general their definition involves the following common properties: They are short sets of nucleotides (50–100 bp in length), often repeated in tandem, which work in concert with the other promoter elements to increase the efficiency of transcription of an associated gene as much as 100–1000 fold. A remarkable feature of enhancer elements is their relative position- and orientation-independence. For example, the SV40 72-bp repeat element will activate transcription from a promoter when located in either orientation, 5' or 3' to the cap site, and several kilobases away. Enhancer elements can generally act on heterologous genes as well as on the natural promoter with which they are associated [6,7]. Finally, a number of enhancer elements show tissue or species specificity, which contributes significantly to the host range of the virus with which they are associated [8–10]. In this review, we discuss the possibility that enhancer specificity not only contributes to the target tissue in which the virus is active, but also frequently dictates the disease potential of the viral agent.

Enhancer Elements in the Host Range of Papovaviruses

The enhancer elements of both simian virus 40 (SV40) and polyoma are located between the early and late coding sequences [11]. This transcriptional regulatory sequence is absolutely required for early gene expression in vivo [3–5], and thus is important for manifestation of the viral lytic cycle and tranformation of nonpermissive cells. At the transcriptional level, the enhancer elements of both SV40 and polyoma virus are somewhat promiscuous and function at reasonable levels in most cells except undifferentiated cells ([12–15]; see below). For example, when the SV40 early promoter and enhancer are linked to the bacterial chloramphenicol acetyltransferase (CAT) gene and transfected into monkey kidney cells and mouse cells, the efficiency of enhancer-dependent gene expression is about fivefold greater in the monkey kidney cell line [8]. The decreased efficiency of SV40 early transcription in mouse cells, and thus T-antigen expression, is not, however, the sole determinant of the restricted host cell specificity of SV40. Graessman has demonstrated in microinjection experiments that although elevation of DNA template number allows production of late viral protein VP1, the block to SV40 DNA replication, and thus to the full lytic cycle, is not observed in mouse cells [16]. Similarly, using an inducible metallothionein promoter linked to the SV40 early gene, Gluzman et al. (personal communication) have demonstrated that an increase in the level of T-antigen by >10 fold in mouse cells does not result in efficient SV40 DNA replication. These results suggest that although the level of early gene expression controlled by these enhancer elements may be maximal in the cell that serves as the lytic host for the virus, the enhancer is not the principal determinant of host range in all cases. It seems more likely that replication-specific proteins interact with the SV40 replication origin in monkey kidney cells and with the polyoma viral replication origin in mouse cells, thus permitting full expression of the lytic cycle.

In contrast, there are situations in which a papovavirus enhancer element appears specifically to determine the ability of a virus to be expressed in a particular host cell. One of these mentioned above is the expression of polyoma virus EC mutants in embryonal carcinoma cells. Although the wild-type virus is both unable to express its early gene product efficiently or to replicate in undifferentiated cells, mutants in the enhancer region (Py-EC mutants) overcome both of these restrictions [12–15]. The polyoma virus enhancer element contributes both to the expression of early viral RNA (which encodes the replication-required protein, T-antigen), and to a separate but essential requirement for viral DNA replication [17–18]. Whether the specific role of the enhancer in DNA replication is linked to its ability to stimulate transcription and/or its function in chromatin organization is unclear.

The human papovavirus, JCV, is closely associated with the degenerative neurologic disease, progressive multifocal leukoencephalopathy (PML). The virus replicates efficiently only in human glial cells in tissue culture and has been repeatedly isolated from demyelinated plaques found in the brains of patients with this disease. Recent studies have shown that the JVC enhancer element functions efficiently only in glial cells [19]. Thus, the tissue-specific transcriptional efficiency of the JCV enhancer element directly correlates with the tissue specificity of the disease. Whether or not there are additional restrictions (such as replication associated proteins) that limit the growth of JCV to human glial cells is a question presently under investigation.

Along these lines, Major and his colleagues [20] have recently obtained an SV40-transformed human glial cell line that constitutively produces SV40 T-antigen, and continues to express a number of cell markers specific for glial cells. JCV replicates efficiently in these cells, and its replication appears to be dependent upon the production of JCV T-antigen from the transfected genome rather than upon the endogenous SV40 T-antigen. Although this observation would appear to conflict with recent studies by Li and Kelly [21], who have shown that replication of JCV DNA in vitro can be supported by SV40 T-antigen with 20% the efficiency of SV40 DNA replication in vitro, the present in vitro replication assay scores for only a subset of the determinants required for efficient in vivo replication. For example, there is no benefit seen in vitro after addition of transcriptional control sequences (such as the 21-bp repeats of SV40) to the minimal SV40 origin of replication (Li and Kelly, personal communication). In vivo, addition of these sequences stimulates replication at least fivefold ([22], Peden et al., personal communication). The transformed human glial cell line should be important in determining the role of the enhancer element in the host-range restriction and disease spectra of the JCV virus.

Enhancer Elements and Retrovirus-induced Disease

Several laboratories have demonstrated that the retrovirus enhancer elements function with host-cell specificity and are responsible for the disease spectra of some retroviruses. For example, in one study the 3' LTR of the Friend viral genome was replaced by the analogous sequence from Moloney murine leukemia virus (Mo-MuLV) [23]. When introduced into mice, the recombinant virus produced thymic lymphomas almost exclusively. Conversely, the recombinant in which the Friend viral regulatory elements replaced the corresponding region from Mo-MuLV, induced principally erythroleukemias. Thus, in both cases the disease spectrum corresponded to the viral LTR retained in the recombinant genome. The sequences responsible for the disease specificities have been localized to the U3 region of the LTR, and presumably represent the transcriptional enhancer elements [24].

In experiments designed to investigate the disease specificity of two murine leukemia virus variants, DNA recombinants were constructed between the two viral strains. One strain (T$^+$) replicated efficiently in thymocytes, whereas the other strain (T$^-$) did not [25]. Because the ability of the murine leukemia viruses to replicate efficiently in thymocytes parallels their leukemogenicity, it was predicted that the efficiency of replication of the T$^+$/T$^-$ recombinants might identify segments of the genome responsible for pathogenicity. Infection of mice with a number of T$^+$/T$^-$ reciprocal recombinants indicated that sequences from the LTR of the T$^+$ strain were required for efficient replication in thymocytes. Recently, investigators have precisely localized the critical elements to the direct repeat sequence present in the LTR of the T$^+$ strain [26]. It seems likely that these sequences correspond to an enhancer element with a specificity for the thymocytes. These same investigators have acquired evidence from recombinant viruses to indicate that the high leukemogenic potential of the Gross passage A murine leukemia virus maps to sequences in the viral LTR [27].

Two Rous sarcoma virus (RSV) transformation-defective variants (i.e., viruses that have lost the *src* oncogene) differ substantially in their disease spectra. The transformation-defective (td) Schmidt–Ruppin strain induces thymic lymphomas whereas the td Prague strain causes a number of other diseases (e.g., osteopetrosis) but induces lymphomas only with very low frequency [28]. Differences in the enhancer regions of these closely related RSV variants may contribute to the diseases they cause in chickens. The 3′ terminal region of the Prague strain contains three distinct enhancer domains which comprise two functional enhancer elements. Two of these domains, located in the U3 region of the 3′ long terminal repeat, are almost identical to those found in the Schmidt–Ruppin strain [29]. The third enhancer region (XSR), located upstream from the LTR, may be important in the activation of downstream genes. This third domain, located outside the LTR, is quite distinct in the two strains of virus, suggesting a possible model for differential activation of cellular genes by either Schmidt–Ruppin or Prague RSV [30]. Although many avian retroviruses lacking oncogenes, such as the Rous associated viruses (RAVs), produce disease by integrating near a proto-oncogene (i.e., promoter or enhancer insertion), RAV-0 does not produce such a disease. Weber and Schaffner [31] have linked the lack of pathogenicity in RAV-0 to the absence of an enhancer.

The human T cell leukemia/lymphoma (HTLV) group of retroviruses appears to cause its pathogenic effect in human T lymphocytes. These viruses include HTLV-I, which induces human T cell leukemia/lymphoma, HTLV-II, which has been associated with hairy cell leukemia, bovine leukemia virus (BLV), and perhaps HTLV-III (lymphadenopathy associated virus, LAV, or AIDs retrovirus), associated with the acquired immune deficiency syndrome or AIDs. It is not yet entirely clear whether the enhancer elements associated with these retroviruses are important determinants of their tissue tropism [32–36]. It has been suggested that virus-encoded trans-acting proteins, the pX or LOR gene products, are responsible for activation of transcription from the viral LTRs. Thus, it is possible that the pX gene product is expressed in a tissue-specific way, and that this protein in turn may further stimulate viral RNA expression in the same cell.

Transgenic Experiments

In addition to tissue culture systems and viral infections of animal models, recent technological advances have allowed investigators to introduce foreign or chimeric constructs into the genomes of mice. Although experiments involving transgenic mice are limited, these studies have repeatedly demonstrated the importance of the enhancer elements (see Chapter 13). Recent experiments suggested that the expression of the SV40 early genes, as evidenced by T-antigen production and tumor formation, occurs specifically in the choroid plexus of the cerebral ventricles [37]. Another study in which the SV40 T-antigen has been linked to regulatory elements of the insulin gene showed that expression of the recombinant transgene was restricted to the predicted tissue, namely, the β cells of the pancreas [38]. When the mouse mammary tumor virus (MMTV) LTR is linked to the *myc* oncogene in transgenic mice, a principal site of disease is in the lactating breast of the pregnant female, suggesting the role of the MMTV enhancer and its associated hormone regulatory sequences [39].

Conclusions

The life cycle of the papovaviruses and retroviruses includes a number of stages such as penetration and uncoating, early and late gene expression, replication, packaging, and release. Any of these stages in the life cycle can affect the relative efficiency of expression of the virus in a particular tissue. Nevertheless, it is clear that one of the most important steps involves expression of "early" viral genes, which for the papovaviruses and retroviruses are generally under the control of viral enhancer elements. As such, these transcriptional regulatory elements (e.g., enhancers and promoters), which show tissue-specific expression, contribute to the disease potential of the viral agent. The linkage of particular transcriptional control sequences to specific genes and the subsequent introduction of these recombinant molecules into animals, either through viral vectors or by transgenic manipulations, should provide an insight into understanding tissue-specific viral diseases.

Acknowledgment

We thank P. Howley and L. Laimins for helpful advice, and M. Priest and Janet Duvall for preparation of the manuscript.

References

1. Khoury G, Gruss P (1983) Enhancer elements. Cell 33:313–314
2. Gluzman Y, Shenk T (eds) (1983) Enhancers and Eukaryotic Gene Expression—Current Communications in Molecular Biology. Cold Spring Harbor Laboratory, Cold Spring Harbor, New York
3. Benoist C, Chambon P (1981) In vivo sequence requirements of the SV40 early promoter region. Nature 290:304–310
4. Gruss P, Dhar R, Khoury G (1981) Simian virus 40 tandem repeated sequences as an element of the early promoter. Proc Natl Acad Sci USA 78:943–947

5. Fromm M, Berg P (1982) Deletion mapping of DNA regions required for SV40 early region promoter function in vivo. J Mol and Appl Genet 1:457–481

6. de Villiers J, Schaffner W (1981) A small segment of polyoma virus DNA enhances the expression of a cloned β-globin gene over a distance of 1400 base pairs. Nucleic Acids Res 9:6251–6254

7. Levinson B, Khoury G, Vande Woude G, Gruss P (1982) Activation of SV40 genome by 72-base pair tandem repeats of Moloney sarcoma virus. Nature 295:568–572

8. Laimins LA, Khoury G, Gorman C, Howard B, Gruss P (1982) Host-specific activation of gene expression by 72 base pair repeats of simian virus 40 and Moloney murine leukemia virus. Proc Natl Acad Sci USA 79:6453–6457

9. de Villiers J, Olson L, Tyndall C, Schaffner W (1982) Transcriptional 'enhancers' from SV40 and polyoma virus show a cell type preference. Nucleic Acids Res 10:7965–7976

10. Kriegler M, Botchan M (1983) Enhanced transformation by a simian virus 40 recombinant virus containing a Harvey murine sarcoma virus long terminal repeat. Mol Cell Biol 3:325–339

11. Tooze J (ed) (1981) DNA Tumor Viruses: Molecular Biology of Tumor Viruses, 2nd Ed. Cold Spring Harbor Laboratory, Cold Spring Harbor, New York

12. Katinka M, Yaniv M, Vasseur M, Blangy D (1980) Expression of polyoma early functions in mouse embryonal carcinoma cells depends on sequence rearrangements in the beginning of the late region. Cell 20:393–399

13. Fujimura FK, Deininger PL, Friedmann T, Linney E (1981) Mutation near the polyoma DNA replication origin permits productive infection of F9 embryonal carcinoma cells. Cell 23:809–814

14. Sekikawa K, Levine A (1981) Isolation and characterization of polyoma host range mutants that replicate in nullipotential embryonal carcinoma cells. Proc Natl Acad Sci USA 78:1100–1104

15. Fujimura FK, Linney E (1982) Polyoma mutants that productively infect F9 embryonal carcinoma cells do not rescue wild-type polyoma in F9 cells. Proc Natl Acad Sci USA 79:1479–1483

16. Graessman A, Graessman M, Mueller C (1981) Regulation of SV40 gene expression. In Klein G, Weinhouse S (ed) Advances in Cancer Research. Academic Press, New York, vol 35, pp 111–146

17. de Villiers J, Schaffner W, Tyndall C, Lupton S, Kamen R (1984) Polyoma virus DNA replication requires an enhancer. Nature 312:242–246

18. Veldman GM, Lupton S, Kamen R (1985) Polyomavirus enhancer contains multiple redundant sequence elements that activate both DNA replication and gene expression. Mol Cell Biol 5:649–658

19. Kenney S, Natarajan V, Strike D, Khoury G, Salzman NP (1984) JC virus enhancer–promoter active in the brain cells. Science 226:1337–1339

20. Major EO, Miller AE, Mourrain P, Traub RG, De Widt E, Sever J (1985) Establishment of a line of human fetal glial cells that supports JC virus multiplication. Proc Natl Acad Sci USA 82:1257–1261

21. Li JJ, Kelly TJ (1985) Simian virus 40 DNA replication in vitro: Specificity of initiation and evidence for bidirectional replication. Mol Cell Biol 5:1238–1246

22. Bergsma DJ, Olive DM, Hartzell SW, Subramanian KN (1982) Territorial limits and functional anatomy of the simian virus 40 replication origin. Proc Natl Acad Sci USA 79:381–385

23. Chatis P, Holland C, Hartley J, Rowe W, Hopkins N (1983) Role of the 3' end of the genome in determining the disease specificity of Friend and Moloney murine leukemia viruses. Proc Natl Acad Sci USA 80:4408–4411

24. Chatis P, Holland CA, Silver JE, Frederickson TN, Hopkins N, Hartley JW (1984) A 3' end fragment encompassing the transcriptional enhancers of nondefective Friend virus confers erythroleukemogenicity on Moloney leukemia virus. J Virol 52:248–254

25. DesGroseillers L, Rassart E, Jolicoeur P (1983) Thymotropism of murine leukemia virus is conferred by its long terminal repeat. Proc Natl Acad Sci USA 80:4203–4207

26. DesGroseillers L, Jolicoeur P (1984) Mapping the viral sequences conferring leukemogenicity and disease specificity in Moloney and amphotropic murine leukemia viruses. J Virol 52:448–456

27. DesGroseillers L, Villemur R, Jolicoeur P (1983) The high leukemogenic potential of Gross passage A leukemia virus maps in the region of the genome corresponding to the long terminal repeat and to the 3' end of *env*. J Virol 47:24–32

28. Robinson HL, Blais BM, Tsichlis PN, Coffin JM (1982) At least two regions of the viral genome determine the oncogenic potential of avain leukosis viruses. Proc Natl Acad Sci USA 79:1225–1229

29. Luciw PA, Bishop JM, Varmus HE, Capecchi MR (1983) Location and function of retroviral and SV40 sequences that enhance biochemical transformation after microinjection of DNA. Cell 33:705–716

30. Laimins LA, Tsichlis P, Khoury G (1984) Multiple enhancer domains in the 3' terminus of the Prague strain of Rous sarcoma virus. Nucleic Acids Res 12:6427–6442

31. Weber F, Schaffner W (1985) Enhancer activity correlates with the oncogenic potential of avian retroviruses. EMBO J 4:949–956

32. Sodroski JG, Rosen CA, Haseltine WA (1984) Trans-acting transcriptional activation of the long terminal repeat of human T lymphotrophic viruses in infected cells. Science 225:381–385

33. Josephs SF, Wong-Staal F, Manzari V, Gallo RC, Sodroski JG, Trus MD, Perkins D, Patarca R, Haseltine WA (1984) Long terminal repeat structure of an American isolate of type I human T-cell leukemia virus. Virology 139:340–345

34. Rosen C, Sodorski J, Kettman R, Burny A, Haseltine W (1985) Trans-activation of the bovine leukemia virus long terminal repeat in BLV-infected cells. Science 227:320–323

35. Sodroski JG, Rosen C, Wong-Staal F, Salahuddin S, Popovic M, Arya S, Gallo R, Haseltine W (1985) Trans-acting transcriptional regulation of human T-cell leukemia virus type III long terminal repeat. Science 227:171–173

36. Derse D, Caradonna S, Casey J (1985) Bovine leukemia virus long terminal repeat: A cell-type specific promoter. Science 227:317–320

37. Brinster R, Chen H, Messing A, Van Dyke T, Levine A, Palmiter R (1984) Transgenic mice harboring SV40 T-antigen genes develop characteristic tumors. Cell 37:367–379

38. Hanahan D (1985) Heritable formation of pancreatic β-cell tumours in transgenic mice expressing recombinant insulin/simian virus 40 oncogenes. Nature 315:115–122

39. Stewart T, Pattengale P, Leder P (1984) Spontaneous mammary adenocarcinomas in transgenic spontaneous mice that carry and express MTV/*myc* fusion genes. Cell 38:627–637

Oncogenes, Transfection, and Differentiation

CHAPTER 9
Amplification of Proto-oncogenes in Tumorigenesis

J. Michael Bishop

The search for genetic damage in neoplastic cells has become a central theme of contemporary cancer research. Two sorts of damage have been sought: (a) lesions that alter the expression of otherwise normal genes, and (b) lesions that change the functions of gene products. Both have been found, and both may play a role in the genesis of human neoplasia [1]. Here I review the evidence that one often-neglected genetic abnormality—the spontaneous amplification of DNA—may impose a tumorigenic burden on cells by enforcing inordinate gene expression.

A Glance at the Amplification of DNA

Both unicellular and metazoan organisms can amplify regions of their DNA [2]. In some instances, the amplification is scheduled in order to meet a large demand for a specific gene product (for example, during the course of embryogenesis). In other instances, the amplification is unscheduled and comes to view only because it accidentally confers a selective advantage on the cell in which it has occurred. Scheduled amplification usually affects a single gene or a small set of genes. By contrast, unscheduled amplification may encompass vast domains of DNA—in excess of 1000-kb in mammalian cells.

It is unscheduled amplification of DNA that concerns us here. The phenomenon was first encountered during exploration of why leukemia cells become resistant to cytotoxic drugs [3]. The replication of DNA in mammalian cells occasionally goes awry, so that a region of chromosome begins to multiply in increments of roughly five. The regions of amplification are apparently distributed at random throughout the genome; the frequency of amplification at any given locus on a chromosome is very low (perhaps once in every millionth cell division); and the mechanism of amplification remains in dispute. But the cell bearing the amplified DNA will become apparent only if the amplification serves some purpose: for example, resistance to drugs, or the competitive advantage for growth that seems inherent to the cancer cell.

Amplification of DNA in mammalian cells gives rise to two karyotypic abnormalities, known as double-minute chromosomes (DMs) and homogeneously staining regions (HSRs). The genesis of these structures has not been decisively explicated, but some observers suggest that DMs arise when amplified DNA is expelled from its site of origin on a chromosome, and that DMs in turn engender HSRs by reentering chromosomes at diverse positions. DMs and HSRs were discovered within the past decade and are now familiar signatures for the presence of amplified DNA. They are not inevitable companions of amplification, however, particularly if the extent of amplification is relatively limited (e.g., 5–10 fold).

DMs multiply in concert with the remainder of chromosomal DNA, but they have no centromere and thus distribute themselves haphazardly when the cell divides. There is no guarantee that both daughter cells will receive at least one DM: Cell lineages can arise that no longer possess the amplified DNA. By contrast, HSRs are passengers within chromosomes and are therefore a more stable component of the cell's genetic dowry, but they too can be lost spontaneously by mechanisms unknown. When cells retain DMs or HSRs, we can assume that the amplified DNA they embody represents a selective advantage.

Tumor cells and cells growing indefinitely in culture remain the sole known provenance of unscheduled DNA amplification. There is as yet no evidence that the phenomenon occurs in normal cells with finite life-spans. It is therefore possible that the widely touted plasticity of the tumor cell genome represents the soil for amplification of DNA that would otherwise not occur. The possibility has direct bearing on the role of amplification in tumorigenesis, a point to which we will return later.

Identifying Amplified Genes in Tumor Cells

The context in which unscheduled amplification of DNA was discovered in mammalian cells gave rise to a bias that the phenomenon would be found only in tumor cells established in culture, perhaps only in those cells that had been exposed to chemotherapy. We now know that this bias is incorrect. A substantial variety of human tumors have shown karyotypic and/or molecular evidence of amplified DNA in advance of any therapy. The mere presence of the amplified DNA is provocative: Because it has survived countless generations of cell division, it must endow the tumor cells that bear it with vital properties. What then are the genes whose amplification confers these properties? How are these genes to be found amid the vast morass of amplified DNA? Two strategies have been deployed to answer these questions:

1. The physical properties of DMs and of chromosomes bearing HSRs permit their isolation, and the tricks of recombinant DNA can be used in turn to ferret out the amplified DNA [4,5]. By these means, amplified DNA has been purified from several human and rodent tumors. Isolation of the amplified DNA is but a first step, however, because it does not by itself define genes within the vast domain of amplification, and it provides no indication as to which of the genes is responsible for the biological impact of the amplification.

2. Vertebrate cells contain more than a dozen genes that can be transduced into retroviruses to give oncogenes [6]. These cellular *proto-oncogenes* are thought to exemplify the genetic keyboard on which various carcinogens might play. The thought has been sustained by the discovery that point mutations, chromosomal translocations, and amplifications of DNA found in diverse tumors can all affect proto-oncogenes first identified because of their transduction into retroviral genomes [6]. Table 9.1 summarizes the proto-oncogenes found to date in the amplified DNA of tumor cells. The list includes several proto-oncogenes also afflicted by other forms of genetic damage (c-*myc* and c-*ras* prominent among them), and it includes two proto-oncogenes newly identified by the study of amplified DNA (N-*myc* and L-*myc*). The amplification of proto-oncogenes has taken either of two patterns: (a) as a

Table 9.1. Amplification of proto-oncogenes in tumor cells.

Proto-oncogene	Tumor	Reference
c-*abl*	K562 (chronic myelogenous leukemia)	18
c-*erb*-B	A431 (epidermoid carcinoma)	23, 24, 25
	Glioblastomas	26
c-*myb*	Adenocarcinoma of colon	27
	Acute myelogenous leukemia	28
c-*myc*	HL60 (promyelocytic leukemia)	29, 30
	COLO 320 (APUDoma)	31
	Small-cell carcinoma of lung	15, 32
	SEWA (osteosarcoma)	33
	SKBR-3 (carcinoma of breast)	34
	Gastric adenocarcinoma	17
L-*myc*	Carcinoma of lung	J. Minna, pers. commun.
N-*myc*	Neuroblastoma	11, 12
	Carcinoma of lung	16
	Retinoblastoma	21
Ki-*ras*	Carcinoma of lung	32
	Adrenocortical tumor	5

sporadic feature of diverse tumors (e.g., c-*myc* in occasional samples of a variety of tumors); or (b) as a frequent feature of particular tumors (e.g., N-*myc* in neuroblastoma). The repertoire of cellular genes that may contribute to tumorigenesis extends well beyond the identified proto-oncogenes of retroviruses, so we expect the list of genes affected by amplification in tumor cells to grow and ramify with time. But the remarkable frequency with which amplification of DNA in tumor cells has affected proto-oncogenes that were first recognized in another setting adds credence to the view that these genes must figure in the genesis of the neoplastic phenotype. It strains reason to explain all this by coincidence [7].

The Pathogenic Burden of Gene Amplification

Amplified genes are expressed in inordinate amounts because production of their mRNA is increased by the surfeit of DNA template. Direct measurements have also shown that production of proteins encoded by amplified genes is increased [5,8,9].

Intuition argues that excesses of at least some gene products might fuel neoplastic growth [10]. Experiment follows suit: Several of the known proto-oncogenes elicit aspects of the neoplastic phenotype when they are implanted into cells in culture and expressed in unusual abundance [6]. It is possible, of course, that amplification might also inflict structural damage, such as point mutations or topographical rearrangements, but no evidence for this possibility has surfaced as yet. It is therefore presumed that trouble arises because amplification enforces inordinate expression of the overgrown genes, which in turn imposes a pathogenic burden on the cell. There is no direct evidence to support this presumption, but provocative circumstantial evidence has emerged from the study of several types of tumors:

1. A gene dubbed N-*myc* because of its kinship with the proto-oncogene c-*myc* was discovered by exploring the amplified DNA in cells derived from human neuroblastoma [11,12]. Amplification and the consequent increased expression of N-*myc* appears to be one of the factors that evokes progression of human neuroblastoma to advanced malignancy [13,14]. The amplification of N-*myc* may also have practical implications, because it apparently marks those neuroblastomas whose course will be rapid and refractory to therapy ([13], and personal communication from R. Seeger). Here is a cameo example of how our burgeoning knowledge of oncogenes might be brought to bear on the care of the cancer patient.

2. Aggressive variants of small cell carcinoma of the human lung often display amplification of c-*myc*, N-*myc*, or a third related gene known as L-*myc* ([15,16], and personal communication from J. Minna). These findings suggest not only that amplification of a *myc* gene accompanies and perhaps contributes to the phenotypic progression of a malignancy, but also that

the structurally kindred genes c-*myc*, N-*myc*, and L-*myc* share related physiological functions.

3. Three samples of human gastric adenocarcinoma displayed amplification of c-*myc* [17]. The amplified gene was expressed at a high level in a rapidly growing and poorly differentiated tumor, at a much lower level in a slowly growing and more differentiated tumor.
4. A transplantable osteosarcoma (SEWA) originally induced in a mouse by polyoma virus carries amplified c-*myc* [8]. When SEWA cells are propagated in cell culture, their tumorigenicity and the amplification of c-*myc* diminish in concert (personal communication from G. Levan).

Fitting Gene Amplification into Tumorigenesis

When during the course of tumorigenesis does amplification of proto-oncogenes occur? The examples just given suggest that amplification arose well after tumorigenesis had begun, and there are presently no examples that can be safely interpreted to the contrary. It is therefore possible that unscheduled amplification of proto-oncogenes is never the initial event in tumorigenesis: Amplification may not occur until cells have taken at least some tentative steps toward malignancy (see above).

Whenever it occurs, the amplification of proto-oncogenes can be viewed as one of several genetic lesions that combine to produce the neoplastic phenotype. As examples, consider the following:

1. Amplification of c-*abl* has been found in combination with a translocation that damages the protein encoded by c-*abl* [18,19].
2. Amplification of c-*myc* sometimes accompanies a mutation in N-*ras* in leukemia cells [20].
3. A chromosomal deletion and amplification of c-*myc*, N-*myc*, or L-*myc* can be found together in carcinoma of the lung [15,16].
4. Amplification of N-*myc* has been found in retinoblastomas that are also either hemi- or homozygous for a chromosomal deletion [21].
5. A mutant version of N-*ras* was first isolated from a neuroblastoma cell line [22], although the mutant gene has yet to be found in tumor cells that also contain amplified N-*myc*.

It has been suggested that the abnormal expression of c-*myc* is an "establishment function" in tumorigenesis, that it sustains indefinite growth of cells without eliciting other aspects of the neoplastic phenotype. The findings with gene amplification reviewed here suggest that this view is unjustifiably simplistic. For example, it is difficult to view c-*myc* as an establishment function in carcinoma of the lung, where amplification of the gene has been associated with events late in the progression of the tumor to advanced malignancy. The same argument can be applied to the amplification of N-*myc* in both carcinoma of the lung and neuroblastoma.

Conclusion

Some forms of damage to proto-oncogenes occur sporadically among a wide variety of tumors, whereas others are more consistent in their occurrence [1]. Amplification provides two especially informative examples of consistency: (a) the amplification of N-*myc* in neuroblastomas; and (b) the amplification of c-*myc*, N-*myc*, or L-*myc* in aggressive variants of carcinoma of the lung. It is possible that amplification of proto-oncogenes is an inconsequential anomaly of tumor cells. I doubt that this is so, because amplified DNA must have survived countless generations of tumor cell growth—as if a strong selective advantage were in force. The place of gene amplification in tumorigenesis should become more apparent as the functions of proto-oncogenes and the mechanism of amplification come into better focus.

References

1. Varmus HE (1984) The molecular genetics of cellular oncogenes. Annu Rev Genet 18:553–612
2. Stark GR, Wahl GM (1984) Gene amplification. Ann Rev Biochem 53:447–493
3. Schimke RT (1984) Gene amplification in cultured animal cells. Cell 37:705–713
4. Kanda N, Schreck R, Alt F, Bruns G, Baltimore D, Latt S (1983) Isolation of amplified DNA sequences from IMR-32 human neuroblastoma cells: Facilitation of fluorescence-activated flow sorting of metaphase chromosome. Proc Natl Acad Sci USA 80:4069–4073
5. Schwab M, Alitalo K, Varmus HE, Bishop JM, George D (1983) A cellular oncogene (c-Ki-*ras*) is amplified, over-expressed, and located within karyotypic abnormalities in mouse adrenocortical tumor cells. Nature 303:497–501
6. Bishop JM (1983) Cellular oncogenes and retroviruses. Annu Rev Biochem 52:301–354
7. Duesberg PH (1985) Activated proto-onc genes: Sufficient or necessary for cancer? Science 228:669–677
8. Schwab M, Ramsay G, Alitalo K, Varmus HE, Bishop JM, Martinsson TG, Levan A (1984) Amplification and enhanced expression of the gene c-*myc* in mouse SEWA tumor cells may be related to malignant progression. Nature 315:345–347
9. Ramsay G, Evan GI, Bishop JM (1984) The protein encoded by the human proto-oncogene c-*myc*. Proc Natl Acad Sci USA 81:7742–7746
10. Pall ML (1981) Gene-application model of carcinogenesis. Proc Natl Acad Sci USA 78:2465–2468
11. Kohl NE, Kanda N, Schreck RR, Burns G, Latt SA, Gilbert F, Alt FW (1983) Transposition and amplification of oncogene-related sequences in human neuroblastomas. Cell 35:359–367
12. Schwab M, Alitalo K, Klempnauer K-H, Varmus HE, Bishop JM, Gilbert F, Brodeur G, Goldstein M, Trent J (1983) Amplified DNA with limited homology to the *myc* cellular oncogene. Nature 305:245–248
13. Brodeur GM, Seeger RC, Schwab M, Varmus HE, Bishop JM (1984) Amplification of N-*myc* in untreated human neuroblastomas correlates with advanced disease stage. Science 224:1121–1124

14. Schwab M, Ellison J, Busch M, Rosenau W, Varmus HE, Bishop JM (1984) Enhanced expression of the human gene N-*myc* consequent to amplification of DNA may contribute to malignant progression of neuroblastoma. Proc Natl Acad Sci USA 81:4940–4945

15. Little CD, Nau MM, Carney DN, Gazdar AF, Minna JF (1983) Amplification and expression of the c-*myc* oncogene in human lung cancer cell lines. Nature 306:194–196

16. Nau MM, Carney DN, Battey J, Johnson B, Little C, Gazdar A, Minna JD (1984) Amplification, expression and rearrangement of c-*myc* and N-*myc* oncogenes in human lung cancer. Curr Top Microbiol Immunol 113:172–177

17. Shibuya M, Yokota J, Ueyama Y (1985) Amplification and expression of a cellular oncogene (c-*myc*) in human gastric adenocarcinoma cells. Mol Cell Biol 5:414–418

18. Collins SJ, Groudine M (1983) Rearrangement and amplification of c-*abl* sequences in the human chronic myelogenous leukaemia cell line. Proc Natl Acad Sci USA 80:4813–4817

19. Konopka JB, Watanabe SM, Singer JW, Collins SJ, Witte ON (1985) Cell lines and clinical isolates derived from Ph¹-positive chronic myelogenous leukemia patients express c-*abl* proteins with a common structural alteration. Proc Natl Acad Sci USA 82:1810–1815

20. Murray MJ, Cunningham JM, Parada LF, Dautry F, Lebowitz P, Weinberg RA (1983) The HL-60 transforming sequence: A *ras* oncogene coexisting with altered *myc* genes in hematopoietic tumors. Cell 33:749–757

21. Lee W-H, Murphree AL, Benedict WF (1984) Expression and amplification of the N-*myc* gene in primary retinoblastoma. Nature 309:458–460

22. Shimizu K, Goldfarb M, Perucho M, Wigler M (1983) Isolation and preliminary characterization of the transforming gene of a human neuroblastoma cell line. Proc Natl Acad Sci USA 80:383–387

23. Ullrich A, Coussens L, Hayflick JS, Dull TJ, Gray A, Tam AW, Lee J, Yarden Y, Libermann TA, Schlessinger J, Downward J, Mayes ELV, Whittle N, Waterfield MD, Seeburg PH (1984) Human epidermal growth factor receptor cDNA sequence and aberrant expression of the amplified gene in A431 epidermoid carcinoma cells. Nature 309:418–425

24. Merlino GT, Young-Hua X, Ishii S, Clark AJL (1984) Amplification and enhanced expression of the epidermal growth factor receptor gene in A431 human carcinoma cells. Science 224:417–420

25. Lin CR, Chen WS, Kruiger W, Stolarsky LS, Weber W, Evans RM, Verma IM, Gill GN, Rosenfeld MG (1984) Expression cloning of human EGF receptor complementary DNA: Gene amplification and three related messenger RNA produces in A431 cells. Science 224:843–849

26. Liebermann TA, Nusbaum HR, Razon N, Kris R, Lax I, Soreq H, Whittle N, Waterfield MD (1985) Nature 313:144–147

27. Alitalo K, Winqvist R, Lin CC, de la Chapelle A, Schwab M, Bishop JM (1984) Aberrant expression of an amplified c-myb oncogene in two cell lines from a colon carcinoma. Proc Natl Acad Sci USA 81:4534–4539

28. Pelicci P-G, Lanfrancone L, Brathwaite MD, Wolman SR, Dalla-Favera R (1984) Amplification of N-myc in untreated human neuroblastomas correlates with advanced disease stage. Science 224:1117–1121

29. Collins S, Groudine M (1982) Amplification of endogenous myc-related DNA sequences in a human myeloid leukaemia cell line. Nature 298:679–681

30. Dalla-Favera R, Wong-Staal F, Gallo RC (1982) Onc gene amplification in promyelocytic leukaemia cell line HL-60 and primary leukaemia cells of the same patient. Nature 299:61–63
31. Alitalo K, Schwab M, Lin CC, Varmus HE, Bishop JM (1983) Homogeneously staining chromosomal regions contain amplified copies of an abundantly expressed cellular oncogene (c-myc) in malignant neuroendocrine cells from a human colon carcinoma. Proc Natl Acad Sci USA 80:1707–1711
32. Taya Y, Hosogai K, Hirohashi S, Shimosato Y, Tsuchiya R, Tsuchida N, Fushimi M, Sekiya T, Nishimura S (1984) A novel combination of K-ras and myc amplification accompanied by point mutational activation of K-ras in a human lung cancer. Eur Molec Biol Org J 3:2943–2946
33. Schwab M, Ramsay G, Alitalo K, Varmus HE, Bishop JM, Martinsson T, Levan G, Levan A (1985) Amplification and enhanced expression of the gene c-*myc* in mouse SEWA tumor cells may be related to malignant progression. Nature 315:345–347
34. Kozbor D, Croce C (1984) Amplification of the c-myc oncogene in one of five human breast carcinoma cell lines. Cancer Res 44:438–441

CHAPTER 10
Multistep Scenarios in Tumor Development and the Role of Oncogene Activation by Chromosomal Translocations

GEORGE KLEIN

Few would contest the statement that every somatic cell is regulated by multiple controls. It is therefore not surprising that most tumors evolve by multiple changes [1–3]. Cell proliferation is often dependent on the interaction between specific growth factors and their receptors on the target cell surface. Such receptors usually appear in relation to maturation switches induced by specific signals. After the cell has been determined to enter a certain differentiation lineage, it proceeds by a series of such switches along a programmed pathway. The one-dimensional implication of the word *lineage* is rarely if ever, true in reality. It is more appropriate to regard a lineage as a complex tree that can harbor more than two choices at certain bifurcation points. This explains the widely diversified assortment of markers that is becoming obvious with the rapid progress of monoclonal antibody analysis of hemopoietic and other tumors.

Many neoplastic cells behave as if their maturation has been blocked at a specific point. In at least some of them, maturation can still be induced by an appropriate signal [4–10]. The teratocarcinoma line of Mintz that could be induced to generate all normal mouse tissues, after exposure to the appropriate inductive fetal environment, is the most spectacular example [4]. There is no basic difference between this and less dramatic cases of tumor cell maturation, except insofar as the positioning of the cells within their differentiation program is concerned. The normal progenitor of the teratocarcinoma is a totipotent, early embryonic cell, whereas other tumor cells arise within more narrowly restricted branches of the differentiation tree.

Responsiveness to maturation-inducing signals is not the rule, however, even for the leukemias. The penetrating analysis of Sachs [8] has shown that maturation-inducible cells can give rise to noninducible variants. The relatively well controlled (Ph$_1$)-positive form of chronic granulocytic leukemia

(CGL) in humans ends in the uncontrollable proliferation of the blast crisis, as a rule.

Many solid tumors develop by a series of changes, collected under the general umbrella of *tumor progression*. First formulated by Rous and Beard as "the process whereby tumors change from bad to worse"[11], it was defined by Foulds [3] as the development of tumors by the stepwise change and reassortment of "multiple unit characteristics," and was specified by Furth [1] in relation to various hormone–target interactions. The resulting concept of a multistep evolution is still the most easily generalizable representation of the natural history of cancer. In contrast to the largely descriptive analysis that was available to the pioneers of the field, modern developments in cytogenetics and molecular biology have now permitted a precise definition of some steps involved in tumor development and progression.

Most attention has been focused on the oncogenes, partly for historical reasons and partly because of the relatively easy experimental demonstrability of dominantly acting transforming genes. Such experimentation is often based on rather artificial systems, however, in both main areas of oncogene analysis: viral transduction and DNA transfection. As a rule v-*onc*-carrying viruses are passaged over long periods of time by the investigator, with continuous selection for high tumorigenic activity. This tends to fix multiple changes within the same oncogene sequences that favor transforming and/or tumorigenic activity [12]. DNA transfection of rodent fibroblasts is even more artificial, particularly when immortal, aneuploid lines such as NIH-3T3 are used as targets. The transformation of normal diploid fibroblasts requires the complementary interaction between two or more oncogenes [13].

Tumor development in vivo is a more complex process. Selection acts on the cell, *not* on the virus. This phenomenon favors single changes in multiple genes. Statistical analysis of age-incidence curves suggests that three to four successive mutation-like changes may be responsible for the leukemias, and six to seven may be responsible for carcinomas [14,15]. This must be considered the *minimum,* rather than the actual number, as only the rate-limiting steps will make an impact on the statistical analysis.

Histological, cytogenetic, and experimental studies have also confirmed that the natural history of spontaneously occurring human and animal cancers is usually a multistep process, on the basis of the successive emergence of new subclones that replace their predecessors as a result of some selective growth advantage [2,3,16].

Is there any evidence that some of these steps are due to the activation of cellular oncogenes? The best evidence comes from the field of *chromosomal translocations*. At least in two systems, (a) juxtaposition of the *myc* oncogene to an immunoglobulin locus in human Burkitt's lymphoma and mouse plasmacytoma, and (b) the production of an abnormal fusion protein in Ph_1-positive CGL, genetic transposition leads to the activation of an oncogene. The regularity of this phenomenon suggests that it must play an essential role in the tumorigenic process. Although superficially similar, these two systems must act in a very different way. The evidence will be reviewed briefly in the following sections.

Myc/Immunoglobulin Juxtaposition by Chromosomal Translocation

In studies of the cytogenetics of various tumors, convergent chromosomal findings were made in two entirely different systems—a mouse B cell tumor (plasmacytoma) and a human B cell tumor (Burkitt's lymphoma). Mouse plasmacytomas have one of two important cytogenetic changes, as seen in Figure 10.1 The most frequent or typical translocation involves the reciprocal exchange between the distal portion of chromosome 15 and chromosome 12. The more infrequent or variant translocation involves the same distal band of chromosome 15, which now translocates to chromosome 6. When this translocation was observed [17], it was already known that chromosome 12 carried the immunoglobulin heavy chain (IgH) locus and that chromosome 6 carried the kappa light chain locus. It was also known that B cell tumors with this t(15;6) translocation were kappa producers. It was therefore hypothesized that because these were immunoglobulin-producing tumors, per-

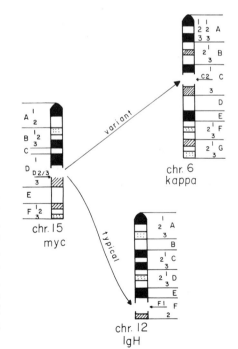

MOUSE (PC)

Figure 10.1 Schematic representation of typical and variant translocations involving c-*myc* on chromosomes 15 in mouse plasmacytomas (MPC). The typical translocation t(15:12) involves the movement of c-*myc* to the immunoglobulin heavy chain locus (IgH) on chromosome 12. The variant translocation t(15:6) involves the movement of c-*myc* to the kappa light-chain locus on chromosome 6.

haps there was an oncogene located on the distal part of chromosome 15, which had come under the influence of a transcriptionally active, chromosomal immunoglobulin (Ig) gene region, leading to activation of the oncogene, and thus resulting in a B cell tumor [18]. This hypothesis was stimulated by the finding of Hayward et al. [19] that in avian B cell lymphomas the insertion of a retrovirus next to the cellular oncogene c-*myc* led to the activation of that oncogene. This hypothesis proved to be correct, because an oncogene was localized right at the breakpoint on chromosome 15 [20,21]. It proved to be c-*myc*, the same oncogene found by Hayward et al. [19] in the promotor insertion model of bursa-derived B cell chicken lymphoma.

The hypothesis actually became more plausible long before the *myc* oncogene was discovered, because a similar set of cytogenetic findings was made on Burkitt's lymphoma cells. In this B cell-derived human tumor, Manolov and Manolova first observed that there was a characteristic marker called 14q+ which was longer than usual because of an extra band on the distal long arm of chromosome 14 [22]. When this was examined more closely, it turned out that the extra piece on 14 was translocated from chromosome 8, i.e., a t(8;14) translocation [23] (Fig. 10.2). The variant translocations were also interesting because in about 10%, each, the same portion of chromosome 8 went to either chromosome 2 or chromosome 22. It was earlier known that the immunoglobulin heavy chain (IgH) locus was localized on chromosome 14; later, it was shown that the IgH locus was located right at the breakpoint. With the use of in situ hybridization techniques, chromosome 2 was later found to carry the κ gene locus at the breakpoint; chromosome 22 was found to carry the λ gene locus, again near the breakpoint. The only three recipient human chromosomes, numbers 14, 2, and 22, therefore contained the three immunoglobulin gene loci, with the donor piece of chromosome 8 always being the same (reviewed in refs. 24 and 25). So again it was postulated that an oncogene is located on chromosome 8 and that, as in the mouse, it should be c-*myc*. Again, this was shown to be true. It is also notable that with the exception of c-*myc*, there are no other homologies between human chromosome 8 and mouse chromosome 15.

The detailed molecular analysis of the c-*myc* oncogene then followed. The *myc* oncogene has three exons. The first exon does not code for protein; it is an unusually long leader sequence. All the coding sequences are found in the second and third exons. The original v-*myc* sequence, which is carried by the avian MC29 acute leukemia virus, also contains only the two coding exons. The most frequent breakpoint of the c-*myc* gene in mouse plasmacytomas is within the c-*myc* gene, and always involves the noncoding exon or its 3' intron. For some time, this observation was considered important because "decapitation" of the gene, as it was called, might have resulted in the removal of the necessary 5' regulatory sequences, resulting in the activation and transcription of the coding exons, which are always found to be highly transcribed in these B cell tumors. In fact, how activation and increased transcription occur has been shown to be more complex, in that

Figure 10.2. Schematic representation of typical and variant translocations involving c-*myc* on chromosome 8 in human Burkitt's lymphoma (BL). The typical translocation (8;14) involves the movement of c-*myc* to the immunoglobulin heavy-chain locus (IgH) on chromosome 14. The variant translocations t(8;2) and t(8;22), involve the movement of c-*myc* to either the kappa light-chain locus on chromosome 2 or to the lambda light-chain locus on chromosome 22.

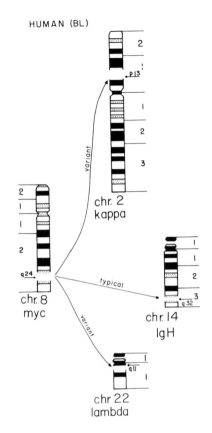

there are many breakpoints that occur outside the *myc* gene on either the 5' or 3' side. In the human situation, the variability of the breakpoint in and around the *myc* gene is even greater than in the mouse. There is, therefore, a wide variety of translocational breakpoints that are associated and compatible with the activation and increased transcription of the c-*myc* gene. Thus the hypothesis that a *single* enhancer and promoter is responsible for the activation of c-*myc* is no longer tenable (reviewed in refs. 21 and 26).

It is interesting, however, that in the mouse plasmacytoma system, the translocated *myc* gene faces the switch α region of the nonproductively rearranged allele of chromosome 12. In the human, the translocated *myc* gene is most commonly found near the switch μ region of chromosome 14. Most of the mouse plasmacytomas produce IgA, whereas most of the human Burkitt's lymphomas make IgM, again with some exceptions. This is, however, a pseudofunctional correlation, because a broken immunoglobulin heavy-chain locus will not make functional immunoglobulin; rather it is the allele on the other chromosome that codes for the functional immunoglobulin protein. Why this pseudofunctional correlation exists is, in itself, an interesting question. It tells us something about the tandem behavior of the two chromosomes in normal DNA rearrangement during B cell differentiation. This and some other "pseudofunctional puzzles" have been reviewed in detail elsewhere [27].

From these initial observations, another set of questions can then be formulated. First, are these *myc* translocations essential for generating B cell tumors? Also, how do the translocations arise? How do they act? What is their position within the chain of progressional events? And, finally, are there functionally analogous translocations in other systems?

In response to these questions, it can be said that the *myc* translocation appears to be an essential step in the genesis of B cell tumors, because in Epstein–Barr virus (EBV)-positive or -negative Burkitt's lymphoma/leukemia, the association of *myc* translocations is 100% at present (provided that we disregard the atypical BJAB cell line) [28]. Although the association of *myc* translocations in mouse plasmacytoma is only 85% (as determined by karyotype analysis), there are *myc* rearrangements in some of the exceptional tumors, which are due to similar but more complex mechanisms [29,30]. It is therefore widely held that these translocations are an essential event in the development of these B cell tumors.

How do *myc* translocations arise? The first question is, why should they be present in both mouse plasmacytoma and Burkitt's lymphoma, which represent different B cell tumors? Mouse plasmacytoma is a mature plasma cell, whereas Burkitt's lymphoma is a less mature B cell, which is detected much earlier in differentiation than a mature plasma cell. The only common denominator between these two diverse B cell tumors is the long preneoplastic latency period. In Burkitt's lymphoma, B cells are stimulated by the combination of Epstein–Barr virus and chronic malaria. In the mouse plasmacytoma, B cells are stimulated by mineral oil-induced granuloma in the peritoneal cavity. These preneoplastic B cells persist for long periods of time and are continuously being stimulated to divide; the risk of genetic aberrations thus increases with each division.

How do the translocations act? Early interpretations of the mechanism whereby the c-*myc*/immunoglobulin juxtaposition contributes to the tumorigenic process included claims of abnormally high transcription rates, abnormal transcript size, changed promoter use, mutations [31], and changed translational control. Each claim has been documented for some tumors either in the Burkitt's lymphoma (BL) or in the closely analogous mouse plasmacytoma (MPC) system, but none was found to apply to all or even most of the tumors. Perhaps there are several ways of activating c-*myc*, and no unifying mechanism exists [32]. Alternatively, the crucial event may be more subtle than a relatively crude quantitative or qualitative change. The current concept is dysregulation. It implies that the immunoglobulin-juxtaposed *myc* gene has become subject to cis control by the constitutively active immunoglobulin region and must therefore behave as if it were part of the immunoglobulin locus itself. The transcription of the normal *myc* gene is associated with competence for cell division [33,34]. In terminally differentiating cells, c-*myc* transcription ceases [9]. Mature plasma cells continue to produce immunoglobulins, however, and an immunoglobulin-juxtaposed *myc* may be similar. This would keep the cells in continuous division. Although there is no direct evidence for this hypothesis, it is supported by comparative studies on the behavior of the translocated *myc* gene and its normal allele in MPCs and BLs. It has been found repeatedly in both systems that the translocated gene is highly transcribed, whereas its normal counterpart is switched off [35–40]. This is compelling evidence for the differential regulation of the two genes.

The Ph$_1$ Chromosome

This translocation shares some features with the *myc*/Ig juxtaposition in BL and MPC but is markedly different in other respects. In the typical t(9;22) translocation, the c-*abl* oncogene, localized on the distal tip of chromosome 9, moves to the Ph$_1$ breakpoint on chromosome 22 [41–43.] Initially, this finding was not compelling because of the large diversity of the variant Ph$_1$ translocations where the distal fragment of chromosome 22 moves to autosomes other than chromosome 9. However, in situ hybridization experiments revealed that the cytogenetically invisible c-*abl*-containing telomeric fragment of chromosome 9 moves regularly to the Ph$_1$ breakpoint on chromosome 22 in these cases as well [44].The transposed c-*abl* gene is activated to produce an abnormally long mRNA [45] which codes for a 210K protein, a combined product of chromosome 22 and chromosome 9-derived sequences, instead of the normal 145K c-*abl* protein [45]. The 210K protein has an abnormal tyrosine kinase activity, like the product of the virally carried (v-*abl*) gene. In contrast to the behavior of the c-*myc* in the BL/MPC, where the translocated c-*myc* is expressed but the normal allele is switched off, Ph$_1$-positive CGL cells express both the translocated gene and the normal c-*abl*. This may reflect an important difference in the mechanism by which the *myc* versus the *abl* translocation contributes to the tumorigenic process. The Ig juxtaposition of the normal c-*myc* gene induces a regulatory change in the production of the normal protein, whereas the Ph$_1$ translocation acts by generating an abnormal fusion protein with an altered function.

Conclusion

Two "experiments of nature" involving oncogene activation by chromosomal translocation were found to contribute to the genesis of neoplastic growth in murine or human B lymphocytes, and in human granulocytic cells, respectively. The regular association of these translocations with the corresponding tumor can be interpreted only as showing that the translocation represents an esssential, rate-limiting step in the tumorigenic process. These two cases already revealed the existence of two fundamentally different mechanisms: Activation of the c-*myc* oncogene by juxtaposition to one of the Ig loci in murine plasmacytoma and human Burkitt's lymphoma acts through the dysregulation (i.e., constitutive activation) of what can be the normal, unchanged gene. In contrast, the transposition of the c-*abl* oncogene to the *bcr* region of chromosome 22 in Ph_1-positive CGL acts by the production of an abnormal fusion protein.

These findings reinforce the conclusion, derived from the more artificial models of viral transduction and DNA transfection, that oncogenes may contribute to the tumorigenic process by qualitative, structural, or quantitative, regulatory changes. As developed elsewhere in more detail [46], the natural development of tumors requires, in all probability, additional changes that involve not only oncogenes, but also *anti-oncogenes* or *suppressor genes* that counteract their effect, and *modulator genes* that influence malignant behavior of the established neoplastic cell.

References

1. Furth J (1953) Conditioned and autonomous neoplasms: A review. Cancer Res 13:477–492
2. Foulds L (1949) Mammary tumours in hybrid mice: Growth and progression of spontaneous tumours. Br J Cancer 3:345
3. Foulds L (1958) The natural history of cancer. J Chron Dis 8:2–37
4. Mintz B, Fleischman RA (1981) Teratocarcinomas and other neoplasms as developmental defects in gene expression. Adv Cancer Res 34:211–273
5. Friend C, Scher W, Holland JG, Sato T (1971) Hemoglobin synthesis in murine virus-induced leukemic cells in vitro: Stimulation of erythroid differentiation by dimethylsulfoxide. Proc Natl Acad Sci USA 68:378–382
6. Andersson LC, Nilsson K, Gahmberg CG (1979) K562—a human erythroleukemic cell line. Int J Cancer 23:143–147
7. Rutherford TR, Clegg JB, Weatherall DJ (1979) K562 human leukaemic cells synthesise embryonic haemoglobin in response to haemin. Nature 280:164–165
8. Sachs L (1984) The reversibility of neoplastic transformation: Regulation of clonal growth and differentiation in hematopoiesis and the normalization of myeloid leukemic cells. Adv Viral Oncol 4:307–329
9. Reitsma PH, Rothberg PG, Astrin SM, Trial J, Bar-Shavit Z, Hall A, Teitelbaum SL, Kahn AJ (1983) Regulation of myc-gene expression in HL-60 leukaemia cells by a vitamin D metabolite. Nature 306:492–494
10. Olsson I, Gullberg U, Ivhed I, Nilsson K (1983) Induction of differentiation of the human histiocytic lymphoma cell line U-937 by 1α, 25-dihydroxycholecalciferol. Cancer Res 43:5862–5867
11. Rous P, Beard JW (1935) The progression to carcinoma of virus. induced rabbit papillomas (Shope). J Exp Med 62:523–548
12. Temin HJ (1984) Do we understand the genetic mechanism of oncogenesis? J Cell Physiol [Suppl] 3:1–11

13. Land H, Parada LF, Weinberg RA (1983) Tumorigenic conversion of primary embryo fibroblasts requires at least two cooperating oncogenes. Nature 304:596–602

14. Farber E, Cameron R (1980) The sequential analysis of cancer development. Adv Cancer Res 31:125–226

15. Knudson A (1973) Mutation and human cancer. Adv Cancer Res 17:317–352

16. Nowell PC (1976) The clonal evolution of tumor cell populations. Science 194:23–28

17. Ohno S, Babonits M, Wiener F, Spira J, Klein G, Potter M (1979) Nonrandom chromosome changes involving the Ig gene-carrying chromosomes 12 and 6 in pristane-induced mouse plasmacytomas. Cell 18:1001–1007

18. Klein G (1981) The role of gene dosage and genetic transpositions in carcinogenesis. Nature 294:313—318

19. Hayward WS, Neel BG and Astrin SM (1981) Activation of a cellular onc gene by promoter insertion in ALV-induced lymphoid leukosis. Nature 290:475–480

20. Klein G (1983) Specific chromosomal translocations and the genesis of B-cell-derived tumors in mice and men. Cell 32:311–315

21. Perry RP (1983) Consequences of myc invasion of immunoglobulin loci: Facts and speculation. Cell 33:647–649

22. Manolov G, Manolova Y (1972) Marker band in one chromosome 14 from Burkitt lymphomas. Nature 237:33–34

23. Zech L, Haglund U, Nilsson K, Klein G (1976) Characteristic chromosomal abnormalities in biopsies and lymphoid cell lines from patients with Burkitt and non Burkitt lymphomas. Int J Cancer 17:47–56

24. Klein G, Lenoir G (1982) Translocations involving Ig-locus-carrying chromosomes: A model for genetic transposition in carcinogenesis. Adv Cancer Res 37:381–387

25. Croce CM, Tsujimoto Y, Erikson J, Nowell P (1984) Chromosomal translocations and B cell neoplasia. Lab Invest 51:258–267

26. Cory S. Activation of cellular oncogenes in hemopoietic cells by chromosome translocation. Adv Cancer Res (in press)

27. Klein G, Klein E (1985) Myc/Ig juxtaposition by chromosomal translocations: Some new insights, puzzles and paradoxes. Immunol Today 6:208–215

28. Klein G, Klein E (1984) Oncogene activation and tumor progression. Commentary. Carcinogen 5:429–435

29. Fahrlander PD, Sümegi J, Yang JQ, Wiener F, Marcu KB, Klein G (1985) Activation of the c-myc oncogene by the immunoglobulin heavy chain gene enhancer after multiple switch region-mediated chromosome rearrangements in a murine plasmacytoma. Proc Natl Acad Sci USA 82:3746–3750

30. Corcoran LM, Cory S, Adams JM (1985) Transposition of the immunoglobulin heavy chain enhancer to the myc oncogene in a murine plasmacytoma. Cell 40:71–79

31. Rabbits TH, Forester A, Hamlyn P, Baer R (1984) Effect of somatic mutation within translocated c-myc genes in Burkitt's lymphoma. Nature 309:592–597

32. Robertson M (1984) Oncogene activation. Message of myc in context. Nature 309:585–587

33. Kelly K, Cochran BN, Stiles CD, Leder P (1983) Cell-specific regulation of the c-myc gene by lymphocyte mitogens and platelet-derived growth factor. Cell 35:603–610

34 Campisi J, Gray HE, Pardee AB, Dean M, Sonnenschein GE (1984) Cell-cycle control of c-myc but not c-ras expression is lost following chemical transformation. Cell 36:241–247

35. Croce CM, Klein G (1985) Chromosome translocations and human cancer. Scient Am 252:54–60

36. Adams JM, Gerondakis S, Webb S, Corcoran LM, Cory S (1983) Cellular myc oncogene is altered by chromosome translocation to and immunoglobulin locus in murine plasmacytomas and is rearranged similarly in human Burkitt lymphomas. Proc Natl Acad Sci USA 80:1982–1986

37. Erikson J, Ar-Rushdi A, Drwinga HL, Nowell PC, Croce CM (1983) Transcriptional activation of the translocated c-myc oncogene in Burkitt lymphoma. Proc Natl Acad Sci USA 80:820–824

38. Harris LJ, Land RB, Marcu KB (1982) Non-immunoglobulin-associated DNA rearrangements in mouse plasmacytomas. Proc Natl Acad Sci USA 79:4175–4179

39. Stanton LW, Watt R, Marcu KB (1983) Translocation, breakage and truncated transcripts of c-myc oncogene in murine plasmacytomas. Nature 303:401–406

40. Leder P, Battey J, Lenoir G, Moulding C, Murphy W, Potter H, Stewart T, Taub R (1983) Translocations among antibody genes in human cancer. Science 222:765–771

41. Heisterkamp N, Groffen J, Stephenson JR, Spurr NK, Goodfellow PN, Solomon E, Carritt B, Bodmer WF (1982) Chromosomal localization of human cellular homologues of two viral oncogenes. Nature 299:747–749

42. De Klein A, Geurts van Kessl A, Grosveld G, Bartram CR, Hagemeijer A, Bootsma D, Spurr NK, Heisterkamp N, Groffen J, Stephenson JR (1982) A cellular oncogene is translocated to the Philadelphia chromosome in chronic myelocytic leukaemia. Nature 300:765–767

43. Groffen J, Stephenson JR, Heisterkamp N, De Klein A, Bartram CR, Grosveld G (1984) Philadelphia chromosomal breakpoints are clustered within a limited region, bcr, on chromosome 22. Cell 36:93–99

44. Bartram CR, de Klein A, Hagemeijer A, von Agthoven T, geurts van Kessel A, Bootsma D, Grosveld G, Ferguson-Smith MA, Davies T, Stone M, Heisterkamp N, Stephenson JR, Groffen J (1983) Translocation of c-abl oncogene correlates with the presence of a Philadelphia chromosome in a chronic myelocytic leukaemia. Nature 306:277–280

45. Konopka JB, Watanabe SM, Witte ON (1984) An alteration of the human c-abl protein in K562 leukemia cells unmasks associated tyrosine kinase activity. Cell 37:1035–1042

46. Klein G, Klein E (1985) Evolution of tumors and the impact of molecular oncology. Nature 315:190–195

CHAPTER 11
Oncogene Activation by Chromosome Translocation

PETER C. NOWELL AND CARLO M. CROCE

There is increasing evidence that many of the nonrandom chromosomal alterations found in neoplastic cells represent mechanisms by which human proto-oncogenes are "activated" and play an important role in carcinogenesis. Most tumors have karyotypic abnormalities [1,2]. In any given tumor, all the cells frequently show the same, or related, alteration, indicating that the cytogenetic change is conferring a selective advantage on the cells carrying it. As various nonrandom karyotypic alterations have been recognized in malignant tumors, specific translocations have been among the most common. Recent studies of such rearrangements, particularly in human leukemia and lymphoma, have begun to indicate the specific oncogenes involved and certain of the mechanisms by which their function may be critically altered. In the following sections, current findings and hypotheses will be briefly reviewed, emphasizing our own investigations. The field is moving rapidly, and significant additional data will undoubtedly be available by the time this summary is published.

Lymphocytic Tumors

The first, and most extensive, evidence concerning "activation" of a human proto-oncogene by chromosomal translocation has come from studies of Burkitt's lymphoma. These studies are now being extended to other human B cell and T cell neoplasms. The Burkitt tumor will be considered first.

In approximately 75% of the cases of Burkitt's lymphoma, there is a reciprocal translocation between chromosomes 8 and 14 (Fig. 11.1). In the remaining 25% of the cases, there are variant translocations involving chromosomes 8 and 22 or chromosomes 2 and 8. In all instances, the breakpoint

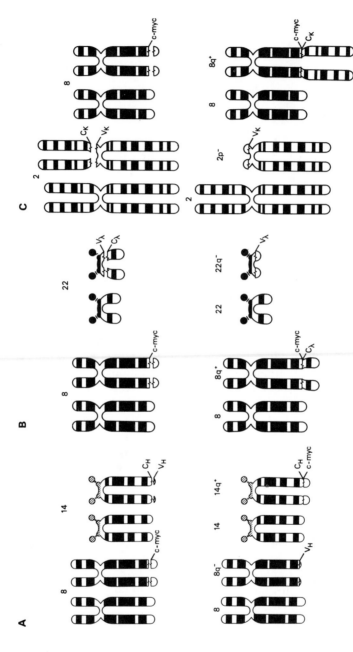

Figure 11.1. Diagram of the three chromosomal translocations (A, B, C) seen in Burkitt's lymphoma that result in the juxtaposition of the *c-myc* oncogene with an immunoglobulin gene locus (C_H, C-λ, C-κ). (From ref. 3.)

on chromosome 8 is the same, at band q24 in the terminal portion of the long arm of the chromosome. The genes for human immunoglobulin heavy and light chains have been mapped to the regions of chromosomes 14, 22, and 2 involved in these translocations: the immunoglobulin heavy chain locus, to band 14q32; the λ light chain genes, to band 22q11; and the κ light-chain genes, to band 2p11. At the same time, the human homologue of the retroviral v-*myc* oncogene was mapped to the terminal portion of 8q, suggesting that both the immunoglobulin genes and the c-*myc* proto-oncogene might have important roles in the development of the Burkitt tumor [3].

Using a combination of cytogenetic and molecular genetic techniques, several laboratories have now confirmed this hypothesis [3–5]. It has been found that in each of these translocations a transcriptionally active and rearranged immunoglobulin gene is brought into juxtaposition with the c-*myc* gene, resulting in deregulation of the oncogene. In the case of the common t(8;14) translocation, the immunoglobulin heavy chain locus is split and the c-*myc* oncogene, with or without structural alteration, is brought into a "head-to-head" association with it. In the variant translocations, the c-*myc* gene, usually without structural alteration, remains on chromosome 8, and the κ or λ immunoglobulin gene is translocated to the 3' end of the oncogene [3–5] (see Fig. 11.1).

Detailed studies of these associations are still ongoing, but it appears that in these translocations, the c-*myc* proto-oncogene comes under the influence of enhancers in or adjacent to the immunoglobulin loci, resulting in deregulation in expression of the *myc* gene and a presumed critical role in the altered growth of the neoplastic B cells. The c-*myc* gene product is a nucleoprotein, apparently having a normal role in growth regulation of lymphocytes and other cells. Studies of the translocated c-*myc* gene, on the 14q+ chromosome of Burkitt tumor cells, indicate that it can still be regulated in an appropriate cellular background (e.g., a fibroblast), but, when under the influence of an enhancing element in a B cell at the proper stage of differentiation, it does not respond to normal regulatory mechanisms [3].These findings concerning the c-*myc* gene in Burkitt's lymphoma, with similar observations in tumors of rats and mice, have provided strong evidence for tumorigenic effects of a structurally unaltered, but deregulated, oncogene, and are also providing important insights into specific "windows of differentiation" in which different oncogenes may function.

Other B Cell Tumors

A significant proportion of other human lymphomas and chronic B cell leukemias have characteristic translocations [1,2] that involve the terminal portion of 14q (band q32) and either the long arm of chromosome 11 (q13) or the long arm of chromosome 18 (q21). It seemed likely that these rearrangements might involve the immunoglobulin heavy chain locus, as in the t(8;14) translocation of Burkitt's lymphoma, but there was no candidate oncogene mapped to the relevant regions of 11q and 18q.

Recently, we have used neoplastic cells from patients with B cell tumors with either the t(11;14)(q13;q32) or the t(14;18)(q32;q21) translocations to molecularly clone the breakpoints involved [6,7]. DNA probes flanking the breakpoints on chromosomes 11 and 18 were then used to detect rearrangements of the homologous sequences in B cell lymphomas and leukemias carrying these translocations. The breakpoints in these cases were found to be clustered within short segments either on chromosome 11 or chromosome 18, and the breakpoint on chromosome 14 was consistently 5' to the constant region of the immunoglobulin heavy chain gene. These findings suggested that unknown oncogenes, located on chromosomes 11 and 18, might be involved in these rearrangements in a manner analogous to the c-*myc* gene in the t(8;14) translocation. We have suggested the names *bcl*-1 and *bcl*-2, respectively, for these two putative new oncogenes, and their characterization is currently under way.

The findings to date indicate that in both cases the translocation is into the *J* region of the immunoglobulin locus on chromosome 14. Furthermore, in the breakpoint regions of the translocated sequences of chromosomes 11 and 18, we have identified several short signal sequences typically used in V–D–J joining during normal rearrangement within the heavy chain locus. This finding indicated that following chromosome breakage, the probability of occurrence of the specific t(11;14) or t(14;18) translocations might increase when the recombinase system normally involved in immunoglobulin gene rearrangement erroneously utilized the chromosome 11 or chromosome 18 signal sequence [6–7]. We have also recently used the *bcl*-2 gene probe to identify a 6-kb RNA transcript, and determined that levels of this transcript are tenfold higher in leukemic B cells with the t(14;18) translocation than in neoplastic B cells without it, suggesting deregulation of the *bcl*-2 gene [7]. These preliminary data are consistent with an oncogene activation mechanism similar to that of the Burkitt lymphoma, but more data are needed concerning the identity of the *bcl*-1 and *bcl*-2 genes and the nature of their presumed altered function.

In addition to these ongoing studies of chromosomal translocations involving the immunoglobulin heavy chain locus as an "activating" site in various B cell tumors, there are several other nonrandom translocations that have been described in acute lymphocytic leukemia (ALL) that are just beginning to be investigated at the molecular level. In some case of ALL, there is a t(9;22) translocation with breakpoints in the same regions as the typical t(9;22)(q32;q11) rearrangement that produces the Philadelphia chromosome in chronic myelogenous leukemia (CML; see below) [1,2].

Preliminary data suggest that the chromosome 22 breakpoint in Ph-positive ALL may involve the immunoglobulin λ chain locus, which is not the case in CML, and the details of this rearrangement in ALL, including involvement of the c-*abl* oncogene from chromosome 9, need clarification [8]. Similarly, it has recently been suggested that a characteristic t(1;19)(q21;23) translocation associated with pre-B cell acute leukemia might involve the insulin receptor gene that has recently been mapped to 19p13 [9]. Because this gene has homologies with both the c-*src* and c-*erb*B oncogenes, the possibility of altered insulin receptor function having an oncogenic effect warrants exploration. There is also a group of cases of childhood ALL characterized by a t(4;11)(q21;q23) translocation involving the terminal portion of the long arm of chromosome 11 to which the c-*ets* oncogene has recently been mapped [10]. Because several other neoplasms (e.g., acute monocytic leukemia, Ewing's tumor) have also been shown to have translocations involving this site [1,2], the possibility of altered structure or function of the c-*ets* oncogene is now being actively investigated.

T Cell Tumors

Consistent chromosomal rearrangements are less common in human T cell neoplasms than in B cell leukemias and lymphomas. Several recent findings, however, have suggested that approaches similar to those used for the study of oncogenes involved in B cell tumors may now be feasible for the investigation, at the molecular level, of T cell neoplasia. The genes coding for several of the chains of the T cell antigen receptor have now been mapped, and in at least some T cell tumors, translocations involving these sites may result in "activation" of proto-oncogenes in a manner analogous to the role of immunoglobulin genes in B cell tumors [11–14]. The gene for the α subunit of the T cell receptor has been mapped to band 14q11, a site frequently involved in translocations and inversions in T cell tumors; the β chain gene has been mapped to the terminal portion of 7q (7q32–35); and other subunit genes may be located at 7p13 and 14q32.

It appears that all four of these sites are unusually fragile in human T cells, perhaps reflecting the effects of somatic recombination [13,14], but their possible "activating" role in neoplasia and the putative oncogenes that might be involved have not yet been defined. This is particularly true of the 14q32 region, where the various chromosomal rearrangements noted thus far in T cell tumors [11] suggest that "activating" genes (immunoglobulin gene? T cell receptor gene?), as well as one or more proto-oncogenes, may all be located within this short chromosomal segment, a phenomenon similar to the q11 region of chromosome 22.

In addition to inversions and translocations involving 14q11 and 14q32, a nonrandom rearrangement that has been described in T cell leukemia [15] involves a translocation from the short arm of chromosome 11 (11p13) to the proximal portion (q11) of chromosome 14. No known oncogene is implicated, but our analysis of this rearrangement indicates that it does split the α chain of the T cell receptor, with the constant portion translocated to chromosome 11 and the variable portion remaining on chromosome 14 [16]. If studies of other nonrandom alterations in T cell neoplasia show similar involvement of T cell receptor genes, this will provide the potential for investigating proto-oncogenes important in T cell tumors, and mechanisms for their activation.

Nonlymphocytic Leukemias

Among nonlymphocytic leukemias, the only chromosomal translocation that has been extensively studied with respect to oncogene involvement is the characteristic t(9;22)(q34;q11) rearrangement that produces the Philadelphia chromosome in most typical cases of chronic myelogenous leukemia (CML) [1,2]. Mapping of the chromosomal breakpoints in this translocation has indicated that the c-*abl* proto-oncogene is regularly translocated from its normal site on chromosome 9 to a very restricted region of chromosome 22 located between the λ immunoglobulin gene and the c-*sis* oncogene. The latter two genes do not appear to be significantly involved in the rearrangement, and the critical segment of chromosome 22 has been termed the *breakpoint cluster region* or BCR [17].

Recent studies of the c-*abl* oncogene in CML have demonstrated the production of a novel 8-kb mRNA, larger than the normal transcript. Cloning of the joining region on chromosome 22 has shown that the altered RNA is in fact a hybrid molecule containing both BCR and abl sequences [18]. Furthermore, this transcript appears to code for an abnormal protein having higher molecular weight and also tyrosine kinase activity, which has not been demonstrated with the normal product. Because some receptors for cellular growth factors are tyrosine kinases, it has been suggested that the altered BCR–abl product, if it proves to be a cell surface protein, might represent a hybrid receptor with altered growth regulatory effects [18]. Unlike the Burkitt tumor translocations, where the *myc* gene product appears unaltered, the recent findings in CML suggest that this may be a circumstance in which a tumorigenic chromosomal translocation results in the production of a modified protein encoded by an oncogene.

Other Nonlymphocytic Leukemias

A number of other specific chromosomal translocations have been shown to be associated, with various degrees of consistency, with different forms of acute nonlymphocytic leukemia. In several instances, the possible involvement of known oncogenes (or possible activation mechanisms) has been suggested, but none has yet been adequately documented.

Nearly every case of acute promyelocytic leukemia (APL) is characterized by a t(15;17)(q22;q21) translocation [1,2]. The c-*erb*A oncogene has been mapped closely proximal to the breakpoint on chromosome 17, suggesting its possible involvement [19]. If true, this would indicate that "activation" of the oncogene resulted from the juxtaposition with a critical sequence on chromosome 15, but a candidate activating sequence or mechanism has not been defined.

A similar situation obtains with the t(8;21)(q22;q22) translocation that characterizes a subgroup of cases of acute myelogenous leukemia (AML). The c-*mos* oncogene is proximal to the breakpoint on chromosome 8, but there is no evidence of its involvement [2]. Studies are underway to attempt to identify and characterize the involved DNA sequence at band 8q22, as well as a possible activating sequence on chromosome 21. We recently studied an atypical case of acute leukemia characterized by an unusual t(17;21)(q21;q22) translocation in which the breakpoints on chromosomes 17 and 21 appeared to be identical, respectively, with those of the t(15;17) of APL and the t(8;21) of AML [19]. In this case, as in APL, the *erb*A oncogene was closely proximal to the breakpoint on chromosome 17, and this suggested that perhaps the activating sequence in this rearrangement, and in AML, was derived from chromosome 21.

Finally, in one form of acute myelomonocytic leukemia (AMMoL) there is a characteristic inversion of chromosome 16 or a t(16;16) translocation involving bands p13 and q22. Involvement of the metallothionein genes in band 16q22 has been demonstrated, with the suggestion that they might activate a proto-oncogene from the 16p13 region [20]. At present, all of these various suggestions concerning oncogene involvement in the acute nonlymphocytic leukemias remain in the realm of speculation, but definitive results should be soon forthcoming.

Conclusions

The findings of the past several years concerning the oncogenes associated with the chromosome translocations of Burkitt's lymphoma and CML have clearly demonstrated that this type of somatic genetic rearrangement represents one mechanism by which gene function can be significantly altered and contribute to the development of human malignancies. The findings in these two neoplasms have also demonstrated that the mechanism of "activation" of an oncogene, following translocation, may be different in different tumors. In the Burkitt's lymphoma, a usually unaltered c-*myc* oncogene appears to be deregulated by its juxtaposition to a rearranged and transcriptionally active immunoglobulin gene. In CML, altered function of the c-*abl* oncogene appears to result from a new gene product produced from a "hybrid" gene formed in the translocation event.

Preliminary data on other lymphocytic neoplasms suggest that the circumstances in these tumors may be more analogous to the Burkitt lymphoma than to CML. The new putative oncogenes *bcl*-1 and *bcl*-2 that are translocated into juxtaposition with the immunoglobulin heavy chain locus in various lymphomas and leukemias may well be deregulated in the same fashion as the c-*myc* gene in the Burkitt tumor. In T cell leukemias, genes coding for the subunits of the T cell receptor, and having many properties in common with the immunoglobulin genes, may similarly activate as yet undefined oncogenes in these neoplasms.

In the acute nonlymphocytic leukemias, the effects of chromosomal translocation are currently much less clear. No known oncogene has been specifically demonstrated to be involved, nor has a putative new oncogene been cloned. It has also not yet been demonstrated whether oncogene "activation" is the result of juxtaposition with another transcriptionally active gene functioning at a particular stage of hemic cell differentiation, or whether the activating process results in a fused gene product analogous to CML, or whether a mechanism as yet completely undefined is involved.

The past several years have seen remarkable progress in our ability to explain how certain chromosomal translocations can confer selective advantage on neoplastic cells. It is clear that this approach to the identification of oncogenes, and of mechanisms by which their function may be critically altered, can be of great value in unraveling molecular mechanisms of oncogenesis. The next few years should prove equally exciting.

References

1. Yunis J (1983) The chromosomal basis of human neoplasia. Science 221:227
2. Rowley JD (1984) Biological implications of consistent chromosome rearrangements in leukemia and lymphoma. Cancer Res 44:3159–3168
3. Croce CM, Nowell PC (1985) Molecular basis of human B cell neoplasia. Blood 65:1–7
4. Leder P, Battey J, Lenoir G, Moulding C, Murphy W, Potter H, Stewart T, Taub R (1983) Translocations among antibody genes in human cancer. Science 222:765
5. Rabbitts TH, Foster A, Hamlyn P, Baer R (1984) Effect of somatic mutation within translocated c-myc genes in Burkitt's lymphoma. Nature 309:592
6. Tsujimoto Y, Jaffe E, Cossman J, Gorham J, Nowell PC, Croce CM (1985) Clustering of breakpoints on chromosome 11 in human B-cell neoplasms with the t(11;14) chromosome translocation. Nature 315:340–343
7. Tsujimoto Y, Cossman J, Jaffe E, Croce CM (1985) Involvement of the bcl-2 gene in human follicular lymphoma. Science 228:1440–1443
8. Erikson J, Griffin CG, ar-Rushdi A, Valtieri M, Hoxie J, Finan J, Emanuel BS, Rovera G, Nowell PC, Croce CM (1986) Heterogeniety of Chromosome 22 breakpoint in PH-positive acute lymphocytic leukemia. Proc Acad Natl Sci USA 83:1807–1811
9. Yang-Feng TL, Francke U, Ullrich A (1985) Gene for human insulin receptor: Localization to site on chromosome 19 involved in pre-B-cell leukemia. Science 228:728–731
10. deTaisne C, Gegonne A, Stehelin D, Bernheim A, Berger R (1984) Chromosomal localization of the human proto-oncogene c-ets. Nature 310:581–583
11. Hecht F, Morgan R, Hecht BK-M, Smith SD (1984) Common region on chromosome 14 in T-cell leukemia and lymphoma. Science 226:1445–1446
12. Croce CM, Isobe M, Palumbo A, Puck J, Ming J, Tweardy D, Erikson J, Davis M, Rovera G (1985) Gene for alpha-chain of human T-cell receptor: Location on chromosome 14 region involved in T-cell neoplasms. Science 227:1044–1047
13. Isobe M, Emanuel BS, Erikson J, Nowell PC, Croce CM (1985) Location of gene for beta subunit of human T cell receptor at band 7q35, a region prone to rearrangements in T cells. Science 228:580

14. Morton CC, Duby AD, Eddy RL, Shows TB, Seidman JG (1985) Genes for beta chain of human T-cell antigen receptor map to regions of chromosomal rearrangement in T cells. Science 228:582–585

15. Williams DL, Look AT, Melvin SL, Roberson PK, Dahl G, Terri Flake, Stass S (1984) New chromosomal translocations correlate with specific immunophenotypes of childhood acute lymphoblastic leukemia. Cell 36:101–109

16. Erikson J, Williams DL, Finan J, Nowell PC, Croce CM (1985) Locus of the alpha chain of the T cell receptor is split by chromosome translocation in T cell leukemias. Science 229:784–786

17. Groffen J, Stephenson JR, Heistercamp N, DeKlein A, Barton CR, Grosveld G (1984) Philadelphia chromosomal breakpoints are clustered within a limited region, bcr, in chromosome 22. Cell 36:93–99

18. Shtivelman E, Lifshitz B, Gale RP, Canaani E (1985) Fused transcript of *abl* and *bcr* genes in chronic myelogenous leukaemia. Nature 315:550–554

19. Dayton AI, Selden JR, Laws G, Dorney DJ, Finan J, Tripputi P, Emanuel BS, Rovera G, Nowell PC, Croce CM (1984) A human c-*erb*A oncogene homologue is closely proximal to the chromosome 17 breakpoint in acute promyelocytic leukemia. Proc Acad Natl Sci USA 81:4495

20. LeBeau MM, Diaz MO, Karin M, Rowley JD (1985) Metallothionein gene cluster is split by chromosome 16 rearrangements in myelomonocytic leukaemia. Nature 313:709–711

CHAPTER 12
Oncogenes, Growth Factors, and Receptors

STEVEN R. TRONICK AND STUART A. AARONSON

Acutely transforming retroviruses are among the most carcinogenic agents known. Because of this property, they have been studied extensively [1]. The model systems provided by these viruses have proven to be valuable sources of information not only on the mechanisms by which normal cells become malignant, but also on how they grow and differentiate. The acutely transforming retroviruses contain genes, termed *oncogenes,* that are responsible for causing tumors in animals and inducing the malignant phenotype in cultured cells. Viral oncogenes have counterparts in normal cells, termed *proto-oncogenes* [2], and detailed analyses of the corresponding viral and cellular sequences have revealed that viral oncogenes arose as a result of the acquisition (by as yet ill-defined recombinational events) of portions of proto-oncogenes by replication-competent retroviruses [2]. These findings have implied the existence of a subset of normal cellular genes with oncogenic potential.

   The integrity of proto-oncogene sequences has been carefully maintained during evolution. For example, some mammalian oncogenes have homologues in organisms as distantly related as yeasts [3,4]. From this, investigators have inferred that proto-oncogenes must play critical roles in basic normal cellular processes such as proliferation and differentiation. As summarized in this review, investigations of the structure and function of these cellular genes as well as their altered viral counterparts are extremely rewarding for deciphering the mechanisms by which naturally occurring tumors arise.

Viral Oncogenes

Nineteen distinct viral oncogenes have been described, and many of their encoded transforming protein products have been identified [5]. The physiological functions of only three of these proteins are known thus far; however, this knowledge has provided important insights about how oncogenes and proto-oncogenes work in both normal and malignant cells. The viral oncogenes are cataloged in Table 12.1. Certain oncogenes share such properties as enzymatic activities, subcellular location, or sequence similarities, and thus can be grouped accordingly [5,6].

The first class includes one member, v-*sis*, whose gene product (as will be discussed in detail shortly) is closely related to a growth factor polypeptide. The second and most populous class comprises genes whose products phosphorylate tyrosine, serine, or threonine residues in proteins, the majority possessing the first activity. All of these transforming proteins share significant amino acid sequence similarities. The v-*erb*B and v-*fms* oncogenes (see Oncogenes and Growth Factor Receptors, below) are related to growth factor receptors. The proteins encoded by members of this class, known alternatively as the *src* or tyrosine kinase family, have been found in cytoplasmic or membrane, but not in nuclear subcellular compartments.

The *ras* family of oncogenes, the third class, encode similarly sized proteins (21,000 daltons) that bind guanine nucleotides, catalyze the hydrolysis of GTP, and phosphorylate proteins. These proteins share similar enzymatic properties and even sequence similarities with the class of proteins termed *G proteins*. G proteins can be involved in signal transduction, as in the adenylate cyclase system [7]; in protein synthesis (elongation factors) [8]; and even in interactions with cellular structural components (tubulin) [9].

The fourth class of viral oncogenes encode proteins that are present in the cell nucleus usually within its matrix [5]. Some of these nuclear oncogene products have been found to bind double-stranded DNA and are thus suspected of influencing transcription.

Oncogenes and Growth Factors

A diverse array of proteins are mitogens for many different cell types [10]. These growth factors exert their mitogenic effects by first binding to specific receptors on the cell surface. In the case of platelet-derived growth factor (PDGF), epidermal growth factor (EGF), and insulin and insulinlike growth factor (IGF-I), this binding leads to activation of intrinsic receptor protein kinase activity [11, 40]. The subsequent biochemical events that next lead to mitogenesis are in large part yet to be deciphered.

The first direct link between oncogenes and growth factor-mediated pro-

Table 12.1. Viral oncogenes.

Oncogene	Virus	Species of origin	Known function(s)	Localization
sis	SSV	woolly monkey	growth factor analogue (PDGF)	cytoplasm, plasma membrane
abl	Ab-MuLV	mouse	protein tyrosine kinase	plasma membrane
erbB	AEV-ES4/AEV-H	chicken	growth factor receptor analogue (EGF receptor)	plasma membrane
fes/fps	FeSV/FSV	cat, chicken	protein tyrosine kinase	cytoplasm
fgr	GR-FeSV	cat	protein tyrosine kinase	?
fms	SM-FeSV	cat	protein tyrosine kinase growth factor receptor analogue (CSF-1 receptor)	plasma membrane
ros	UR2	chicken	protein tyrosine kinase	?
src	RSV	chicken	protein tyrosine kinase	plasma membrane
yes	Y73	chicken	protein tyrosine kinase	?
mos	Mo-MSV	mouse	protein serine–threonine kinase	?
raf/mil	MSV3611/MH2	mouse/chicken	protein serine–threonine kinase	?
H-ras	Ha-MSV/BALB-MSV	rat/mouse	GTP binding, GTPase, autokinase	plasma membrane
K-ras	Ki-MSV	rat	GTP binding, GTPase, autokinase	plasma membrane
fos	FBJ-MSV	mouse	?	nucleus
myb	AMV/E26	chicken	?	nucleus
myc	MC29	chicken	?	nucleus
ski	SKV770	chicken	?	nucleus
erbA	AEV	chicken	?	cytoplasm
ets	E26	chicken	?	nucleus
rel	REV-T	chicken	?	cytoplasm

liferative pathways was established when it was determined that the amino acid sequence of one of the two major polypeptide chains of human PDGF is highly similar to that predicted for the transforming protein encoded by the v-*sis* oncogene [12,13]. The v-*sis* product undergoes rapid disulfide-linked dimer formation and further processing of its N and C termini, yielding a molecule analogous to biologically active PDGF [14]. Moreover, the v-*sis* gene product has been shown to bind PDGF receptors, to stimulate tyrosine phosphorylation of the receptor [15], and to act as a potent mitogen specific for connective tissue cells that possess PDGF receptors [16]. Strong evidence that its transforming activity is mediated directly by its interaction with the PDGF receptor derives from studies demonstrating that v-*sis* transforming activity is also specific to those cell types possessing PDGF receptors [15].

Molecular cloning and nucleotide sequence analysis of the *sis* proto-oncogene from the normal human cellular genome demonstrated that it codes for the PDGF-2 polypeptide chain [17, 41, 42]. When the normal human *sis*/PDGF-2 coding sequence was linked to a powerful promoter, it caused transformation of fibroblasts in culture [17]. In fact, this construct was as potent as the original retrovirus containing the v-*sis* oncogene. These findings established that transcriptional activation of the gene coding this normal growth factor can transform a cell responsive to its growth-promoting actions. The abnormal expression of *sis* in some human spontaneous malignancies of connective tissue origin [18] suggests that *sis*/PDGF-2 may very well play a role in the neoplastic process.

Oncogenes and Growth Factor Receptors

The link between *sis* and PDGF was rapidly followed by the discovery that the next component in the growth factor-mediated proliferation pathway, that is, the growth factor receptor, could also be related to an oncogene product. Amino acid sequences of some peptides derived from the EGF receptor were found to exhibit close similarity with portions of the sequence deduced for the v-*erb* oncogene tranforming protein [19]. Analysis of EGF receptor cDNA clones revealed that v-*erb* represents a short portion of the extracellular domain, the entire transmembrane segment, and all but the C-terminal cytoplasmic region of the EGF receptor. Available evidence suggests that a truncated growth factor receptor, such as the v-*erb*B transforming protein, functions as a constitutively "turned-on" signal for cell proliferation.

A second example of relatedness between growth factor receptors and oncogene products has recently been provided [20]. Studies on the v-*fms* transforming protein, a member of the *src* family, showed that transformation by this oncogene requires the expression of its product at the cell surface. Furthermore, the deduced amino acid sequence of v-*fms* contains a transmembrane binding domain characteristic of known growth factor receptors. The tissue specificity of c-*fms* proto-oncogene expression was found to coincide with that of a receptor for a murine macrophage growth factor designated colony stimulating factor-1 (CSF-1) [20]. Antibodies prepared against a v-*fms*-coded polypeptide could be shown to react specifically with purified murine CSF-1 receptor [20], and it was further demonstrated that the CSF-1 receptor underwent phosphorylation on tyrosine in response to addition of CSF-1. Thus, it is very likely that the CSF-1 receptor is encoded by the c-*fms* proto-oncogene.

Cellular homologues of other oncogenes also appear to be related to growth factor receptors. The predicted amino acid sequence of the cytoplasmic tyrosine kinase domain of the v-*ros* oncogene of the UR-2 avian sarcoma virus is most closely related to the human insulin receptor [21]. Furthermore, the v-*ros* sequence contains a stretch of hydrophobic amino acids that may act as a membrane spanning domain [22]. Sequence analysis of the human *ros* proto-oncogene will be required to determine just how closely it resembles the insulin receptor gene.

A transforming gene *(neu)* isolated from a rat neuroblastoma cell line was found to be expressed on the cell surface and to be antigenically related to the EGF receptor [23]. Independently, human genes amplified in a breast carcinoma [24] and a salivary gland malignancy [25] were identified, and shown by sequence analysis to be the same and related to but distinct from the EGF receptor. It is likely that *neu* and this human gene, designated *erb*B-2 or MAC [24,25], are the same because their chromosomal localization in human cells is coincident [26]. Comparison of the coding sequence of this new human EGF-related oncogene with that of the EGF receptor shows close overall structural organization; the highest degree of predicted amino acid sequence similarity between the two is located within their cytoplasmic tyrosine kinase domains.

It is worthwhile to reexamine Table 12.1 and note that there exists a large family of retroviral oncogenes that either encode tyrosine protein kinases or are related to such enzymes on the basis of their nucleotide sequences. With the exceptions of v-*erb*B, v-*fms,* and v-*ros* discussed above, none of the other family members possesses the extracellular or transmembrane binding domain, hallmarks of growth factor receptors. However, it is known that some are membrane associated whereas others are cytoplasmic. Because retroviral oncogenes represent, in the majority of cases, truncated cellular genes, it is possible that their normal cellular homologues function as growth factor receptors. Alternatively, it is possible that some or all of these molecules function in growth factor activated pathways in positions downstream from the ligand–receptor interaction.

Oncogenes and the Second Messenger System

A variety of hormones and neurotransmitters bind to cell surface receptors. The outcome of these interactions, which can be either positive or negative, affects adenylate cyclase activity [7]. The cAMP-dependent protein kinases and phosphoprotein kinases then come into play in a manner that depends upon the new levels of adenylate cyclase activity. The G proteins interact with receptors and adenylate cyclase either to stimulate or to inhibit activity of the latter. Analogous pathways are involved in signal transduction in olfactory reception and photoreception, although adenylate cyclase is not involved in the latter pathway [27].

Ras oncogene products share some structural and functional properties with G proteins, suggesting that these oncogenes may play critical roles in another pathway central to the regulation of growth. The predicted amino acid sequences of *ras* proteins have been extraordinarily highly conserved in species as diverse as human and yeasts [3,4]. *Ras* proteins of yeast (reviewed in ref. 5) have been reported to act either directly or indirectly as controlling elements of adenylate cyclase, but evidence to date has failed to implicate *ras* p21 as a regulatory component of adenylate cyclase in mammalian cells. Thus, the function of *ras* proteins still remains to be elucidated.

Oncogenes Coding for Nuclear Proteins

Indirect evidence exists that some oncogene products function more distally in pathways activated by growth factors that lead to DNA synthesis. Thus, several viral oncogene products are localized within the cell nucleus (Table 12.1) and, in addition, some possess the ability to bind DNA [5,6]. Recent studies have indicated that certain of these genes are rapidly mobilized following growth factor stimulation of quiescent cells. In particular, c-*fos* appears to be transcriptionally activated within minutes and increases dramatically following growth factor addition [28,29]. Others, such as c-*myc* and c-*myb*, are also affected but not nearly as rapidly or markedly. There is also some evidence that nuclear oncogenes may act to affect the transcription of other genes required for growth and cell division.

Transformation by Viral Oncogenes and Normal Growth Control Pathways

Advances in techniques for culturing hematopoietic cells have made it possible to expand certain populations of hematopoietic cells by use of various growth factors. Such culture systems have been utilized to great advantage by investigators to study the effects of retroviral transforming genes on hematopoietic cells that normally respond to specific physiological regulators. Infection of normal fetal liver mast cells with Abelson murine leukemia virus (Ab-MuLV), which contains the *abl* oncogene, malignantly transforms the cells and also releases them from their dependence on the growth factor, interleukin-3 (IL-3) [30]. These Ab-MuLV-transformed cells do not release detectable levels of mitogens for IL-3-dependent cells, IL-3 expression could not be detected in such cells, and antibody against IL-3 had no effect on the growth of the transformants. Thus, it appears that an autocrine mechanism is not involved in the Ab-MuLV abrogation of mast cell IL-3 requirement. Similarly, Ab-MuLV transformation of a murine myeloid cell line that is absolutely dependent on the presence of IL-3 or granulocyte–macrophage colony stimulating factor (GM-CSF) for proliferation and survival resulted in growth of the cells in the absence of these factors [31]. The malignant cells did not synthesize detectable levels of either IL-3 or GM-CSF. Genetically engineered constructs of the v-*myc* gene can also convert murine hematopoietic cell lines dependent on either IL-2 or IL-3 to factor independence, presumably by also short-circuiting the growth factor pathways upon which the cells are normally dependent [32].

In contrast, chicken cells transformed by either v-*myc* or v-*myb* retain dependence upon growth factors. Superinfection with avian viruses containing *src*-related oncogenes releases the cells from growth factor requirements apparently by an autocrine mechanism [33] involving synthesis of a substance closely related to chicken myelomonocytic growth factor [34].

Oncogenes can also short-circuit growth factor-dependent pathways required for epithelial cell proliferation. One well-characterized epithelial model system involves clonal lines of mouse epidermal keratinocytes (BALB/MK) cells that are dependent upon EGF for their growth. This system is particularly relevant, because epithelial cells are the most common targets for neoplastic alterations in vivo. Infection of BALB/MK-2 cells with viruses representing two major oncogene families (*src* and *ras*) frees them from their dependence on EGF [35]. None of the oncogenes utilized in these studies encodes a protein that shares structural properties with EGF. Moreover, EGF-related factors were not found to be detectably released by virus-altered BALK/MK cells, suggesting that such factors do not play a direct role in the growth alterations induced by oncogenes of the *src* and *ras* families. If so, these oncogenes must act by circumventing normal EGF control of epithelial cell replication.

Oncogenes and Human Cancers

One of the lessons learned from studying retroviral oncogenes and their cellular homologues is that the number of cellular genes capable of acquiring transforming properties when incorporated within retrovirus genomes must be rather limited. Thus, several viruses isolated from the same or different species have been shown to have captured identical *onc* sequences [2,5,36]. These observations would not be expected if any of the thousands of cellular genes could acquire neoplastic properties when captured by a retrovirus. Thus, if the neoplastic process involved in spontaneous cancers were to utilize similar mechanisms, investigators might not have to confront the entire array of cellular genes in their efforts to decipher the processes leading to malignancy in human cells.

This view was greatly strengthened when it was shown that DNAs from some human tumor cells could confer the neoplastic phenotype to susceptible assay cells by DNA-mediated gene transfer techniques, and that the genes involved were actually related to retroviral oncogenes (reviewed in ref. 36). The transforming genes most frequently detected by DNA transfection assays are members of the *ras* family and include a *ras* gene (N-*ras*) that has thus far been able to elude capture by a retrovirus. It has been further demonstrated that *ras* proto-oncogenes most commonly acquire malignant properties by single point mutations affecting either the 12th or 61st codons within the 189 amino acid coding sequence of the p21 molecule [36]. Activated *ras* oncogenes have been detected in a wide variety of tumor types. The frequency of their detection ranges from 10% to 20% of human primary malignancies. The demonstration of *ras* oncogenes in malignant but not normal tissues of the same patient has provided strong evidence that the activating lesions are somatic in origin, powerfully selected within the tumor cell population, and thus, very likely to play an important role in the development of those tumors in which *ras* oncogenes are present.

Other lines of investigation have provided independent evidence implicating oncogenes in human malignancy. It has long been known that nonrandom chromosomal aberrations occur in a number of cancers, the majority affecting hematopoietic tissues. These chromosomal translocations are of importance because of their specific association with morphologically well-defined subtypes of leukemias and lymphomas. The best-characterized tumors in which specific chromosomal translocations occur are Burkitt's lymphoma (BL) and chronic myelogenous leukemia (CML). Advances in cytogenetics and the availability of human oncogene probes made it possible to show that proto-oncogenes (*myc* in BL, *abl* in CML) are present at the breakpoints in virtually all cases of these tumors that have been examined (reviewed in ref. 36). The molecular consequences of *myc* gene translocations have been detailed and are thought to result in inappropriate expression of *myc*. In contrast, the *abl* translocation involves recombination between a gene of unknown function *(bcr)* and c-*abl* so as to replace the N terminus of c-*abl* with *bcr* sequences [37,38]. There is evidence that this DNA rearrangement results in both an altered *abl* transcript and protein. An increasingly fruitful approach to studying the involvement of oncogenes in human tumors and in identifying new oncogenes is the molecular cloning of sequences involved in specific translocations known to occur in other hematopoietic malignancies [43].

In addition to chromosomal translocations, human tumor cells frequently display karyotypic markers of gene amplification such as double-minute chromosomes and homogeneously staining regions. Investigators have demonstrated the presence of amplified oncogenes such as c-*myc* in colon and small-cell lung carcinoma cells [36], N-*myc* in retinoblastomas and neuroblastomas, and *erb*B in squamous carcinomas. In neuroblastomas, amplification of N-*myc* appears to be involved in tumor progression, because only those tumors with a more aggressive phenotype appear to exhibit this particular DNA rearrangement.

Implications

Research advances summarized in this chapter have implicated what appears to be a rather limited set of cellular genes as major targets for genetic alterations that can lead normal cells along the pathway to malignancy. When viewed from the perspective that normal cell growth and development may involve the interactions among literally thousands of cellular genes and their products, these recent findings offer hope that defining the mechanisms of neoplastic transformation has become much more amenable to solution.

It has long been known from clinical observations that cancer is very likely a multistep process. Such evidence strongly argues against the concept that neoplastic transformation may be accounted for by a single event, such a point mutation activating a *ras* oncogene. The number of genetic alterations needed for neoplastic transformation remains to be elucidated. In cell culture systems, transfection of primary rat embryo fibroblasts with two oncogenes, *ras* and *myc* or *ras* and E1A, an early gene of the adenovirus genome, appears sufficient to induce the malignant phenotype, whereas either transforming gene alone cannot do so [6]. In human cell culture systems, it has been possible to induce neoplastic transformation of human epidermal keratinocytes by the sequential addition of viruses containing different oncogenes [39]. Such transformants achieved frank malignancy and could induce progressively growing squamous cell carcinomas in athymic nude mice, whereas uninfected keratinocytes or keratinocytes infected with only one virus could not induce tumors. Such findings suggest that the number of discrete steps in the neoplastic pathway may be relatively few, and that there may be complementing groups of oncogenes.

There is a great likelihood that not all of the proto-oncogenes capable of being activated as oncogenes have been discovered. Moreover, genetically predisposing states certainly exist but probably involve recessive alterations whose effects are more subtle than that of the germ-line transmission of an activated oncogene. Finally, the few genes that have already been identified as targets of the neoplastic process are highly conserved and in their altered state are almost certainly involved in essential growth processes. Nonetheless, the repeated detection of the same small set of oncogenes, as well as knowledge that they often seem to act within the proliferative pathways triggered by growth factors and their receptors, gives an important conceptual framework toward understanding the neoplastic process. Hopefully, the knowledge being gained will eventually lead to insights into approaches that will be useful in the diagnosis and treatment of human cancer.

References

1. Weiss RA, Teich N, Varmus H (1982) Molecular biology of tumor viruses. *In* Coffin J (ed) RNA Tumor Viruses, 2nd Ed. Cold Spring Harbor Laboratory, Cold Spring Harbor, New York
2. Bishop JM (1983) Cellular oncogenes and retroviruses. Annu Rev Biochem 52:301–354
3. Wigler M (1984) Genes in *S. cerevisiae* encoding proteins with domains homologous to the mammalian *ras* proteins. Cell 36:607–612.
4. Dhar R, Neito A, Koller R, DeFeo-Jones D, Scolnick EM (1984) Nucleotide sequence of two *ras*H-related genes isolated from the yeast *Saccharomyces cerevisiae*. Nucleic Acids Res 12:3611–3618
5. Bishop JM (1985) Viral oncogenes. Cell 42:23–38
6. Weinberg RA (1985) The action of oncogenes in the cytoplasm and nucleus. Science 230:770–776
7. Gilman AG (1984) G proteins and dual control of adenylate cyclase. Cell 36:577–579
8. Miller DL, Hachmann J, Weissbach H (1971) The reactions of the sulfhydryl groups on the elongation factors Tu and Ts. Arch Biochem Biophys 144:115–121
9. Carlier M-E (1982) Guanosine-5'-triphosphate hydrolysis and tubulin polymerization. Mol Cell Biochem 47:97–113
10. James R, Bradshaw RA (1984) Polypeptide growth factors. Annu Rev Biochem 53:259–292
11. Hunter T, Cooper JA (1985) Protein–tyrosine kinases. Annu Rev Biochem 54:897–930
12. Doolittle RF, Hunkapiller MV, Hood LE, Devare SG, Robbins KC, Aaronson SA, Antoniades HN (1983) Simian sarcoma virus *onc* gene, v-*sis*, is derived from the gene (or genes) encoding a platelet-derived growth factor. Science 221:275–277
13. Waterfield MD, Scrace GT, Whittle N, Stroobant P, Johnsson A, Wasteson A, Westermark B, Heldin CH, Huang JS, Deuel TF (1983). Platelet-derived growth factor is structurally related to the putative transforming protein p28^{sis} of simian sarcoma virus. Nature 304:35–39
14. Robbins KC, Antoniades, HN, Devare SG, Hunkapiller MV, Aaronson SA (1983) Structural and immunological similarities between simian sarcoma virus gene product(s) and human platelet-derived growth factor. Nature 305:605–608.
15. Leal F, Williams LT, Robbins KC, Aaronson SA (1985) Evidence that the v-*sis* gene product transforms by interaction with the receptor for platelet-derived growth factor. Science 230:327–330
16. Stiles CD (1983) The molecular biology of platelet-derived growth factor. Cell 33:653–655
17. Igarashi H, Gazit A, Chiu I-M, Srinivasan A, Yaniv A, Tronick SR, Robbins KC, Aaronson SA (1985) Normal human *sis*/PDGF-2 gene expression induces cellular transformation. *In* Feramisco J, Ozanne B, Stiles C (eds) Cancer Cells, Vol 3: Growth Factors and Transformation, Cold Spring Harbor Laboratories, Cold Spring Harbor, New York, pp 159–166
18. Eva A, Robbins KC, Andersen PR, Srinivasan A, Tronick SR, Reddy EP, Ellmore NW, Galen AT, Lautenberger JA, Papas TS, Westin EH, Wong-Staal F, Gallo RC, Aaronson SA (1982) Cellular genes analogous to retroviral *onc* genes are transcribed in human tumor cells. Nature 295:116–119

19. Downward J, Yarden Y, Mayes E, Scrace G, Totty N, Stockwell P, Ullrich A, Schlessinger J, Waterfield MD (1984) Close similarity of epidermal growth factor receptor and v-*erb*B oncogene protein sequences. Nature 307:521–527

20. Sherr CJ, Rettenmier CW, Sacca R, Roussel MF, Look AT, Stanley ER (1985) The c-*fms* proto-oncogene product is related to the receptor for the mononuclear phagocyte growth factor, CSF-1. Cell 41:665–676

21. Ullrich A, Bell JR, Chen EY, Herrera R, Petruzzelli LM, Dull TJ, Gray A, Coussens L, Liao YC, Tsubokawa M, et al. (1985) Human insulin receptor and its relationship to the tyrosine kinase family of oncogenes. Nature 313:756–761

22. Neckameyer WS, Wang LH (1985) Nucleotide sequence of avian sarcoma virus UR2 and comparison of its transforming gene with other members of the tyrosine protein kinase oncogene family. J. Virol 53:879–884

23. Schechter AL, Stern DF, Vaidyanathan L, Decker SJ, Drebin JA, Green MI, Weinberg RA (1984) The *neu* oncogene: An *erb*B-related gene encoding a 185,000-Mr tumour antigen. Nature 312:513–516

24. King CR, Kraus MH, Aaronson SA (1985) Amplification of a novel v-*erb*B-related gene in a human mammary carcinoma. Science 229:974–976

25. Semba K, Kamata N, Toyoshima K, Yamamoto T (1985) A v-*erb*B-related protooncogene, c-*erb*B-2, is distinct from the c-*erb*B-1/epidermal growth factor-receptor gene and is amplified in a human salivary gland adenocarcinoma. Proc Natl Acad Sci USA 82:6497–6501

26. Coussens LC, Yang-Feng TL, Liao Y-C, Chen E, Gray A, McGrath J, Seeburg PH, Libermann KTA, Schlessinger J, Francke U, Levinson A, Ullrich A (1985) Tyrosine kinase receptor with extensive homology to EGF receptor shares chromosomal location with *neu* oncogene. Science 230:1132–1139

27. Shepherd GM (1985) Olfactory transduction: Welcome whiff of biochemistry. Nature 316:214–215

28. Greenberg ME, Ziff EB (1984) Stimulation of 3T3 cells induces transcription of c-*fos* proto-oncogene. Nature 311:433–438

29. Kelly K, Cochran BH, Stiles CD, Leder P (1983) Cell-specific regulation of the c-*myc* gene by lymphocyte mitogens and platelet-derived growth factor. Cell 35:603–610

30. Pierce JH, Di Fiore PP, Aaronson SA, Potter M, Pumphrey J, Scott A, Ihle JN (1985) Neoplastic transformation of mast cells by Abelson-MuLV: Abrogation of IL-3 dependence by a nonautocrine mechanism. Cell 41:685–693

31. Cook WD, Metcalf D, Nicola NA, Burgess AW, Walker F (1985) Malignant transformation of a growth factor-dependent myeloid cell line by Abelson virus without evidence of an autocrine mechanism. Cell 41:677–683

32. Rapp UR, Cleveland JL, Brightman K, Scott A, Ihle JN (1985) Abrogation of IL-3 and IL-2 dependence by recombinant murine retroviruses expressing v-*myc* oncogenes. Nature 317:434–438

33. Sporn MB, Roberts AB (1985) Autocrine growth factors and cancer. Nature 313:745–747

34. Adkins B, Leutz A, Graf T (1984) Autocrine growth induced by *src*-related oncogenes in transformed chicken myeloid cells. Cell 39:439–445

35. Weissman B, Aaronson SA (1985) Members of the src and ras oncogene families supplant the EGF requirement of BALB/MK-2 keratinocytes and induce distinct alterations in their terminal differentiation program. Mol Cell Biol 5:3386–3396

36. Varmus HE (1984) The molecular genetics of cellular oncogenes. Annu Rev Genet 18:553–612
37. Shtivelman E, Lifshitz B, Gale RP, Canaani E (1985) Fused transcript of *abl* and *bcr* genes in chronic myelogenous leukaemia. Nature 315:550–554
38. Stam K, Heisterkamp N, Grosveld G, de Klein A, Verma RS, Coleman M, Dosik H, Groffen J (1985) Evidence of a new chimeric *bcr*/c-*abl* mRNA in patients with chronic myelocytic leukemia and the Philadelphia chromosome. N Eng J Med 313:1429–1433
39. Rhim JS, Jay G, Arnstein P, Price FM, Sanford KK, Aaronson SA (1985) Neoplastic transformation of human epidermal keratinocytes by AD 12-SV40 and Kirsten sarcoma viruses. Science 227:1250–1252
40. Heldin C-H, Westermark B (1984) Growth factors: Mechanisms of action and relation to oncogenes. Cell 37:9–20
41. Johnsson A, Heldin C-H, Wasteson Å, Westermark B, Deuel TF, Huang HS, Seeburg PH, Gray A, Ullrich A, Scrace G, Stroobant P, Waterfield MD (1984) The c-*sis* gene encodes a precursor of the B chain of platelet derived growth factor. EMBO J 3:921–928
42. Josephs SF, Guo C, Ratner L, Wong-Staal F (1984) Human proto oncogene nucleotide sequence corresponds to the transforming gene of simian sarcoma virus. Science 223:487–491
43. Croce CM, Nowell PC (1986) Molecular genetics of human B cell neoplasia. Adv Immunol 38:245–274

CHAPTER 13
Specificity of Viral Gene Controlling Elements in Transgenic Mice

Heiner Westphal

Among the many facets of viral pathogenesis, one of particular interest to the molecular biologist is the ability of the virus to adapt its mechanisms for gene control to specific host environments. Although we may be accustomed to analyze viral gene regulation in cultures of cells that are permissive for a given virus or suitable for transfection with cloned gene constructs, in the living organism the virus associates itself with cells that may be quite different from those in the Petri dish. Why does one virus single out the brain, another the respiratory system, a third one the intestinal tract? This complex problem of viral tissue tropism has been the topic of previous contributions to this series (see *Concepts in Viral Pathogenesis,* Vol I, Chapters 11, 15, and 19). New clues have come from studies of viral gene controlling elements and their role in the selection of target tissues in the animal. These viral control sequences, also called *enhancers* are thought to interact with specific host factors to regulate the transcription of genes that are under their influence. The reader is referred to Chapter 8 in this volume for details. Here, the focus will be on the expression in transgenic mice of genes that are under the control of viral enhancers.

Transgenic is a term now commonly used for mice carrying exogenous DNA that has been integrated in the genome and is usually transmitted to progeny through the germ line. The first transgenic mouse strains ever generated in the laboratory carried retroviral DNA, which integrated after infection of embryos with Moloney murine leukemia virus [1]. Packaged into retrovirus vectors, genes other than those of retroviruses should also be transferable by this method. However, the gene transfer system used by most laboratories today is based on Graessmann's manual microinjection technique [2]. The gene construct of choice is injected via a fine glass capillary into the male or the female pronucleus of the early mouse embryo. The embryo is placed into the oviduct of a pseudopregnant foster animal and brought to term. Animals found to have the new gene integrated as part of their genome usually pass it on as a Mendelian trait to progeny. Many strains of transgenic mice carrying a wide variety of gene constructs have been obtained by this method (see ref. 3 for a recent review). Among the facts that have been gathered from studying these mice, the following are important with respect to the topic of this chapter:

1. Gene expression is often faithfully directed to the correct target tissue, irrespective of the integration site of the transferred gene.
2. A number of chimeric genes containing a 5' flanking region of gene A and the coding sequence of gene B have been found to express gene product B in the target tissue determined by gene A.
3. In the few instances where this has been measured, tissue specificity of expression coincides with correct temporal control, i.e., the expression of a transferred gene occurs at the predicted time of development.

The following two examples may illustrate these facts: In both cases, chimeric gene constructs were used, and a number of transgenic strains were obtained, each presumed to contain the inserted gene at a different chromosomal site of integration. The first study [4] was performed with fusion genes in which the 5' flanking sequences of the elastase gene were required to direct expression of the human growth hormone gene to pancreatic acinar cells, the normal site of elastase gene expression. The second study involved work of our own laboratory [5]. A 409-bp 5' flanking sequence of the mouse αA lens crystallin gene, fused to the bacterial gene encoding chloramphenicol acetyltransferase (CAT), was inserted into the mouse genome. CAT expression occurred selectively in lens fibers and epithelia, the only tissue known to contain α-crystallin. Moreover, both CAT activity and α-crystallin became detectable in embryo eyes at approximately day 12.5 of development, suggesting that the spatial and temporal control of the fusion gene coincided with that of the genuine αA crystallin gene.

Success with fusion genes such as those described in these examples set the stage for examining tissue specificities of viral gene control elements in transgenic mice. Extensive work on the cell tropisim of viruses in vitro or on their disease specificities in vivo has focused attention on viral enhancer sequences as likely determinants of tissue specificity. For example [6], when enhancer sequences of a murine leukemia virus were exchanged for those of polyoma virus, the recombinant virus, although still able to replicate in cell culture, completely lost its leukemogenicity. This and other studies (see Chapter 8) ascribe to viral enhancers a pivotal role in disease specificity and suggest a gene activation mechanism that is based on the recognition of enhancers by factors specific to certain target tissues in the living organism.

Transgenic mice constitute an ideal experimental system to determine the tissue specificity of viral enhancers. A chimeric gene, consisting of a given viral enhancer and promoter sequence fused to an easy-to-score marker gene such as CAT, for example, can be constructed. Once inserted in the mouse germ line, this gene will be present in every conceivable cell type of the developing organism. When various tissues of the animal are screened for marker gene expression, a pattern should emerge that is specific for the enhancer controlling the marker gene. In our own studies [5], we have used the construct of Rous sarcoma virus pRSV–CAT [7] for such a screening procedure. The 5′ end of the gene construct consists of a long terminal repeat (LTR) sequence of Rous sarcoma virus (RSV). This sequence contains viral enhancer and promoter sequences. Fused to the RSV LTR are the bacterial CAT gene sequences. The 3′ end of the chimeric gene is formed by SV40 sequences containing signals for RNA splicing and polyadenylation. We inserted the pRSV–CAT gene into mouse embryos and obtained several strains of transgenic mice. CAT enzyme activity was detected almost exclusively in muscle and in connective tissue. Tumors caused by Rous sarcoma virus in birds and rodents originate in corresponding tissues [8]. Therefore, the fact that the gene controlling elements contained in the RSV LTR direct marker gene expression to the same tissues that are the site of diseases caused by the intact Rous sarcoma virus can be used as the strongest argument so far for the key role of these elements in viral disease specificity.

Two other recent studies also deal with tissue specificities exerted by viral gene controlling sequences in transgenic mice. The first of these studies [9] reports on choroid plexus tumors in the brain of transgenic mice carrying early SV40 genes and cites evidence suggesting that the SV40 enhancer region is responsible for this remarkable tissue tropism. In the second study [10], insertion of a fusion gene containing the LTR of mouse mammary tumor virus 5′ to the mouse *myc* gene led to expression of this gene and to tumor formation in breast tissue of some of the resulting transgenic animals.

If the three examples cited above are any guide, one may expect the palette of tissue tropisms detected in this type of experiment to be no less colorful than that of extant viral enhancers. There is also plenty of room for surprises. For instance, complex patterns of expression in several tissues or organs and at more than one stage of development may emerge, and one may begin to reflect on possible pathways of differentiation that link such sites of expression in time and space. Or one may detect that control sequences of seemingly unrelated viruses generate similar patterns of gene expression in the animal, pointing to common denominators in the evolution of the viruses so compared.

With several constellations of viral enhancers and their corresponding target tissues in the animal already established, the question as to the nature of host factors that are thought to be involved in enhancer-mediated gene expression becomes more pressing. New and powerful methods are available [11] to purify such trans-acting factors that bind specifically to the regulatory sequences of eukaryotic genes. Whereas the virus uses these factors for its own purpose, their functions in the animal are unknown. Viral enhancer-mediated gene expression is influenced by the state of differentiation of a cell, and viruses can adapt to changes in differentiation by altering the enhancer sequence (for details, see Chapters 6 and 8, this volume). It is attractive to speculate, therefore, that our studies on the control of viral gene controlling elements will ultimately lead to new insights into mechanisms that govern tissue differentiation in vivo.

References

1. Jaenisch R, Jähner D, Nobis P, Simon I, Löhler J, Harbers K, Grotkopp D (1981) Chromosomal position and activation of retroviral genomes inserted into the germ line of mice. Cell 24:519–529
2. Graessmann A, Graessmann M, Mueller C (1980) Microinjection of early SV40 DNA fragments and T antigen. Methods Enzymol 65:816–825
3. Gordon JW, Ruddle FH (1985) DNA-mediated genetic transformation of mouse embryos and bone marrow—a review. Gene 33:121–136
4. Ornitz DM, Palmiter RD, Hammer RE, Brinster RL, Swift GH, MacDonald RJ (1985) Specific expression of an elastase–human growth hormone fusion gene in pancreatic acinar cells of transgenic mice. Nature 313:600–612
5. Westphal H. Overbeek PA, Khillan JS, Chepelinsky AB, Schmidt A, Mahon KA, Bernstein KE, Piatigorsky J, de Crombruggne B (1985) Promoter sequences of murine αA crystallin, murine α2(1) collagen or of avian sarcoma virus genes linked to the bacterial chloramphenicol acetyl transferase (CAT) gene direct tissue specific patterns of CAT expression in transgenic mice. Cold Spring Harbor Symp Quant Biol 50:411–416
6. Davis B, Linney E, Fan H (1985) Suppression of leukaemia virus pathogenicity by polyoma virus enhancers. Nature 314:550–553
7. Gorman CM, Merlino GT, Willingham MC, Pastan I, ward BH (1982) The Rous sarcoma virus long terminal repeat is a strong promoter when introduced into a variety of eukaryotic cells by DNA-mediated transfection. Proc Natl Acad Sci USA 79:6777–6781
8. Purchase JT, Burmester BR (1978) Neoplastic diseases. Leukosis/sarcoma group. In Hofstad MS (eds) Disease of Poultry. Iowa State University Press, Ames, pp 418–468
9. Brinster RL, Chen HY, Messing A, van Dyke T, Levine AJ, Palmiter RD (1984) Transgenic mice harboring SV40 T-antigen genes develop characteristic brain tumors. Cell 37:367–379
10. Stewart TA, Pattengale PK, Leder P (1984) Spontaneous mammary adenocarcinomas in transgenic mice that carry and express MTV/myc fusion genes. Cell 38:627–637
11. Wu C (1985) An exonculease protection assay for specific DNA-binding proteins in crude nuclear extracts. Nature 317:84–87

Viral Tropism and Entry into Cells

CHAPTER 14
Isolation of Cellular Receptors for Viruses

RICHARD L. CROWELL AND K-H. LEE HSU

The inital event in the life cycle of a virus is the attachment to specific receptors on the cell surface [1,2]. In many cases, this interaction is a major determinant of virus tropism in pathogenesis [3,4]. Virus attachment is mediated by specialized structures, referred to as virion attachment proteins (VAP), occurring in multiple copies on the virion. For some viruses, the local environmental conditions may cause a conformational change in the VAP to permit attachment and entry to virions into cells. Cellular receptors with polar groups complementary to the VAP occur as multiple copies in the plasma membrane of human and animal cells. These receptors may serve to bind both viruses and hormones and may be comprised of either multiple subunits, as the nicotinic acetylcholine receptor [5], or a single polypeptide chain, as the β-adrenergic hormone receptor [6]. Because virions have a relatively large mass, as compared to hormones, virions may occupy multiple cellular receptor sites (CRS), whereas an isolated VAP such as the adenovirus fiber would combine with only one site [1]. Furthermore, the receptors are embedded in the lipid matrix of the plasma membrane, whose composition may influence the reactivity of the receptors for binding and processing virions [7]. In general, receptors for viruses are glycoproteins (or occasionally glycolipids or sialyloligosaccharides) that show high-affinity binding for a given virus and demonstrate saturability and specificity [1,8].

The results of numerous studies of virus–cell interactions have revealed that different viruses utilize different specialized cell structures as receptors (Table 14.1). Apparently, the different viruses have evolved to take advantage of these different sites, which may be strategically located for facilitating entry into the cell [32].

A significant but little studied area of inquiry is the definition of the number of qualitatively different types of cellular receptors for a given virus, that

Table 14.1. Cellular organelles or molecules proposed as receptors for viruses.

Virus	Organelle or molecule	Approx. mol. wt.	Cell type	Reference
Enveloped				
Semliki Forest	H-2K and H-2D	44	multiple	9,10
Sindbis	Catecholaminergic neurotransmitter (?)	(?)	skeletal muscle	11
Epstein–Barr	Complement receptor 2	145	B lymphocytes	12,13
	Class II antigen	60	B lymphocytes	14
LAV/HTLV-III[a]	CD4 (T4) antigen	55	T lymphocytes	15,16
Murine leukemia (Rauscher)	(?)	10	fibroblasts	17
Rabies	Acetylcholine receptor	292	skeletal muscle	18,19
Vesicular stomatitis	Phospho- or glycolipid	<10	fibroblasts	20,21
Lactate dehydrogenase	Ia antigen	30	macrophages	22,23
Influenza A	glycophorin A (MN agn.) sialyloligosaccharides	84	erythrocytes	24,25
Sendai	gangliosides	<10	multiple	26
Vaccinia	epidermal growth factor receptor	170	L Cells	55
Nonenveloped				
Reovirus T3	β-adrenergic	67	neurons, lymphs L cells	27
Encephalomyocarditis	glycophorin A	84	erythrocytes	28
group B coxsackievirus	Rp-a (?)	49.5	HeLa	29
Rhinovirus	(?)	90	HeLa	30
Adenovirus	(?)	42	HeLa	31

[a]LAV/HTLV-III, lymphadenopathy virus / human T cell leukemia/lymphoma III virus.

may serve to facilitate natural infection of cells. For most viruses, nonspecific binding of virions to cells may occur, but these sites are nonsaturable and usually do not lead to productive infection [2]. Certain enveloped viruses (influenza, Sendai, vesicular stomatitis, etc.) bind to sialoglycoproteins, phospholipids, or glycosides which are widely distributed in nature and may be considered to be nonspecific receptors [21]. However, some of these molecules show specificity for selecting virus variants from populations of virions [20,25,26]. The binding and internalization of flaviviruses by cells is facilitated by neutralizing antibodies, which serve as ligand bridges to the Fc receptors [33]. Other studies have revealed that transfection of viruses can be accomplished by surrounding virions with liposomal membranes of a specific composition [7,34,35]. Uncoating of the genomes in receptorless cells for some viruses may not require anything more specific than the degradation of the viral coat by lysosomal enzymes [32]. On the other hand, the encapsulation of intact poliovirions in liposomes does not facilitate an efficient mechanism of virus infection of receptorless cells, as compared to the delivery of virus RNA in liposomes [36]. Undoubtedly, the binding of specific receptors to the poliovirions is needed to accomplish efficient disassembly of the virions [37].

It is likely that more than one specific type of receptor can serve to bind and facilitate virus uncoating [9,10,11,38]. For example, hemagglutination-positive, host-range variants of the group B coxsackieviruses probably acquired a second VAP for infection of RD cells, because these variant viruses showed unidirectional receptor competition with parental virus strains for attachment to HeLa cells [38]. In addition, a poliovirus host-range variant that has been selected for growth in murine cells (lacking receptors for prototype virus) may have acquired a second VAP for attachment to mouse cells, because the variant virus also grows in human cells [39]. An analogous situation may exist for the three serotypes of reoviruses, which share a receptor on L cells [40], but have distinct receptor-mediated histotropisms in mice [4]. Whether more than one specific receptor serves to facilitate infection of Semliki Forest virus [9,10] or rabies virus [11] remains to be determined. Perhaps the use of monoclonal or monospecific antibodies with specificity for blocking a given cellular receptor [30,41–43] will help to provide direct evidence for the role of a second receptor in initiating infection by host-range virus variants.

Isolation of Cellular Receptors for Viruses

The identification and isolation of the many cellular receptors for the different human and animal viruses involves a number of problems. First of all, the receptors are present in very low quantities, ranging from 10^4 to 10^5 per cell [1]. They are intrinsic membrane proteins and require detergents for solubilization [44,45], in the presence of inhibitors of proteolytic or hydrolytic enzymes [46]. Once in solution, it is necessary that the reactive site keeps its native conformation to permit detection of its functional activity, and assays may need to be developed to measure the soluble receptor [47]. Because virus and cellular receptor interactions are multivalent, the apparent affinity in a soluble system will be less than that provided by membranes, making the detection of soluble CRS difficult [48]. Finally, the availability of large amounts of purified virions or virion proteins that are stable in the detergent preparations represents another obstacle. At present, no alternative to a trial-and-error approach is available for selection of appropriate detergents and conditions that will accomplish the task. The methods and reagents used for the purification of hormone receptors are informative and serve as useful models for study of receptors for viruses [5,6,49,50]. The purification of membrane receptors has been facilitated by binding detergent-solubilized membrane preparations to specific ligands immobilized on inert surfaces, as in affinity chromatography. The choice of ligand includes purified virions, isolated viral proteins, and anti-receptor antibodies.

Use of Purified Virions or Viral Proteins

A recent report has described the use of free virions, as the affinity surface, to bind the detergent-solubilized receptors. A coxsackievirus–receptor complex (VRC) was purified on sucrose gradients by methods normally used for virus purification [29]. This strategy was made possible because the coxsackievirus B3 (CB3)–cellular receptor bond was stable in a detergent mixture of deoxycholate (DOC), Triton X-100 (TX), and sodium dodecylsulfate (SDS). The VRC was identified by its sedimentation coefficient (140S), which was less than that of virions (155S). The use of sequential differential and gradient centrifugations of the VRC, followed by iodination of the purified preparation (30,000-fold purification), permitted the identification of a single receptor protein by SDS–polyacrylamide gel electrophoresis (SDS–PAGE) and autoradiography. The protein, designated Rp-a (MW, 49.5K) was dissociated from the VRC and found to be capable of combining with CB3 and CB1, but not with poliovirus T1, providing evidence of virus-receptor specificity.

Application of affinity purification of cellular receptors by use of virions offers two advantages. First, if a receptor is purified bound to its natural ligand, the capacity of the receptor to bind virus will be retained after purification, because the combining site is protected from denaturation. Second, routine virus purification methods can be used. In the adenovirus system, isolation of cellular receptors from KB [51] and HeLa cells [31] has been accomplished by specific adsorption of receptor proteins to the fiber antigen (VAP of adenovirus), penton base-fiber, or virions that were immobilized on inert surfaces. However, no direct evidence was provided to show that the designated protein(s) bound virus or fibers.

Use of Anti-receptor Antibodies

Modern immunological techniques offer useful strategies for isolation and identification of receptors. Antibodies directed against cellular receptors for viruses provide specific, high-affinity probes for these molecules. However, the preparation of receptor-specific antibodies has been difficult [41–43,52,53]. The major limitation has been the lack of purified receptor proteins for use as immunogens. To surmount this limitation, a panel of monoclonal antibodies is raised against the entire cell or crude membrane preparations and screened for blocking activity against specific receptors. Unfortunately, only one monoclonal antibody from thousands of screened hybridoma fluids is obtained, and this antibody may not be capable of immunoprecipitating the receptor protein from detergent-solubilized membrane preparations [41,42]. Fortunately, an IgG class antibody obtained by Colonno et al. [52] had a high affinity for the major group receptor for human rhinoviruses. This antibody was coupled to Affi-gel and selected a receptor protein of 90 kilodaltons from a DOC-solubilized HeLa cell membrane preparation [30].

Monoclonal antibodies prepared for reasons other than for use in virology have been helpful in identifying receptors for viruses. Two groups of investigators [15,16] have reported that monoclonal antibodies against the T4 antigen of lymphocytes, specifically inhibited cell infection by lymphadenopathy associated virus / human T cell leukemia/lymphoma virus (LAV/HTLV-III) and suggested that T4 is an essential component of the receptor for this virus. The T4 antigen is a surface glycoprotein of 62 kilodaltons and is specifically expressed on a subset of T lymphocytes that are target cells for LAV. In a similar way, monoclonal antibodies with specificity for a complement receptor protein of 145 kilodaltons on B lymphocytes have identified this protein as the receptor for EBV [12,13]. Alternatively, a class II antigen has been considered to be the receptor for EBV [14].

The acetylcholine receptor (AChR) has been suggested to be the receptor for rabies virus [18], although there is reservation regarding this suggestion because rabies virus can infect cells in the absence of AChR [19]. Nevertheless, monoclonal antibody to the α subunit of AChR blocked attachment of rabies virus to cultured muscle cells. In addition, immunoblotting results showed that rabies virus proteins, as well as the monoclonal antibody, bound specifically to two membrane proteins of 43 kilodaltons (α subunit) and 110 kilodaltons (undetermined).

Use of Anti-idiotype Antibodies

Anti-idiotype antibodies raised against antibodies to VAP may mimic the conformation of the VAP and therefore recognize the cell surface receptor for that virus [43]. This approach has been used successfully for isolation of the cellular receptor for reovirus type 3 [27]. The outer capsid protein of reovirus type 3, the hemagglutinin (HA), governs virus attachment to target cell surfaces. Because anti-HA antibodies expressed a restricted set of idiotypes, anti-idiotypic antibodies made in rabbits were shown to mimic HA binding and also trigger cellular responses in a similar manner to the viral HA. The reovirus receptor was isolated from a murine thymoma cell line R1.1 by utilizing the rabbit anti-idiotype antibodies. After surface iodination, the membrane proteins were solubilized with TX-100 and Nonidet P-40, the anti-idiotype antibodies were incubated with solubilized membranes, and the antibody–protein complexes were precipitated with Sepharose–protein A. The receptor isolated was a glycoprotein with a molecular weight of 67 kilodaltons and a pI of 5.9. Immunoblot analysis demonstrated that the 67-kilodalton glycoprotein was the only surface structure recognized by both reovirus type 3 and the anti-receptor antibody. Subsequently, the reovirus receptor was reported to be structurally similar to the mammalian β-adrenergic receptor [54].

In summary, a large number of different components of the plasma membrane of cells have been identified as receptor molecules that serve both for cellular functions and for attachment of viruses. Only recently has progress been made in their isolation and purification. In most reports of receptor isolation, evidence of purity and specificity was given. However, the binding affinity between virus and solubilized receptor is usually much lower than the high binding affinity betwen virus and intact cells. Whether the isolated protein is in fact the receptor on cells requires further functional characterization for final proof. Also, the isolated viral binding protein may be only one of the subunits comprising a multicomponent receptor complex. Reconstitution of the isolated protein(s) as an active receptor complex in lipid vesicles is an important step for demonstration of receptor function. In addition, amino acid sequencing of the viral receptor protein(s) may help to determine its identity and to facilitate molecular cloning of receptor genes. Finally, the complete characterization of purified receptors will permit a better understanding of their roles in the economy of the cell and their participation in regulating viral tropism in pathogenesis.

References

1. Lonberg-Holm K, Philipson L (eds) (1981) Receptors and Recognition, series B, vol 8: Virus Receptors, part 2. Chapman and Hall, London.
2. Tardieu M, Epstein RL, Weiner HL (1982) Interaction of viruses with cell surface receptors. Int Rev Cytol 80:27–61
3. Crowell RL, Landau BJ (1983) Receptors in the initiation of picornavirus infections. In Fraenkel-Conrat H and Wagner RR (eds) Comparative Virology, vol 18, pp 1–42
4. Sharpe AH, Fields BN (1985) Pathogenesis of viral infections. Basic concepts derived from the reovirus model. N Engl J Med 312:486–497

5. Changeux JP, Devillers-Thiery A, Chemouilli P (1984) Acetylcholine receptor: An allosteric protein. Science 225:1334–1345.
6. Cerione RA, Strulovici B, Benovic JL, Lefkowitz RJ, Caron MG (1983) Pure β-adrenergic receptor: The single polypeptide confers catecholamine responsiveness to adenylate cyclase. Nature 306:562–566
7. Kielian MC, Helenius A (1984). Role of cholesterol in fusion of Semliki Forest virus with membranes. J Virol 52:281–283
8. Crowell RL (1966) Specific cell-surface alteration by enteroviruses as reflected by viral-attachment interference. J. Bacteriol 91:198–204
9. Helenius A, Morein B, Fries E, Simons K, Robinson P, Schirrmacher V, Terhorst C, Strominger JL (1978) Human (HLA-A and HLA-B) and murine (H-2K and H-2D) histocompatibility antigens are cell surface receptors for Semliki Forest virus. Proc Natl Acad Sci USA 75:3846–3850
10. Oldstone MBA, Tishon A, Dutko FJ, Kennedy SIT, Holland JJ, Lampert PJ (1980) Does the major histocompatibility complex serve as specific receptor for Semliki Forest virus. J Virol 34:256–265
11. Tignor GH, Smith AL, Shope RE (1984) Utilization of host proteins as virus receptors. In Notkins AL, Oldstone MBA (eds) Concepts in Viral Pathogenesis. Springer-Verlag, New York, pp 109–116
12. Fingeroth JD, Weis JJ, Teddler TF, Strominger JL, Biro A, Fearon DT (1984) Epstein–Barr virus receptor of human B lymphocytes is the C3d receptor CR2. Proc Natl Acad Sci USA 81:4510–4514
13. Nemerow GR, Wolfert R, McNaughton ME, Cooper NR (1985) Identification and characterization of the Epstein–Barr virus receptor on human B lymphocytes and its relationship to the C3d complement receptor (CR2). J Virol 55:347–351
14. Reisert PS, Spiro RC, Townsend PL, Stanford SA, Sairenji T, Humphreys RE (1985). Functional association of class II antigens with cell surface binding of Epstein–Barr virus. J. Immunol 134:3776–3780
15. Dalgleish AG, Beverley PCL, Clapham PR, Crawford DH, Greaves MF, Weiss RA (1984). The CD4 (T4) antigen is an essential component of the receptor for the AIDS retrovirus. Nature 312:763–767
16. Klatzmann D, Champagne E, Chamaret S, Gruest J, Guetard D, Hercend T, Gluckman J-C, Montagnier L (1984) T-lymphocyte T4 molecule behaves as the receptor for human retrovirus LAV. Nature 312:767–768
17. Landen B, Fox CF (1980) Isolation of BP gp 70, a fibroblast receptor for the envelope antigen of Rascher murine leukemia virus. Proc Natl Acad Sci USA 77:4988–4992
18. Burrage TG, Tignor GH, Smith AL (1985) Rabies virus binding at neuromuscular junctions. Virus Res 2:273–289
19. Wunner WH, Reagan KJ, Koprowski, H (1984) Characterization of saturable binding sites for rabies virus. J Virol 50:691–697
20. Schlegel R, Wade M (1985) Biologically active peptides of the vesicular stomatitis virus glycoprotein. J Virol 53:319–323
21. Bailey CA, Miller DK, Lenard J (1984) Effects of DEAE–dextran on infection and hemolysis by VSV. Evidence that nonspecific electrostatic interactions mediate effective binding of VSV to cells. Virology 133:111–118
22. Inada T, Mims CA (1984) Mouse Ia antigens are receptors for lactate dehydrogenase virus. Nature 309:59–61

23. Kowalchyk K, Plagemann PGW (1985) Cell surface receptors for lactate dehydrogenase-elevating virus on subpopulation of macrophages. Virus Res 2:211–229

24. Marchesi VT, Furthmayr H, Tiomita M (1976) The red cell membrane. Annu Rev Biochem 45:667–698

25. Higa HH, Rogers GN, Paulson JC (1985) Influenza virus hemagglutinins differentiate between receptor determinants bearing N-acetyl-, N-glycollyl-, and N, O-diacetylneuraminic acids. Virology 144:279–282

26. Markwell MAK, Fredman P, Svennerholm L (1984) Receptor ganglioside content of three hosts for Sendai virus. Biochem Biophys Acta 775:7–16

27. Co MS, Gaulton GN, Fields BN, Greene MI (1985) Isolation and biochemical characterization of the mammalian reovirus type 3 cell-surface receptor. Proc Natl Acad Sci USA 82:1494–1498

28. Burness ATH, Pardoe IU (1983) A sialoglycopeptide from human erythrocytes with receptor-like properties for encephalomyocarditis and influenza viruses. J Gen Virol 64:1137–1148

29. Mapoles JE, Krah DL, Crowell RL (1985) Purification of a HeLa cell receptor protein for group B coxsackieviruses. J Virol 55:560–566

30. Tomassini JE, Colonno RJ (1986) Isolation of the receptor protein involved in attachment of human rhinoviruses. J Virol 58:290–295

31. Svensson U, Persson R, Everitt E (1981) Virus–receptor interaction in the adenovirus system. I. Identification of virion attachment proteins of the HeLa cell plasma membrane. J Virol 38:70–81

32. Pastan IH, Willingham MC (1981) Journey to the center of the cell: Role of the receptosome. Science 214:504–509

33. Gollins SW, Porterfield JS (1984) Flavivirus infection enhancement in macrophages: Radioactive and biological studies on the effect of antibody on viral fate. J Gen Virol 65:1261–1272

34. Shimura H, Kimura G (1985) Phospholipid liposomes enhance the infectivity of purified simian virus 40 virions. Virology 144:268–272

35. Faller DV, Baltimore D (1984) Liposome encapsulation of retrovirus allows efficient superinfection of resistant cell lines. J Virol 49:269–272

36. Wilson T, Paphadjopoulos D, Taber R (1979) The introduction of poliovirus RNA into cells via lipid vesicles (liposomes). Cell 17:77–84

37. Crowell RL, Siak JS (1978) Receptor for group B coxsackieviruses: Characterization and extraction from HeLa cell plasma membranes. In Pollard M (ed) Perspect Virol 10:39–53

38. Reagan KJ, Goldberg B, Crowell RL (1984) Altered receptor specificity of coxsackievirus B3 after growth in rhabdomyosarcoma cells. J Virol 49:635–640

39. Racaniello VR (1984) Poliovirus type II produced from cloned cDNA is infectious in mice. Virus Res 1:669–675

40. Lee PWK, Hayes EC, Joklik WK (1981) Protein sigma-1 is the reovirus cell attachment protein. Virology 108:156–163

41. Campbell BA, Cords CE (1983) Monoclonal antibodies that inhibit attachment of group B coxsackieviruses. J Virol 48:561–564

42. Minor PD, Pipkin PA, Hockley D, Schild GC, Almond JW (1984) Monoclonal antibodies which block cellular receptors of poliovirus. Virus Res 1:203–212

43. Gaulton GN, Co MS, Royer HD, Greene MI (1985) Anti-idiotypic antibodies as probes of cell surface receptors. Mol Cell Biochem 65:5–21

44. Hjelmeland LM, Chrambach A (1984) Solubilization of functional membrane proteins. Methods Enzymol 104:305–318
45. Reynolds JA (1981) Solubilization and characterization of membrane proteins. *In* Jacobs S, Cuatrecasas P (eds) Receptors and Recognition, series B, vol 11:Membrane Receptors: Methods for Purification and Characterization. Chapman and Hall, London, pp 33–59
46. Umezawa K, Aoyagi T (1984) Elimination of protein degradation by use of protease inhibitiors. *In* Venter JC, Harrison LC (eds) Receptor Biochemistry and Methodology, vol 2. Liss Inc, New York, pp 139–148
47. Krah DL, Crowell RL (1982) A solid-phase assay of solubilized HeLa cells plasma membrane receptors for binding group B coxsackieviruses and polioviruses. Virology 118:148–156
48. Philipson L, Everitt E, Lonberg-Holm K (1976) Molecular aspects of virus-receptor interaction in the adenovirus system. *In* Beers RF Jr, Basset EG (eds) Cell Membrane Receptors for Viruses, Antigens and Antibodies, Polypeptide Hormones, and Small Molecules. Raven Press, New York, pp 203–216
49. Jacobs S, Cuatrecasas P (1981) Affinity chromatography for membrane receptor purification. *In* Jacobs S, Cuatrecasas P (eds) Receptors and Recognition, series B, vol 11: Membrane Receptors: Methods for Purification and Characterization. Chapman and Hall, London, pp. 61–86
50. Wilchek M, Miron T, Kohn J (1984) Affinity chromatography. Methods Enzymol 104:3–55
51. Hennache B, Boulanger P (1977) Biochemical study of KB-cell receptor for adenovirus. Biochem J 166:237–247
52. Colonno RJ, Callahan PL, Long WL (1986) Isolation of a monoclonal antibody which blocks attachment of the major group of human rhinoviruses. J Virol 57:7–12
53. Crowell RL, Field AK, Schleif WA, Long WL, Colonno RJ, Mapoles JE, Emini EA (1986) A monoclonal antibody that inhibits infection of HeLa and rhabdomyosarcoma cells by selected enteroviruses through receptor blockade. J Virol 57:438–445
54. Co MS, Gaulton GN, Tominaga A, Homcy CJ, Fields BN, Greene MI (1985) Structural similarities between the mammalian β-adrenergic and reovirus type 3 receptors. Proc Natl Acad Sci USA 82:5315–5318
55. Eppstein DA, Marsh YV, Schreiber AB, Newman SR, Todaro GJ, Nestor JJ, Jr (1985) Epidermal growth factor receptor occupancy inhibits vaccinia virus infection. Nature 318:663–665

CHAPTER 15
Viral Receptors Serving Host Functions

MAN SUNG CO, BERNARD N. FIELDS, AND MARK I. GREENE

Viral receptors are structures on the cell surface that are recognized by virus particles and provide specific points of entry into the host cells. These structures can be polypeptides, glycoproteins, or glycolipids. Many investigators have speculated that viruses utilize specific cell surface structures that serve host functions for binding and entry. In several bacterial systems, it has been demonstrated that bacteriophages bind and subsequently infect cells through receptors that are involved in normal host functions. These include the binding of λ phage to the maltose receptor [1] and the recognition of the ferrichrome binding site by bacteriophage φ 80 [2].

A number of investigators have studied features of the interaction between the animal viruses and their hosts [3]. One goal has been to identify the receptor molecules on cell surfaces that are recognized by animal viruses. Using biochemical and immunological techniques, investigators have identified structures with specificity for a number of viruses [4–8,15–20,22–24,26–27]. A few of these are related to molecules associated with normal cellular functions.

Anti-receptor Antibody Development

Work in our laboratory has centered on the utilization of anti-idiotypic (anti-receptor) antibodies in the study of virus receptors [9]. We have developed a monoclonal antibody [10] and a monospecific polyclonal antibody [11] specific for a cell surface receptor for the mammalian reovirus type 3. A panel of monoclonal antibodies to the hemagglutinin (sigma 1) protein of reovirus serotype 3 (HA3) was initially prepared. One monoclonal antibody (9BG5) binds to the neutralizing region of the hemagglutinin (the reovirus hemagglutinin directs tissue binding, tropism, and susceptibility to injury; reviewed in ref. 9). Rabbit anti-idiotypic antibodies prepared to BALB/c anti-HA3 antibodies specifically block the binding of the purified hemagglutinin protein to 9BG5 monoclonal antibody [11]. More recently, a syngeneic monoclonal anti-idiotype (87.92.6) has been prepared, that binds to the 9BG5 immuno-globulin and that displays properties identical to those seen with xenogeneic anti-idiotypic antibodies [10]. The development of syngeneic monoclonal anti-idiotype has permitted the preparation of large amounts of highly purified specific antibodies. We have demonstrated that monoclonal anti-idiotype and a subset of the polyclonal anti-idiotype represent an effective "internal image" for the reovirus hemagglutinin domain that interacts with the cellular receptor for the virus [10–13]. The evidence supporting this claim can be summarized as follows:

1. Anti-idiotype binding parallels the cellular tropism of reovirus type 3.
2. The anti-idiotype and reovirus have similar biological effects in limiting concanavalin A-induced stimulation of murine lymphocytes.
3. The anti-idiotype and reovirus inhibit the host-cell DNA synthesis in a serotype-specific manner ([14], and unpublished).
4. Anti-idiotype specifically inhibits reovirus type 3 binding to target cells and can prevent viral infection of neurons in vitro (unpublished).
5. Anti-idiotype prevents isolated HA protein from binding to monoclonal anti-HA antibodies.
6. Anti-idiotype specifically stimulates both T and B cell immunity to reovirus type 3 in vivo.

Reovirus Receptor

Using the monospecific polyclonal antibody, we have isolated a protein (MW, 67,000; pI, 5.8–6.0) present on L cells, a mouse thymoma line (R1.1), and a rat neuroblastoma line (B104) [15]. We have further demonstrated that this isolated protein is the only membrane protein recognized by the reovirus particles in immunoblotting experiments. The receptor is a glycoprotein containing sialic acid residues that can be removed by neuraminidase. We have also found that this protein shares structural and biochemical features with the mammalian β-adrenergic receptor [16]. A structural comparison of the reovirus receptor and an affinity-purified β-adrenergic receptor on a two-dimensional gel (isoelectric focusing followed by sodium dodecyl sulfate–polyacrylamide gel electrophoresis, SDS–PAGE) has indicated that the two molecules are very similar, showing indistinguishable molecular weights and isoelectric points.

Trypsin digests of the two proteins also display identical peptide patterns. We have further demonstrated that the purified reovirus receptor is capable of binding the β-antagonists iodohydroxybenzylpindolol and iodocyanoal-prenolol. The binding is saturable and can be blocked by alprenolol and propranolol. All these results suggest that mammalian reovirus type 3 is capable of utilizing the β-adrenergic receptor on the host cell for binding and entry and subsequent replication.

Adrenergic receptors are membrane receptors for the catecholamines norepinephrine and epinephrine. Binding of ligands to the β-adrenergic receptor stimulates the membrane-bound enzyme adenylate cyclase and leads to the intracellular accumulation of cAMP. With the identification of the β-adrenergic receptor as a receptor for reovirus type 3, it should become feasible to determine the role of the virus in its interaction with a hormonal receptor and to determine its role in pathogenesis.

A few other viruses have been reported to bind to receptors associated with known cellular functions. Epstein–Barr virus (EBV) has a strong tropism for B lymphocytes. An association has been found between the EBV receptor and the complement C3d receptor on human lymphoid cells [17,18]. There is a direct correlation between the expression of these two receptors. Two-color fluorescence staining of EBV receptor and C3 receptor showed complete overlapping of green and red fluorescence. Capping of the EBV receptor induced cocapping of the C3 receptor and vice versa. Furthermore, the kinetic pattern of EBV receptor capping was identical with that of C3 receptor capping. The identity of the C3d receptor as the EBV receptor was further confirmed recently with the use of monoclonal antibody [19] and monospecific polyclonal antibody [20]. However, there are also reports indicating that the C3 receptor may be distinct from the EBV receptor [21].

Recently, Inada and Mims demonstrated that mouse Ia antigens are receptors for lactate dehydrogenase virus (LDV) [22]. LDV replicates exclusively in a restricted set of macrophages, leading to elevation of plasma lactate dehydrogenase and viremia. Using immunofluroescence techniques, these investigators showed that the percent of infected macrophages was the same as the percent expressing Ia antigens. Infection of I-A$^+$ peritoneal macrophages was blocked when cells were treated simultaneously with monoclonal antibody to I-A and I-E. LDV infectivity was inactivated when the virus was treated with purified rat glycoprotein homologous to mouse I-A and I-E antigens. These results indicate that the receptors for LDV may be the I-A and I-E histocompatibility antigens. However, another possibility is that class II antigens help deliver the virus. Lentz et al. has proposed that the acetylcholine receptor may function as the rabies virus receptor [23]. These investigators found that the rabies viruses are present on mouse diaphragms and on cultured chick myotubes in a distribution coinciding with that of the acetylcholine receptor. They demonstrated that pretreatment of the myotubes with α-bungarotoxin and d-tubocurarine reduced the number of myotubes that became infected with rabies virus. These findings together suggest that acetylcholine receptors may serve as receptors for rabies virus.

Helenius et al. demonstrated that the human HLA-A and HLA-B and mouse H-2K and H-2D histocompatibility antigens are cell surface receptors for Semliki Forest virus (SFV) [24]. Isolated HLA-A and HLA-B antigens reconstituted in lipid vesicles inhibit the binding of viral proteins to human cells. They also showed that viral coat protein inhibit the complement-dependent cytotoxicity of antibodies directed against H-2K and H-2D antigens in mouse cells. They further demonstrated that complexes formed between viral spike proteins and HLA-A and HLA-B antigens or H2-K and H2-D antigens can be isolated from the cell surface via affinity chromatography or immunoprecipitation. However, Oldstone et al. [25] later demonstrated that a cell's possession of histocompatibility antigens is not a requirement for SFV infection. He showed that SFV replicates in H-2-negative murine lymphoblastoid cells, and thus concluded that the MHC antigens are not specific receptors for SFV.

Dalgleish et al. [26] and Klatzmann et al. [27] independently reported that the retrovirus human T cell lymphotropic virus / lymphadenopathy associated virus (HTLV-III/LAV) may utilize the T4 (CD4) antigen as its receptor. Klatzmann et al. [27] further demonstrated that preincubation of $T4^+$ lymphocytes with monoclonal antibodies directed at the T4 glycoprotein blocked cell infectivity by HTLV-III/LAV.

The presence or absence of specific cell surface receptors can influence the host range and tissue tropism of viruses. For example, selective infection of Ia^+ macrophages by LDV may have an important effect on the immunological capability of infected mice. The binding of rabies virus to acetylcholine receptors, which are present in high density at the neuromuscular junction, would provide a mechanism whereby the virus could be locally concentrated at sites in proximity to peripheral nerves, facilitating subsequent uptake and transfer to the central nervous system. Binding of reovirus type 3 to the β-adrenergic receptors in the brain may explain the neurotropism of this virus. The understanding of the B lymphocyte tropism of EBV and the T-helper lymphocyte tropism of HTLV-III/LAV may also provide directions for the treatment of these viral infections. Viral interactions with physiologic receptors may also provide an explanation of metabolic disorders caused by noncytopathic viurses or autoantibodies. The etiology of non-insulin-dependent diabetes mellitus may illustrate such an effect of lymphocytic choriomeningitis virus [28].

References

1. Hazelbauer GL (1975) Role of the receptor for bacteriophage lambda in the functioning of the maltose chemoreceptor of *Escherichia coli*. J Bacteriol 124:119–126
2. Wayne R, Neilands JB (1975) Evidence for common binding sites for ferrichrome compounds and bacteriophage φ80 in the cell envelope of *Escherichia coli*. J Bacteriol 121:497–503
3. Notkins AL, Oldstone MBA (1984) Concepts in Viral Pathogenesis, vol 1. Springer-Verlag, New York

4. Schaffar-Deshayes L, Choppin J, Levy JP (1981) Lymploid cell surface receptor for Moloney leukemia virus envelope glycoprotein gp71. II. Isolation of the receptor. J Immunol 126:2352–2354
5. Markwell MA, Svennerholm L, Paulson JC (1981) Specific gangliosides function as host cell receptors for Sendai virus. Proc Natl Acad Sci USA 78:5406–5410
6. Niman HL, Elder JH (1980) Molecular dissection of Rauscher virus gp70 by using monoclonal antibodies: Localization of acquired sequences of related envelope gene recombinants. Proc Natl Acad Sci USA 77:4524–4528
7. Svensson U, Persson R, Everitt E (1981) Virus–receptor interaction in the adenovirus system. I. Identification of virion attachment proteins of the HeLa cell plasma membrane. J Virol 38:70–81
8. Robinson PJ, Hunsmann G, Schneider J, Schirramacher M. (1980) Possible cell surface receptor for Friend murine leukemia virus isolated with viral envelope glycoprotein complexes. J Virol 36:291–294
9. Fields BN, Greene MI (1982) Genetic and molecular mechanisms of viral pathogenesis: Implications for prevention and treatment. Nature 300:19–23
10. Noseworthy JH, Fields BN, Dichter MA, Sobotoka C, Pizer E, Perry LL, Nepom JT, Greene MI (1983) Cell receptors for the mammalian reovirus. I. Syngeneic monoclonal anti-idiotypic antibody identifies a cell surface receptor for reovirus. J Immunol 131:2533–2538
11. Nepom JT, Weiner HL, Dichter MA, Tardieu M, Spriggs DR, Gramm CF, Powers ML, Fields BN, Greene MI (1982) Identification of a hemagglutinin-specific idiotype associated with reovirus recognition shared by lymphoid and neural cells. J Exp Med 155:155–167
12. Kauffman RS, Noseworthy JH, Nepom JT, Finberg R, Fields BN, Greene MI (1983) Cell receptors for the mammalian reovirus. II. Monoclonal anti-idiotypic antibody blocks viral binding to cells. J Immunol 131:2539–2541
13. Sharpe AH, Gaulton, GN, McDade KK, Fields, FN, Greene MI (1984) Syngeneic monoclonal antiidiotype can induce cellular immunity to reovirus. J Exp Med 160:1195–1205
14. Sharpe AH, Fields BN (1981) Reovirus inhibition of cellular DNA synthesis: Role of the S1 gene. J Virol 38:389–392
15. Co MS, Gaulton GN, Fields BN, Greene MI (1985) Isolation and biochemical characterization of the mammalian reovirus type 3 cell-surface receptor. Proc Natl Acad Sci USA 82:1494–1498
16. Co MS, Gaulton GN, Tominaga A, Homcy CJ, Fields BN, Greene MI (1985) Structural similarities between the mammalian beta-adrenergic and reovirus type 3 receptors. Proc Natl Acad Sci USA 82:5315–5318
17. Jondal M, Klein G, Oldstone MBA, Bokish V, Yeenof E (1976) Surface markers on human B and T lymphoctyes. VIII. Association between complement and Epstein–Barr virus receptors on human lymphoid cells. Scand J Immunol 5:401–410
18. Yeenof E, Eklein G, Jondal M, Oldstone MBA (1976) Surface markers on human B- and T-lymphocytes. IX. Two-color immunofluorescence studies on the association between EBV receptors and complement receptors on the surface of lymphoid cell lines. Int J Cancer 17:693–700
19. Fingeroth JD, Weis JJ, Tedder JF, Strominger JL, Biro PA, Fearon DT (1984) Epstein–Barr virus receptor of human B lymphoctyes is the C3d receptor CR2. Proc Natl Acad Sci USA 81:4510–4514

20. Frade R, Barel M, Ehlin-Henriksson B, Klein G (1985) gp140, the C3d receptor of human B lymphoctyes, is also the Epstein–Barr virus receptor. Proc Natl Acad Sci USA 82:1490–1493
21. Wells A, Koide N, Stein H, Gerdes J, Klein G (1983) The Epstein–Barr virus receptor is distinct from the C3 receptor. J Gen Virol 64:449–453
22. Inada T, Mims CA (1984) Mouse Ia antigens are receptors for lactate dehydrogenase virus. Nature 309:59–61
23. Lentz TL, Burrage TG, Smith AL, Crick J, Tignor GH (1982) Is the acetylcholine receptor a rabies virus receptor? Science 215:182–184
24. Helenius A, Morein B, Fries E, Simons K, Robinson P, Schirrmacher V, Terhorst C, Strominger JL (1978) Human (HLA-A and HLA-B) and murine (H2-K and H2-D) histocompatibility antigens are cell surface receptors for Semliki Forest virus. Proc Natl Acad Sci USA 75:3846–3850
25. Oldstone MBA, Tishon A, Dutko FJ, Kennedy SIT, Holland JJ, Lampert PW (1980) Does the major histocompatibility complex serve as a specific receptor for Semliki Forest virus? J Virol 34:256–265
26. Dalgleish AG, Beverley PCL, Clapham PR, Crawford DH, Greaves MF, Weiss RA (1984) The CD4 (T4) antigen is an essential component of the receptor for the AIDS retrovirus. Nature 312:763–767
27. Klatzmann D, Champagne E, Chamaret S, Gruest J, Guetard D, Hercned T, Gluckman JC, Montagnier L (1984) T-lymphocyte T4 molecule behaves as the receptor for human retrovirus LAV. Nature 312:767–768
28. Oldstone, MBA, Rodriguez M, Daughaday WH, Lampert PW (1984) Viral perturbation of endocrine function: Disordered cell function leads to disturbed homeostasis and disease. Nature 307:278–281

CHAPTER 16
Genetics of Retrovirus Tumorigenicity

Douglas R. Lowy

Retroviruses are the causative agents of a wide variety of malignant and degenerative diseases under natural and experimental conditions. These viruses may be oncogenic, whether or not they contain a viral oncogene. However, the presence of a viral oncogene alters and enhances their tumorigenicity. Retroviruses have been used experimentally to study the pathogenesis of tumor induction because they reproducibly induce similar disease in animals with a similar genetic background. Molecular genetic techniques have now provided investigators with powerful tools to define the role of specific viral sequences in the induction of disease as well as to probe the mechanisms of pathogenesis in greater detail. Current studies indicate that both regulatory elements and protein coding sequences play important roles in determining the disease-inducing capacity of these viruses.

Retroviruses as Tumor-inducing Agents

Members of this virus family have been identified as the presumptive etiologic agents of the acquired immunodeficiency syndrome (AIDS) in humans, lymphoid lymphomas in humans and animals, hematopoietic tumors in rodents and birds, and a variety of sarcomas and solid tumors in various animal species (reviewed in refs. 1 and 2). This brief review will discuss selected genetic aspects of tumor induction by these viruses.

Earlier studies defined the pathogenic range of these viruses, demonstrated the importance of high virus titers for disease induction, and indicated that some viruses were pathogenic only for newborn animals whereas others induced disease in adults as well. The capacity to clone infectious retroviral genomes molecularly represents an important advance of direct relevance to viral genetics. Using the cloned genomes, investigators have analyzed the role of specific sequences by studying the consequences of introducing defined mutations in these genomes and by characterizing chimeric viruses that have been constructed by making recombinants in vitro between two closely related viral genomes that differ in their disease-inducing capacity. As is the case with many other viruses, regulatory sequences and protein coding sequences each serve important functions in determining retroviral oncogenicity.

Virus Replication and Genetics

The retrovirus replication cycle (reviewed in refs. 1 and 2) has several unusual features that are relevant to their oncogenicity. Because most retroviruses are not cytocidal for the cells they productively infect, they readily establish chronic infection of susceptible cells. Although the viral genome in virions is composed of single-stranded RNA, it is replicated intracellularly through a double-stranded DNA intermediate that is integrated efficiently into the host-cell DNA. Integration of the viral genome places replication of the viral DNA under the same regulation as that of host DNA; this feature insures that the viral DNA will be transmitted in a stable and efficient manner to progeny cells. Reverse transcription and integration of the viral genome are both mediated largely by the protein product of the viral *pol* gene. Retroviral DNA integration seems to be relatively random within the host genome, but many tumors are associated with viral integration at specific sites within the host genome, suggesting that site-specific integration represents an important pathogenetic mechanism in these tumors.

Three different viral genes encode the virion proteins that are required for replication of progeny virus. Their 5'-to-3' order is (a) *gag,* which specifies the core proteins; (b) *pol,* whose product is required for reverse transcription and viral DNA integration; and (c) *env,* which specifies the envelope glycoprotein. Some retroviruses, such as the human retroviruses, encode nonvirion proteins that function in *trans* as transcriptional enhancers which exert a major effect on viral replication by directly regulating the rate of RNA transcription. A third important class of genes found in some retroviruses are the viral oncogenes (v-*onc*); these cell-derived genes encode proteins that induce proliferation of susceptible cells. The v-*onc* are important determinants of oncogenicity, but they are not essential for virus replication.

In addition to these protein coding sequences, retrovirus genomes contain several classes of *cis*-acting regulatory elements. These include enhancers of RNA transcription, signals for initiation and termination of RNA transcription, sequences that participate directly in replication of the viral genome, and signals that permit the viral RNA to be packaged (pseudotyped) by the virion proteins. Except for the packaging signals, these regulatory elements are present in the long terminal repeat (LTR), which are viral sequences that are the same at both ends of the viral DNA genome. These viral control elements can affect the transcriptional activity of host genes located near integrated viral genomes in addition to serving as the major transcriptional regulatory elements for the viral genes. This virally mediated enhanced expression of specific host genes is believed to contribute to the pathogenesis of those tumors associated with site-specific integration of the viral genome.

The relative plasticity of retroviral genomes is also relevant to their disease-inducing capacity. These viruses readily undergo recombination in vivo with related retroviral sequences or, less frequently with nonviral sequences (such as c-*onc*), which can lead to the formation of retroviruses with novel biological properties. Such recombinational events may lead to genomes that are replication defective because they have lost one or more of the genes that encode virion proteins. These defective genomes will retain their infectivity if the lesion accounting for their defectiveness does not affect the *cis*-regulatory elements, because the missing proteins can be supplied by a replication-competent (helper) viral genome. The *cis*-acting elements of the defective genome, together with the virion proteins supplied by the helper virus, enable the defective genome to be reverse-transcribed, integrated, expressed, and pseudotyped normally. The presence of a v-*onc* is usually associated with deletion of some sequences that encode virion proteins, presumably because a large increase in the length of the viral genome would grossly impair the efficiency with which the genome could be packaged into virions. Because preparations of defective viruses must contain a helper virus, pathogenicity studies are carried out in the presence of a relatively nonpathogenic helper virus to insure that the disease studied is induced by the defective virus and not by the helper virus.

Disease Induction by Retroviruses that Do Not Contain Oncogenes

Lymphomas and leukemias account for the vast majority of tumors induced by this class of retroviruses (reviewed in ref. 1). Most of these viruses are replication competent. The tumors they induce usually develop after a relatively long latent period (several months), which implies that infection represents only one of several steps in tumor induction by these agents. The genetics of disease induction has been studied most completely in the avian and murine systems. We shall discuss avian leukosis viruses (ALV), murine leukemia viruses (MuLV), and the spleen focus forming virus (SFFV). In these viruses, the *cis*-regulatory sequences, especially the enhancer region of the viral LTR, along with protein coding sequences independently play determining roles in viral tumorigenicity (reviewed in ref. 3).

Inoculation of most strains of chickens with ALV results in the induction of bursal (B cell) lymphomas in many animals as well as a low incidence of other tumors and osteopetrosis, which is a nonmalignant proliferative osteoblastic disease. In the vast majority of the bursal lymphomas, the viral genome is integrated near the cellular *myc* proto-oncogene, activating this gene by virtue of sequences in the viral LTR. The type of LTR markedly affects the pathogenicity of the virus. An endogenous virus, Rous associated virus (RAV-0), is nononcogenic, largely because its enchancer element is less efficient than that of ALV. An in vivo-derived recombinant virus (NTRE-7) that replaces the RAV-0 LTR plus 0.2 kb of contiguous upstream sequences of RAV-0 with the homologous region of ALV replicates as well in vivo as ALV and is now tumorigenic. However, the incidence of lymphomas with NTRE-7 is much lower than with ALV, indicating that sequences outside the LTR somehow affect the "targeting" of the virus. With the use of a virus that induces a high incidence of osteopetrosis, sequences outside the LTR have been shown to confer the ability to induce osteopetrosis, whereas sequences in or near the LTR influence the time of onset and severity of disease [4].

In the murine system the role of enhancer sequences in determining the target tissue of MuLV has been demonstrated most dramatically by making recombinants between a virus that induces thymic leukemia (Moloney [Mo] MuLV) and one that induces splenic erythroleukemia (Friend [F] MuLV). The difference in the type of disease induced by the two viruses resides in the LTR (and/or sequences just upstream from it), because a chimeric viral genome whose LTRs (and sequences just upstream from it) are derived from Mo-MuLV but whose other sequences are derived from F-MuLV induces Mo-MuLV-type (thymic) disease, whereas a viral genome composed of the opposite construction induces F-MuLV-type (splenic) disease [5]. The importance of the LTR enhancer sequences in conferring leukemogenic potential has been shown directly with recombinants between a nonleukemogenic and a highly leukemogenic MuLV [6].

Sequences encoding virion proteins have also been shown to play a significant pathogenic role. The *env* gene product, which determines the viral host range via its interaction with host cell receptors, is of particular importance. Recombination with endogenous *env* sequences appears to be required for disease induction by several replication-competent MuLV; this topic has recently been reviewed in these pages and will not be considered here [7]. The *env* protein of SFFV is a special case. This protein in SFFV is truncated and does not function as a virion protein. However, its presence in cells induces proliferation of erythroid precursor cells in vitro and in vivo. One variant SFFV induces a polycythemic disease, whereas another induces anemia. Subtle differences in the 3' end of *env* account for these different phenotypes [8].

gag and *pol* sequences may also play an ancillary role in disease induction by v-*onc*-viruses [9,10]. It is not yet clear whether these latter differences can be accounted for at the level of viral nucleic acid (RNA stability, recombination, etc.) or protein.

Host genes also play a significant role in disease induction (reviewed in ref. 1). Most appear either to interfere with viral replication directly (such as by interference as a result of endogenous viral expression in DBA/2 mice) or to act at the level of immune response (as with the *H-2* haplotype). A particularly interesting example has been noted in chickens, where the same

ALV that induces bursal lymphomas in most strains induces erythroblastosis in an unusual strain. The susceptibility to erythroblastosis, which is associated with the viral genome's insertion next to and activation of the c-*erv*B locus (the epidermal growth factor [EGF] receptor), results from a dominant host gene [11]. The molecular basis for this striking difference in target cell has not been elucidated.

Disease Induction by Viruses that Contain Oncogenes

v-*onc* represent altered versions of cellular proto-oncogenes (c-*onc*). Several types of structural alterations have been noted, including point mutations, deletion of some amino acid coding sequences, and fusion genes of which a retroviral gene (usually part of *gag*) provides the N terminus and the *onc* sequences encode the C terminus of a fusion protein.

The disease induced by these viruses usually occurs after a shorter latent period (days to weeks) than those induced by viruses without v-*onc*. In most instances, tumor induction appears to depend principally on the particular v-*onc* in the virus. The in vitro transforming activity of different mutants or variants generally correlates quite well with their disease-inducing capacity. To the extent that enhancer elements affect the level of v-*onc* expression, it may be expected that these *cis*-acting sequences will also contribute significantly to disease induction, but only limited attention has been given to this question. Most studies of mutant v-*onc* have been limited to tissue culture analysis, but we shall refer primarily to those instances where disease induction has been tested. Rous sarcoma virus (RSV), Harvey murine sarcoma virus (Ha-MuSV), MC29 avian myelocytomatosis virus, and avian erythroblastosis virus (AEV) will serve as examples.

The *src* gene of RSV induces solid tissue sarcomas in susceptible birds. Its protein product can also alter the proliferation of other stem cells, such as erythropoietic and myelocytic precursors [12,13]. The v-*src* encoded protein (pp60$^{\text{v-src}}$) functions at the plasma membrane and possesses a tyrosyl kinase activity that appears to be necessary, but not sufficient, for cell transformation in vitro and tumor induction in vivo. The protein encoded by its proto-oncogene (c-*src*) has a much lower tyrosyl kinase activity, which correlates with its apparent inability to be directly tumorigenic. Certain mutations at either the N terminus or the C terminus of c-*src* are sufficient to increase tyrosyl kinase activity and transforming activity [14]. Conversely, mutations in v-*src* that interfere with tyrosyl kinase activity or membrane association lower the transforming activity of the gene [15]. Mutation of a tyrosyl residue in pp60$^{\text{v-src}}$ that is autophosphorylated by the *src* kinase (a mutation that does not detectably alter kinase activity) leads to the interesting phenotype that mouse cells transformed by the mutant gene induce tumors only in immunodeficient mice, whereas cells transformed by the wild-type gene induce tumors in immunocompetent syngeneic animals [16].

Studies of Harvey murine sarcoma virus (Ha-MuSV) have demonstrated that single point mutations in an *onc* gene can markedly alter the disease-inducing capacity of the gene [17]. Compared with the c-*ras*H proto-oncogene, Ha-MuSV v-*ras*H contains two point mutations, located at amino acids 12 and 59. Either mutation by itself increases the in vitro transforming activity of the proto-oncogene and lowers the GTPase activity of this GTP binding protein [18]. By making recombinants between the c-*ras*H proto-oncogene and v-*ras*H, it has been possible to test the tumorigenicity of mutants that are isogenic except for encoding the four possible combinations of these two amino acid residues. Newborn mice inoculated with the wild-type virus induce fibrosarcomas and splenic erythroblastosis in less than one month in virtually all animals. Similar amounts of the virus that contains the proto-oncogene sequences induce a rare fibrosarcoma without splenic disease after a latent period of a few months. As is true of wild-type virus, the viruses with mutations only at amino acid 12 or only at amino acid 59 induce fibrosarcomas in almost all animals; however, the mean latent period is longer (about two months), and these animals do not usually develop splenic disease. The results indicate that a single point mutation in this *ras* gene is sufficient to confer the high tumorigenic phenotype, but both mutations together give rise to an even more oncogenic virus. Using a somewhat different construction, other investigators have also shown that a human c-*ras* gene activated by a point mutation at amino acid 12 can give rise to a highly oncogenic *ras*-encoding transforming retrovirus [19].

It has also turned out that mutations in noncoding sequences outside the LTR may influence the disease-inducing capacity of Ha-MuSV [20]. When a 1.7-kb noncoding segment downstream from the v-*ras*H gene was deleted, the resulting virus surprisingly induced lymphoid lymphoma (without fibrosarcomas or splenic disease) with a latent period of two to three months. This change in target cell was correlated with the virus-inducing smaller foci in vitro and lower levels of v-*ras* protein. Preliminary results suggest that these noncoding sequences contain an enhancer element (distinct from the enhancer in the LTR); this enhancer apparently exerts an important influence on the oncogenicity of the virus. Whereas this lowering of *ras* gene expression was associated with a change in the target cell of the virus, substitution of about 150 nucleotides just upstream from the v-*ras*H coding sequences with a similar number of sequences from the same region of the rat c-*ras*H proto-oncogene resulted in a virus whose titer in vitro was almost two orders of magnitude greater than that of wild-type v-*ras*H. This virus induced tumors with a shorter latent period than that of the wild type. When the mutant v-*ras*H gene that encoded the normal amino acid 12 but the mutated amino acid 59 was substituted for the wild-type gene in this construction, the resulting virus also had a very high titer; it resembled the wild-type virus pathogenically in that it induced both fibrosarcomas and splenic erythroblastosis, with a latency that was actually shorter than that of the wild-type virus. These results emphasize the importance of noncoding sequences in determining the oncogenicity of even those viruses that contain a v-*onc*.

Studies of the avian MC29 virus and its variant HBI virus also indicate that relatively subtle differences in v-*onc*-containing viruses can have an enormous effect on the target tissue [21]. MC29 and HBI both encode a fusion v-*onc* protein: The N terminus is encoded by *gag* sequences; the C terminus, by the *myc* oncogene. Despite their similar structure, HBI usually causes lymphoid lymphomas, whereas MC29 induces other tumors. DNA sequence analysis has revealed differences between the two viral genomes in LTR, *gag,*and *myc;* genetic analysis will be required to establish the relative importance of these differences to the pathogenicity of the viruses.

Although viruses containing a single oncogene appear to induce tumors in a single step, there is considerable evidence that nonviral tumors generally develop following several changes in the cell (reviewed in ref. 3). A group of retroviruses that may be especially relevant to this model are those viruses that contain two v-*onc*. AEV is the prototype of this group of viruses [3,22]. This virus contains v-*erb*A and v-*erb*B. The v-*erb*B gene is a truncated version of the gene encoding the EGF receptor, whereas the v-*erb*A gene specifies a fusion protein whose N terminus encodes *gag* sequences and C terminus is encoded by sequences derived from c-*erb*A, which is unrelated to *erb*B. As its name implies, AEV induces erythroblastosis in chickens; it can also induce sarcomas. In vitro, it transforms erythroid precursor cells and fibroblasts. Mutagenesis studies have indicated that both v-*onc* are required for full tumorigenicity, but v-*erb*B is the principal transforming gene of the virus. One type of AEV mutant (v-*erb*B$^+$, v-*erb*A$^-$) still transforms erythroblasts and fibroblasts in vitro, and induces sarcomas and a different spectrum of disease from wild-type AEV. In contrast, the converse type of AEV mutant (v-*erb*$^{A+}$, v-*erb*$^{B-}$) is nononcogenic in vivo and is nontransforming in vitro.

Conclusions

The foregoing studies emphasize the importance of protein coding sequences and regulatory sequences in determining the in vivo tumorigenicity of retroviruses. These principles apply both to v-*onc*$^+$ and v-*onc*$^-$ viruses. Tumorigenicity represents the outcome of the complex interaction among these various factors, played out against the genetic and environmental background of the host. In the case of v-*onc*$^+$ viruses, the study of mutations that determine tumorigenicity has direct implications for the tumorigenic capacity of their c-*onc* counterparts. The participation of regulatory elements in determining the target cell of a particular virus provides an explanation as to how the same v-*onc* may induce different tumors. The recognition that regulatory elements function importantly in determining the target tissue of v-*onc*$^-$ viruses also implies that these elements will be important in determining the target tissue of human retrovirus-induced disease and of tumors that are not associated with retrovirus infection. As noted in Nowell and Croce's chapter on chromosomal translocations (Chapter 11), this latter prediction now has a firm experimental basis.

References

1. Weiss R, Teich N, Varmus H, Coffin J (eds) (1982) RNA Tumor Viruses: Molecular Biology of Tumor Viruses, 2nd Ed. Cold Spring Harbor Laboratory, Cold Spring Harbor, New York
2. Weiss R, Teich N, Varmus H, Coffin J (eds) (1985) RNA Tumor Viruses: Molecular Biology of Tumor Viruses, Supplement to 2nd Ed. Cold Spring Harbor Laboratory, Cold Spring Harbor, New York
3. Teich N, Wyke J, Kaplan P (1985) Pathogenesis of retrovirus-induced disease. In Weiss R, Teich N, Varmus H, Coffin J (eds), RNA Tumor Viruses: Molecular Biology of Tumor Viruses. Supplement to 2nd Ed. Cold Spring Harbor Laboratory, Cold Spring Harbor, New York 187–248
4. Shank PR, Schatz PJ, Jensen IM, Tsichlis PN, Coffin JM, Robinson HL (1985) Sequences in the *gag-pol-5'env* region of avian leukosis viruses confer the ability to induce osteopetrosis. Virology 145:94–104
5. Chatis PA, Holland CA, Hartley JW, Rowe WP, Hopkins N (1983) Role of the 3' end of the genome in determining disease specificity of Friend and Moloney murine leukemia viruses. Proc Natl Acad Sci USA 79:4408–4411
6. DesGroseillers L, Jolicoeur P (1984) The tandem direct repeats within the long terminal repeat of murine leukemia viruses are the primary determinant of their leukemogenic potential. J Virol 52:945–952
7. Elder JH (1984) On the role of recombinant retroviruses in murine leukemia. *In* Notkins AL, Oldstone MBA (eds) Concepts in Viral Pathogenesis. Springer-Verlag, New York, pp 86–95
8. Ruscetti S, Wolff L (1985) Biological and biochemical differences between variants of spleen focus-forming virus can be localized to a region containing the 3' end of the envelope gene. J Virol 717–722
9. Holland CA, Hartley JW, Rowe WP, Hopkins N (1985) At least four viral genes contribute to the leukemogenicity of murine retrovirus MCF 247 in AKR mice. J Virol 53:158–165
10. Oliff A, McKinney MD, Agranovsky O (1985) Contribution of the *gag* and *pol* sequences to leukemogenicity of Friend murine leukemia virus. J Virol 54:864–868
11. Robinson HL, Miles BD, Catalano DE, Briles WE, Crittenden, LB (1985) Susceptibility to *erb*B-induced erythroblastosis is a dominant trait of 15_1 chickens. J Virol 55:617–622
12. Anderson SM, Scolnick EM (1983) Construction and isolation of a transforming murine retrovirus containing the *src* gene of Rous sarcoma virus. J Virol 46:594–605
13. Adkins B, Leutz A, Graf T (1984) Autocrine growth induced by *src*-related oncogenes in transformed chicken myeloid cells. Cell 39:439–445
14. Iba H, Takeya T, Cross FR, Hanafusa T, Hanafusa H (1984) Rous sarcoma virus variants that carry the cellular *src* gene instead of the viral *src* gene cannot transform chicken embryo fibroblasts. Proc Natl Acad Sci USA 81:4424–4428
15. Cross FR, Garber EA, Hanafusa H (1985) N-terminal deletions in Rous sarcoma virus p60^{src}: Effects on tyrosine kinase and biological activities and on recombination in tissue culture with the cellular *src* gene. Mol Cell Biol 5:2789–2795
16. Snyder MA, Bishop JM (1984) A mutation at the major phosphotyrosine in pp60^{v-src} alters oncogenic potential. Virology 136:375–386
17. Tambourin PE, Lowy DR (1986) Unpublished observations

18. Temeles GL, Gibbs JB, D'Alonzo JS, Sigal IS, Scolnick EM (1985) Yeast and mammalian *ras* proteins have conserved biochemical properties. Nature 313:700–703
19. Tabin CJ, Weinberg RA (1985) Analysis of viral and somatic activations of the cHa-*ras* gene. J Virol 53:260–265
20. Tambourin PE, Lowy DR (1986) Unpublished observations
21. Smith DR, Vennstrom B, Hayman MJ, Enrietto PJ (1985) Nucleotide sequences of HBI, a novel recombination MC29 derivative with altered pathogenic properties. J Virol 56:969–977
22. Graf T, Beug H (1983) Role of the v-*erb*A and v-*erb*B oncogenes of avian erythroblastosis virus in erythroid cell transformation. Cell 34:7–9

CHAPTER 17
Adenovirus Entry into Cells: Some New Observations on an Old Problem

IRA PASTAN, PREM SETH, DAVID FITZGERALD, AND
MARK WILLINGHAM

With the advent of electron microscopy, it became possible to investigate
the pathway of viral entry into cells. Viruses were visualized bound to the
cell surface prior to entry and within cells after entry. Some investigators
interpreted their findings to indicate that a virus such as adenovirus entered
the cell in a vacuole or vesicle whose formation was induced by the presence
of the virus at the cell surface. Other investigators concluded that viruses
directly crossed the plasma membrane without being transported in a vesicle.

In the past 5–10 years, intensive effort has been devoted to establishing
the pathway by which growth factors, hormones, toxins, and some viruses
enter cells [1–4]. The pathway of this receptor-mediated form of endocytosis
is shown in Figure 17.1. Ligands bind to receptors on the cell surface; receptors are intrinsic proteins of the plasma membrane that move about by
random diffusion [5]. Embedded in the plasma membrane are depressions
termed *clathrin-coated pits*. A typical cultured cell such as a KB cell or a
fibroblast has about 1000 coated pits per cell; the pits make up about 1% of
the cell surface area. When ligand–receptor complexes, and in some cases
receptors not associated with ligands, enter the pits by random diffusion,
they may become fixed or trapped in these pits. In this manner, each pit
comes to contain many different receptors and ligands. Every 20–30 sec, an
uncoated vesicle forms from each pit, so that about 3000 vesicles are formed
per minute. Each vesicle contains all the substances that were previously
concentrated in the pit. The mechanism by which uncoated vesicles form
from coated pits is not fully understood [6,7]. The rate at which endocytic
vesicles form is not dependent on whether or not the pits contain ligands
and receptors; it appears to be a constitutive process. The endocytic vesicles,
which are termed *receptosomes* (or *endosomes*), move along microtubules
until they encounter tubular membranous elements on the trans-face of the

Figure 17.1 A diagrammatic summary of the pathway of receptor-mediated endocytosis.

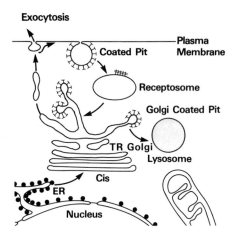

Golgi system. The endocytic vesicles appear to fuse with these elements in a process during which their contents are transferred into the tubular system on the trans-face of the Golgi. Sorting of ligands and receptors occurs in this trans-reticular Golgi (TR Golgi) system. Some ligands and receptors are sent on to lysosomes from the TR Golgi to be degraded; other ligands and receptors are recycled back to the cell surface to be reutilized. This same morphologic pathway is also utilized by soluble materials that do not bind to specific receptors on the surface; this mode of entry has been referred to as *fluid-phase* or *nonabsorptive endocytosis.*

Electron microscopic images published by Dales and co-workers almost twenty years ago clearly show adenovirus entering cells via coated pits [8]. A few minutes thereafter, the virus was observed in vesicles which Dales termed *vacuoles;* these vacuoles are what are now called receptosomes or endosomes. Finally, after approximately 30 minutes, the virus became associated with the nucleus. Images of some of these early steps in adenovirus entry are shown in Figure 17.2. To show that adenovirus enters the cell via the same pathway as growth factors and related molecules, an experiment was performed in which epidermal growth factor (EGF) was attached to colloidal gold and allowed to enter the cell synchronously with adenovirus [9]. EGF could be seen in the same coated pits as adenovirus and subsequently in the same endocytic vesicle as adenovirus. These and similar experiments clearly show that adenovirus enters cells by the pathway of receptor-mediated endocytosis. Soon after adenovirus entry, the virus is found free in the cytosol. Somehow, the virus escapes from the endocytic vesicle.

By electron microscopy we observed that, prior to its exit from the vesicle, the vesicle membrane was distorted and wrapped around the virus [9]. After the virus had escaped into the cytosol, we often observed fragments of membrane close to the virus which could have been remnants of a disrupted vesicle (Figure 17.2D).

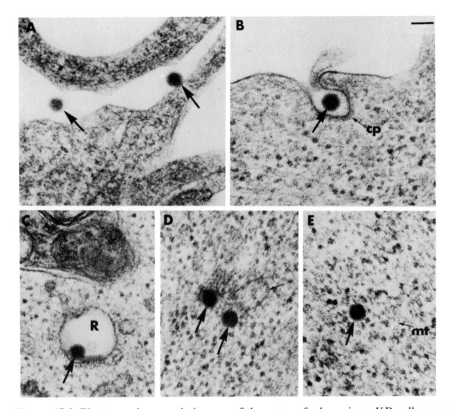

Figure 17.2 Electron microscopic images of the entry of adenovirus. KB cells were incubated with adenovirus at 4°C, then either fixed in glutaraldehyde (**A**) or warmed to 37°C for 1 (**B**), 3 (**C**), 5 (**D**), or 30 minutes prior to fixation. Virus particles bound to the cell surface are shown in (**A**) (arrows), and a virus particle clustered in a coated pit (cp) is shown in (**B**). Endocytosis from coated pits generates intracellular endocytic vesicles (receptosomes, [R] shown in (**C**). These appear to lyse, releasing virus into the cytoplasm and leaving a membrane remnant of the receptosome (small arrow, **D**). Subsequently, free virus particles are found in the cytoplasm (**E**), often near microtubules (mt).
(Bar = 0.1 μm; uranyl acetate–lead citrate counterstains.)

Because EGF entered the cell with adenovirus, we were able to ask if EGF escaped into the cytosol with the virus. One way to do this was to attach EGF to colloidal gold; we found that the complex of EGF with colloidal gold appeared in the cytosol when adenovirus was present, but not when it was absent [9]. This experiment strongly suggested that adenovirus was disrupting the endocytic vesicle. A second way to demonstrate vesicle disruption was to construct an agent that would kill the cell when released in the cytosol. To do this, *Pseudomonas* exotoxin (PE) was coupled to EGF. The EGF–PE conjugate entered the cell bound to the epidermal growth factor receptor. Ordinarily, most of the EGF–PE is carried to lysosomes, where it is degraded.

We found that adenovirus greatly enhanced the ability of EGF–PE to kill cells, and reasoned it did this by allowing more EGF–PE to escape from the vesicle than was possible when EGF–PE was added alone. The enhancement of the toxicity of EGF–PE by adenovirus in KB cells was quite remarkable; up to a 10,000-fold enhancement of EGF–PE toxicity was observed when adenovirus was present. From these experiments it was concluded that adenovirus escaped from the vesicle by disrupting the vesicle membrane.

Mechanism of Vesicle Disruption

One major physiological difference between the extracellular fluid and the interior of receptosomes is the pH; the pH within receptosomes is about pH 5.5 [10]. The addition of weak bases such as chloroquine or methylamine raises the pH within receptosomes to approximately 7 and prevents adenovirus from disrupting the membrane of the receptosome and escaping into the cytosol [16]. These results suggest that the low pH activates a process that enables adenovirus to disrupt the membrane of the receptosome. Membrane lysis could be due to action of a phospholipase whose optimum activity is expressed at pH 5. However, we have been unable to detect any phospholipase activity associated with intact or disrupted adenovirus. Another way membranes might be weakened is by penetration of the membrane lipids by hydrophobic domains of proteins in the adenovirus capsid.

The hydrophobicity of proteins can be studied by measuring their ability to bind to a nonionic detergent such as Triton X-114. We found that the ability of intact adenovirus to bind Triton X-114 is much greater at pH 5 than at pH 7 [11]. When we studied the individual components of the adenovirus capsid we found that the penton base had a very striking pH dependency for binding Triton X-114; the binding of Triton X-114 by fiber and hexon was also pH dependent. These detergent binding studies suggest the following mechanism, depicted in Figure 17.3: At pH 7, the adenovirus fiber binds to a receptor on the cell surface and the virus moves into a coated pit; it is then transferred into a receptosome. Within the receptosome the pH falls; hydrophobic regions on the penton base, fiber, and hexon are exposed and bind to phospholipids; and in this manner the virus becomes intimately associated with the lipids of the vesicle membrane. The vesicle not only has a low pH but is also swollen because of elevated osmotic pressure generated by ion pumps in the vesicle membrane. These pumps maintain a greater concentration of ions within the vesicles than in the surrounding cellular fluid. Receptosomes are fragile [12,13] and could become disrupted when the hydrophobic proteins of the viral capsid penetrate the membrane and cause local weakening. Such a membrane weakening would not occur prior to cellular entry when the virus is attached to the outside of the cell, because the extracellular pH is high (pH 7) and the proteins would not have their hydrophobic residues exposed. Another factor that probably contributes to vesicle lysis is the absence of a stabilizing cytoskeleton attached to the mem-

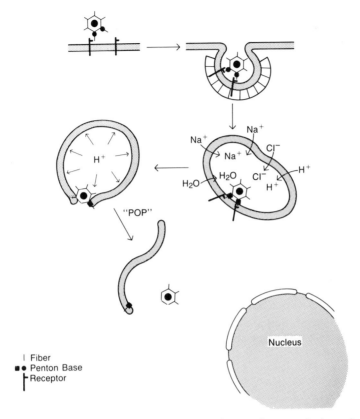

Figure 17.3 A diagrammatic summary of the pathway of entry of adenovirus.

brane of the vesicle. The plasma membrane is intimately associated on its inner surface with actin, myosin, and other cytoskeletal elements. Such a cytoskeletal meshwork is not associated with receptosomes. Although plasma membrane disruption does not occur, it has been possible to demonstrate a direct effect of adenovirus on plasma membrane permeability. When cells with adenovirus attached to their surface are placed at pH 6, the plasma membrane becomes transiently premeable to small molecules such as phosphoryl choline, 2-deoxyglucose, and ^{51}Cr bound to small peptides [14,17].

Other Viruses

A number of other viruses have been shown to enter cells via coated pits and endocytic vesicles. These include enveloped viruses such as Semliki Forest virus and vesicular stomatitis virus [8,15]. These viruses do not disrupt the membrane of the receptosome as they escape into the cytosol. Instead, the membrane of the virus fuses with that of the endocytic vesicle, and the nucleocapsid is released in the cytosol without vesicle disruption. A low pH is also required for this process, and, at pH 5, proteins of the viral membrane have been found to undergo structural changes that cause hydrophobic residues to be exposed [15]. Thus, the low pH of the endocytic vesicle is very important in the entry of at least a few, and perhaps many, different viruses.

References

1. Pastan I, Willingham MC (1983) Receptor-mediated endocytosis: Coated pits, receptosomes, and the Golgi. Trends Biochem Sci 8:250–254
2. Pastan I, Willingham MC (1981) Journey to the center of the cell: Role of the receptosome. Science 214:504–509
3. Helenius A, Mellman I, Wall D, Hubbard A (1983) Endosomes. Trends Biochem Sci 8:245–250
4. Goldstein JL, Anderson RGW, Brown MS (1979) Coated pits, coated vesicles, and receptor-mediated endocytosis. Nature 279:679–685.
5. Maxfield RF, Cheng S-y, Dragsten P, Willingham MC, Pastan I (1981) Binding and mobility of the cell surface receptors for 3,3', 5-triiodo-L-thyronine. Science 211:63–65
6. Willingham MC, Pastan I (1983) Formation of receptosomes from plasma membrane coated pits during endocytosis: Analysis by serial sections with improved membrane labeling and preservation techniques. Proc Natl Acad Sci USA 80:5617–5621
7. Willingham MC, Pastan I (1984) Do coated vesicles exist? Trends Biochem Sci 9:93–94
8. Dales S (1973) Early events in cell–animal virus interactions. Bacteriol Rev 37:103–135
9. FitzGerald DJP, Padmanabhan R, Pastan I, Willingham MC (1983) Adenovirus-induced release of epidermal growth factor and *Pseudomonas* toxin into the cytosol of KB cells during receptor-mediated endocytosis. Cell 32:607–617
10. Tycko B, Maxfield FR (1982) Rapid acidification of endocytic vesicles containing α_2-macroglobulin. Cell 28:643–651
11. Seth P, Willingham MC, Pastan I (1985) Binding of adenovirus and its external proteins to Triton X-114: Dependence on pH. J Biol Chem 260:14431–14434
12. Okada CY, Rechsteiner M (1982) Introduction of macromolecules into cultured mammalian cells by osmotic lysis of pinocytic vesicles. Cell 29:33–41
13. Dickson RB, Beguinot L, Hanover JA, Richert ND, Willingham MC, Pastan I (1983) Isolation and characterization of a highly enriched preparation of receptosomes (endosomes) from a human cell line. Proc Natl Acad Sci USA 80:5335–5339
14. Seth P, Willingham MC, Pastan I (1984) Adenovirus-dependent release of ^{51}Cr from KB cells at an acidic pH. J Biol Chem 259:14350–14353
15. Marsh M, Helenius A (1980) Adsorptive endocytosis of Semliki Forest virus. J Mol Biol 142:439–454
16. Seth P, FitzGerald DJP, Willingham MC, Pastan I (1984) Role of a low-pH environment in adenovirus enhancement of the toxicity of a *Pseudomonas* exotoxin–epidermal growth factor conjugate. J Virol 51:650–655
17. Seth P, Pastan I, Willingham MC (1985) Adenovirus-dependent increase in cell membrane permeability. J Biol Chem 261:9589–9602

CHAPTER 18

Molecular Anatomy of Viral Infection: Study of Viral Nucleic Acid Sequences and Proteins in Whole Body Sections

PETER J. SOUTHERN AND MICHAEL B.A. OLDSTONE

Introduction

There are many examples of acute viral infections that have the potential to develop into prolonged latent or persistent infections. This transition is associated with reduced, or abolished, viral gene expression which probably explains, at least in part, viral evasion of the host immune system [1]. Persistent infections have also been established by use of tissue culture cells in vitro [2], although the molecular details involving defective interfering particles may be substantially different from in vivo persistent infections as a consequence of modified selection pressures. In order to begin to understand the molecular basis of virus-induced disease, it is important to locate virus and viral genetic material within an infected host. Although there are several methods currently available to detect and quantitate viral genes, messages, and proteins during the course of infection, none is ideal. Here we will describe a novel approach using whole animal sections and compare the strengths and weaknesses of this system with other techniques employed for studying viral pathogenesis.

Acute, Latent, and Persistent Infections

There are multiple stages during virus infection from initial contact with the host to final clearance or establishment of stable latent/persistent infections. Natural infections normally occur via the respiratory or intestinal tract, other mucosal membranes, or skin puncture resulting from insect or animal bites. Virus replication may occur at the site of entry; alternatively and/or additionally virus may spread or be transported throughout the body for replication at multiple secondary sites. These secondary sites often reflect tissue tropisms where active virus replication is confined to a particular tissue or even to a certain cell type within a particular tissue. The basis for such tropism remains unclear but may involve both successful penetration of the cell, i.e., recognition of normal cell membrane components that function as receptors [3] for virus infection (see Chapters 14,15) and/or specific enhancer–promoter transcription signals [4,5] (see Chapter 8) which function in an appropriate intracellular environment to allow virus replication. In a wide variety of experimental infections, the time course and severity of virus-induced disease are independent of the virus inoculum, indicating an obligatory link between virus replication, specific cell tropism, and ensuing disease.

Infection of permissive cells normally results in extensive replication of the viral genome, accumulation of viral structural proteins, and release of progeny virus particles. The infected cell may or may not be lysed, depending upon the characteristics of the particular virus, but the infection nonetheless spreads to other susceptible cells. During progression from acute to latent or persistent infection, it is also not clear whether virus must infect semipermissive or nonpermissive cells that do not support virus replication or whether virus variants are produced and attenuate the infection in normally permissive cells. In a true latent infection, there is no detectable infectious virus and no evidence for viral transcription or synthesis of viral specific proteins. The only indication of virus is provided by in situ hybridization with nucleic acid probes to detect (silent) viral genetic information. Molecular analysis of latent infections with herpes simplex virus (HSV) in nerve ganglia cells [6,7] and cytomegalovirus in peripheral blood lymphocytes [8] indicates a restricted transcription of viral RNA with a concomitant absence of late structural proteins. With persistent infections there is continual release of low levels of infectious virus, suggesting that replication and transcription of the viral genome are extensively down-regulated but not absolutely abolished and that viral protein synthesis may be limited. For example, in chronic persistent measles virus infection, subacute sclerosing panencephalitis, there is limited expression [9] or rapid degradation of M protein and both viral glycoproteins [10], whereas in persistent lymphocytic choriomeningitis virus infections of mice (the natural host), there is a dramatic reduction in viral glycoprotein expression as compared to nucleoprotein expression in vivo [(11], see also Figure 18.3) and in vitro.

Different Techniques for Detection of Virus Infection

I)   Cocultivation with Susceptible Cultured Cells In Vitro
Individual infectious particles can theoretically be detected by cocultivation, providing that appropriate cells can be maintained in vitro to allow replication of the virus. This point is best illustrated with recent successes in the isolation of human retroviruses, which were dependent, in large part, on techniques allowing the establishment of human T lymphocytes in culture [12]. Hence, cocultivation is compromised by the need to grow and maintain specialized

cells (i.e., neurons, oligodendrocytes, islets of Langerhans, lymphoctye subsets) over several weeks, and by the requirement that the viral genome remain intact and that replication of this standard virus not be obscured by (defective) interfering particles.

II)   Dot Blot or Gel Electrophoresis with Total Nucleic Acid Extracted from Individual Tissues

The analysis of tissue nucleic acids by dot-blot [13] or gel procedures [14] requires some prior knowledge of the distribution of infectious material throughout the host. It is difficult to imagine sampling all tissues, and, as a consequence, reservoirs of virus infection may pass undetected. Gel electrophoresis can provide size estimates of target nucleic acids that are detected by hybridization with labeled probes, whereas the dot-blot allows fairly precise quantitation of the hybridizing species but gives no indication of the molecular size of these species. The dot-blot procedure is attractive because of simplicity and efficiency and has the advantage that the nucleic acid need not be intact to give clear, meaningful results, whereas degraded nucleic acid is almost always inappropriate for gel analysis. Both gel filters (after transfer of nucleic acids to nitrocellulose or an equivalent nylon support) and dot-blot filters can be hybridized sequentially with multiple, independent, hybridization probes to allow precise comparisons with the same sample rather than using parallel samples which may show artifactual differences.

III)   In Situ Hybridization to Cryostat Sections from Individual Tissues

Thin sections cut with a normal cryostat have been used very successfully for in situ hybridization coupled with light microscopy (see Chapter 36). When the technique is used with ^{3}H or ^{35}S-labeled probes and emulsion autoradiography, it is often possible to identify the cell type(s) showing a hybridization signal. A recently published procedure describing a double detection system [15] (to visualize both protein and nucleic acid on the same section) will be enormously useful in extending the application of the technique as increasing numbers of antibodies become available as specific cell-surface markers. The lower limit of nucleic acid sequence detection differs according to the infection studied and the expertise of the investigator. Usual experience is detection of 20–50 copies of RNA per cell, although detection of about 1–10 copies of measles virus RNA and 1 copy of visna virus RNA per infected cell has been reported [16]. Thus, at best it would be possible to find a single cell in a field of 10^6 or more cells, given adequate scanning facilities. This technique is the most sensitive of the various hybridization procedures and allows detection of viral genetic material in a subset of cells where studies by dot blot or gel electrophoresis using nucleic acid extracted from whole tissue, containing few positive cells, could be negative. However, the procedure requires adaptation for macro-scale analysis and is limited by the work needed to study multiple tissue sources.

IV)   In Situ Hybridization to Whole Body Sections
Recently, we and others have independently described techniques for in situ
hybridization to mouse whole body sections that provide a rapid and reproducible method to detect viral nucleic acid sequences throughout all tissues
[17,18]. This new method clearly provides the opportunity to detect viral
genetic material in unexpected locations because, by cutting sections at various depths, one can sample all major organs (Figure 18.1, panels A, B, C).
This procedure also combines several advantages of the dot-blot and in situ
cell hybridization techniques. The sensitivity is currently about 10 fold less
than that of the dot-blot procedure [17].

V)   Antibody Staining of Tissue Sections
Immunochemical techniques employing polyclonal or monoclonal antibodies
and antibodies to predetermined sequences (peptide antibodies) have been
used for detection of viral antigens in tissue sections at light or electron
microscopic levels. To date, a limitation has been the selection of tissues
for analysis (see Dot Blot or Gel Electrophoresis, above). We have been
able to overcome this by adapting antibody techniques to whole body sections
of animals; these then have a sensitivity of detection in the range of 10–50
ng protein (Figure 18.1, panels D, E). However, screening for viral antigens
cannot be relied upon to provide evidence of virus infection if viral gene
expression has been shut down. In this situation, nucleic acid hybridization
would be required to locate viral genomes.

Synthesis of Hybridization Probes for In Situ Studies

The techniques that have been outlined for hybridization to viral genetic material are critically dependent on the availability of probes of high specific
activity. Typically, viral genomes have been cloned into prokaryotic vectors
for propagation in bacterial cells and convenient DNA purification. Cleavage
with sequence-specific restriction endonucleases allows separation of the
inserted viral sequence from the vector and can also be used to purify selected
regions from the cloned viral genomes prior to labeling in vitro. The nick-
translation reaction has been used extensively with α [^{32}P or ^{35}S] deoxyri-
bonucleotide triphosphates for labeling double-stranded DNA; the reaction
is dependent on the presence of double-stranded DNA, and both DNA strands
are labeled equally [19]. Alternatively, strand-specific probes can be prepared
from recombinant M13 phage genomes [20] or from SP6/T7 polymerase templates [21]. These single-stranded probes increase the specificity of hybridization because label can be exclusively incorporated into the strand that
should hybridize. For example, an antimessage sense probe will efficiently
detect messenger RNA, whereas label in the message-sense in this experiment
will contribute only to nonspecific background. Strand-specific probes have
also been important in unraveling transcription and replication pathways of

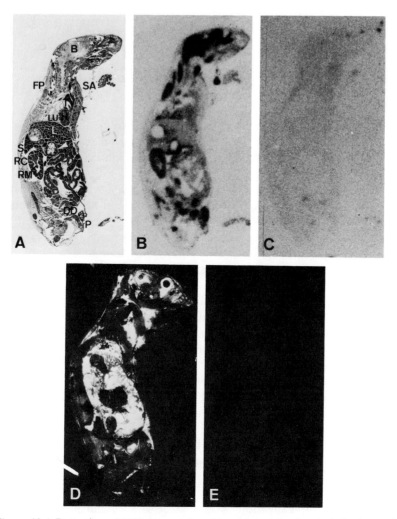

Figure 18.1 Detection of LCMV nucleic acids and proteins in whole body sections of persistently infected mice. (**A** and **B**) Histologic staining and autoradiographic detection of S RNA species in the same 40-μm section from a 12-week-old SWR/J mouse that had been infected at birth with LCMV. Major tissues are identified as follows: B, brain; SA, salivary gland; FP, brown fat pad; T, thymus; H, heart; LU, lung; L, liver; S, spleen; RC, renal cortex; RM, renal medulla; DD, ductus deferens; P, penis. The white arrow indicates the mucosa of the stomach. (**C**) Control autoradiograph of a matched, uninfected SWR/J mouse showing no signal after hybridization with an LCMV-specific probe. (**D** and **E**) LCMV antigen detection with sections from a persistently infected animal and a control, uninfected animal. This technique involves incubation with a guinea pig polyclonal anti-LCMV antiserum followed by iodinated protein A and autoradiography.

single-stranded RNA viruses where both genomic sense and genomic complementary RNAs are predicted during acute infections.

Applications of Whole Body Sectioning and In Situ Hybridization

We have used the whole-body sectioning technique to monitor the distribution of viral genetic material during the establishment and maintenance of persistent lymphocytic choriomeningitis virus (LCMV) infections in mice. Neonatal animals of most mouse strains, when injected within the first 24 hours of life with LCMV by any route, develop normally to reach maturity but retain a lifelong viremia [22]. Initally, extensive viral replication occurs in most major organs, but then, with increasing time, there is a 10^3-to 10^4-fold reduction in the titers of infectious virus that can be recovered from animals. Over the same time period however, there is a dramatic accumulation of LCMV viral nucleic acid sequences within major organs, suggesting that virus particle development has been arrested and that the viral nucleic acid has become trapped at an intracellular stage (Figure 18.2). These studies provide initial evidence for a defective virus state during in vivo infection. LCMV glycoprotein synthesis is reduced when compared with nucleoprotein synthesis in multiple tissues of persistently infected animals. This observation has been derived from whole body sectioning in conjunction with appropriate, specific viral nucleic acid and protein probes (Figure 18.3). The reduction of glycoprotein expression at the cell surface would prevent virus formation by the normal budding mechanism and remove a signal needed for effective antibody and lymphocyte immunologic control of infection. This disordered immunologic surveillance is reflected by viral persistence. Currently, the molecular explanation for reduced glycoprotein synthesis is incomplete, but the "ambisense" character of the viral genome (see Chapter 4) indicates that GP mRNA synthesis requires active replication of the LCMV genome, and there is intensive investigation of the regulatory mechanism(s) for viral replication. We have found that intracellular RNAs from infected tissues of LCMV persistently infected mice contained a complex, heterogeneously sized collection of LCMV RNA species, although the full-length genomic RNAs represent the most abundant individual species (S.J. Francis, M.B.A. Oldstone, and P.J. Southern, unpublished results). It will be necessary to determine whether these subgenomic (deletion) RNAs are causal or coincidental to the persistent infection and whether they may have properties predicted for defective interfering species.

Further, whole body sections have proved useful in other studies. For example, viral genetic material has been detected in unexpected locations in several virus infections. This assay has been particularly informative in following the spread of virus through the host and the clearance of viral materials by immunotherapy [23]. Analyses of virus replication during em-

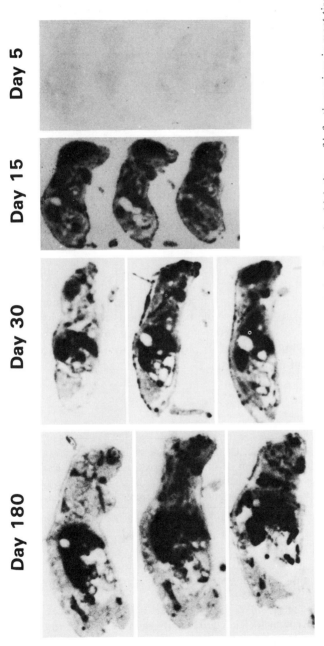

Day 5 **Day 15** **Day 30** **Day 180**

Figure 18.2 Accumulation of viral RNA sequences during the course of LCMV persistent infections in BALB/w mice. All animals were infected at birth with LCMV (60 plaque forming units [PFU] given by intracerebral injection) and then were sacrificed at the indicated times. The in situ hybridization with an S-specific probe was performed as described previously [17]. At day 5, four individual animals showed no detectable hybridization sig- nal, despite having high titers of infectious virus in most tissues (e.g., brain, mean titer at day 5: 6×10^6 PFU/g). By day 15 and onwards, all three animals at each time point showed significant and highly reproducible accumulation of viral nucleic acid. At day 180, the mean titer in brain was 1×10^4 PFU/g. [Reprinted by permission from Nature, vol 312, 5994, pp 556. Copyright © 1984 Macmillan Journals Limited.]

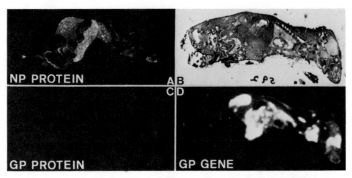

Figure 18.3 Differential expression of LCMV nucleoprotein (NP) and glycoprotein (GP) by use of whole body sections from a persistently infected mouse. (**A**) NP detection with a monospecific antibody to NP. Similar results are seen with both monoclonal and peptide antibody. (**B**) Histologic staining of an adjacent section. (**C**) GP detection with a rabbit antibody to predetermined amino acid sequences (peptide antibody). (**D**) In situ hybridization with a ^{32}P probe from the GP coding region that detects genomic-sized S RNA and the GP mRNA.

bryonic development, virus tropism for various central nervous system transmitter tracts, and expression of a wide variety of host developmental and differentiation gene products have also been performed. The combination of in situ hybridization to mouse whole body sections followed by conventional cryostat sectioning of selected tissues (Chapter 36) for high-resolution microscopic analysis should provide a powerful approach to dissecting the molecular basis of viral pathogenesis.

Acknowledgments

This work was supported in part by USPHS grants NS-12428, AG-04342, and AI-09484 and by USAMRIID Contract C-3013. The findings in this report are not to be construed as an official Department of the Army position unless so designated by other authorized documents. This is publication number 3993-IMM from the Department of Immunology, Scripps Clinic and Research Foundation, La Jolla, California 92037.

References

1. Southern PJ, Oldstone MBA (1986) Medical consequences of persistent virus infections. N Eng J Med 314:359–367
2. Younger JS, Preble OT (1980) Viral persistence: Evolution of viral populations. In Fraenkel-Conrat H, Wagner RR (eds) Comprehensive Virology, vol 16. Plenum Publishing, New York, pp 73–135
3. Lonberg-Holm KL, Philipson L (1984) (eds) Animal Viruses, part 2: Virus Receptors. Chapman and Hall, New York

4. Laimins LA, Gruss P, Pozzatti R, Khoury G (1984) Characterization of enhancer elements in the long terminal repeat of Moloney murine sarcoma virus. J Virol 49:183–189

5. Kenney S, Natarajan V, Strike D, Khoury G, Salzman NP (1984) JC virus enhancer–promoter active in human brain cells. Science 226:1337–1339

6. Puga A, Rosenthal JD, Openshaw H, Notkins AL (1978) Herpes simplex virus DNA and mRNA sequences in acutely and chronically infected trigeminal ganglia of mice. Virology 89:102–111

7. Galloway DA, Fenoglio C, McDougall JK (1982) Limited transcription of herpes simplex genome when latent in human sensory ganglia. J Virol 41:686–691

8. Rice GPA, Schrier RD, Oldstone MBA (1984) Cytomegalovirus infects human lymphocytes and monocytes: Virus expression is restricted to immediate–early gene products. Proc Natl Acad Sci USA 81:6134–6138

9. Hall WW, Choppin PW (1979) Evidence for lack of synthesis of the M polypeptide of measles virus in brain cells in subacute sclerosing penencephalitis. Virology 99:443–447

10. Fujinami RS, Oldstone MBA (1980) Alterations in expression of measles virus polypeptides by antibody: Molecular events in antibody-induced antigenic modulation. J Immunol 125:78–85

11. Oldstone MBA, Buchmeier MJ (1982) Restricted expression of viral glycoprotein in cells of persistently infected mice. Nature 300:360–362

12. Shaw GM, Hahn BH, Arya SK, Groopman JE, Gallo RC, Wong-Staal F (1984) Molecular characterization of human T-cell leukemia (lymphotropic) virus type III in the acquired immune deficiency syndrome. Science 226:1165–1171

13. Kafatos FC, Jones CW, Efstratiadis A (1979) Determination of nucleic acid sequence homologies and relative concentrations by a dot hybridization procedure. Nucleic Acids Res 7:1541–1552

14. Southern EM (1975) Detection of specific sequences among DNA fragments separated by gel electrophoresis. J Mol Biol 98:503–517

15. Brahic M, Haase AT, Cash E (1984) Simultaneous in situ detection of viral RNA and antigens. Proc Natl Acad Sci USA 81:5445–5448

16. Haase AT, Brahic M, Stowring L (1984) Detection of viral nucleic acids by in situ hybridization. In Maramorosch K, Koprowski H (eds) Methods in Virology, vol 7. Academic Press, New York, pp 189–226

17. Southern PJ, Blount P, Oldstone MBA (1984) Analysis of persistent virus infections by in situ hybridization to whole-mouse sections. Nature 312:555–558

18. Dubensky TW, Murphy FA, Villarreal LP (1984) Detection of DNA and RNA virus genomes in organ systems of whole mice: Patterns of mouse organ infection by polyomavirus. J Virol 50:779–783

19. Rigby PWJ, Dieckmann M, Rhodes C, Berg P (1977) Labeling deoxyribonucleic acid to high specific activity in vitro by nick translation with DNA polymerase I. J Mol Biol 113:237–251

20. Hu N-t, Messing J (1982) The making of strand-specific M13 probes. Gene 17:271–277

21. Melton DA, Kreig PA, Rebagliati MR, Maniatis T, Zinn K, Green MR (1984) Efficient in vitro synthesis of biologically active RNA and RNA hybridization probes from plasmids containing a bacteriophage SP6 promoter. Nucleic Acids Res 12:7035–7056

22. Buchmeier MJ, Welsh RM, Dutko FJ, Oldstone MBA (1980) The virology and immunobiology of lymphocytic choriomeningitis virus infection. Adv Immunol 30:275–331
23. Ahmed R, Southern P, Blount P, Byrne J, Oldstone MBA (1985) Viral genes, cytotoxic T lymphocytes and immunity. In Lerner RA, Chanock R, Brown F (eds) Vaccines 85: Molecular and Chemical Basis of Resistance to Parasitic, Bacterial and Viral Disease. Cold Spring Harbor Laboratory, Cold Spring Harbor, New York, pp 125–132

Immune Recognition of Viruses

CHAPTER 19
Mapping Neutralization Domains of Viruses

ECKARD WIMMER, EMILIO A. EMINI, AND
DAVID C. DIAMOND

Introduction

Animal viruses enter the host cell by attachment to a virus-specific host cell
receptor followed by a chain of events that allows the particle to pass through
the plasma membrane. This pathway may involve either the direct fusion of
the viral envelope with the cytoplasmic membrane (paramyxoviruses) or internalization via endocytotic vesicles, called *endosomes* (togaviruses,
myxoviruses, possibly also the picornaviruses). Available evidence suggests
that the virion contained in the endosomes undergoes acid-induced structural
rearrangements that lead to penetration (see ref. 1 for a review).

Antibodies, elicited to surface structures of a virion, can interfere with
the uptake and uncoating of a virus, a process called *neutralization of infectivity*. This process of antibody-mediated neutralization is poorly understood. Considering the complexity of the virion structure and its interaction
with the cell surface, it is safe to assume that several mechanisms are involved
in neutralization, e.g., structural rearrangements of the virion, aggregation,
blockage of the receptor binding site, inhibition of release of the viral genome
from the virion, etc. [2]. It is important to realize, however, that not all
surface protrusions of a virion, although capable of inducing the production
of, and binding to, specific antibodies, participate in virus neutralization. In
other words, antibodies to a surface site A of the virion may bind tightly to
site A without inhibition of infection. Antibodies to a site B, on the other
hand, may bind *and* neutralize the virus. We will call site B a *neutralization
antigenic site* (abbreviated N-Ag). Such an N-Ag can be either a continuous
linear stretch of amino acids, or a noncontinuous, topographical determinant.
Studies with neutralizing monoclonal antibodies (N-mcAb) have shown that
an N-Ag may express several functional conformations, each defined by their

ability to interact with a monospecific immunoglobulin. Each different conformation belonging to an N-Ag we will call a *neutralization epitope* (N-Ep). N-Eps can function independently of each other or may functionally overlap (for definitions used in this paper, see refs. 3, 4). Information concerning the molecular identity of an N-Ag and its N-Eps is of interest for several reasons: First, it can be assumed that an N-Ag is a surface property of a virion and thus reveals some of that virion's architecture. Second, if an N-Ag is also part of an element that interacts with the cell's surface receptor (as is the case with reovirus), the virus's attachment site to the cellular receptor can be analyzed and its significance to viral pathogenicity can be tested [5]. Third, genetic variants of an N-Ag can reveal possible mechanism(s) of neutralization [6]. Fourth, mimicking N-Ags of a virion by chemical oligopeptide synthesis or by expression of genetically engineered viral gene segments may lead to novel vaccines [7].

Various strategies have been followed to map N-Ags of viruses. These involve (a) genetic and serological analyses of reassortant recombinant viruses (for example of myxoviruses or reovirus); (b) the chemical crosslinking of F(ab) fragments of neutralizing monospecific antibodies to surface polypeptides (e.g., with poliovirus, see below); (c) the binding of specific peptides to neutralizing antibodies, or (d) the induction by those peptides of a neutralizing immune response; (e) measurements of the antigenicity of genetically engineered surface proteins; and lastly but most importantly, (f) the analysis of genetic variants resistant to neutralization by monoclonal antibodies.

The scope of this review does not allow us to cover all the viral systems that have been used to search for and characterize N-Ags. We will confine our discussion to the family of Picornaviridae, one of the largest single groups of human and animal pathogens. These viruses are naked, that is, they lack a membrane spiked with glycoproteins. Instead, they form a tightly packed capsid consisiting of 60 copies each of four coat proteins (VP1, VP2, VP3, and VP4) that contains a single strand of genome RNA of positive polarity (see, for example, ref. 8). One or several of these coat proteins carry the N-Ags. Very recently, the crystal structure of two prominent picornavirions has been elucidated: human rhinovirus 14 [9] and poliovirus type 1 (Mahoney) [10], (see also Hogle et al., Chapter 1, in this volume).

Picornaviruses encompass four genera: enteroviruses, rhinoviruses, cardioviruses, and aphthoviruses. Poliovirus (an enterovirus), human rhinovirus 14, and foot-and-mouth disease virus (FMDV, an aphthovirus) are the best-characterized, as the nucleic acid sequences of genomic RNA as well as the genetic map of all known virus-encoded polypeptides have been elucidated. (FMDV genomic RNA contains a poly(C) segment near its 5' end; the nucleotide sequences surrounding the poly (C) remain to be determined.) Moreover, a wealth of information concerning N-Ags is available for polioviruses and FMDV; similar information is just emerging for rhinovirus and for hepatitis A virus (HAV), another enterovirus.

Capsid Structures Involved in Antibody Binding

The first clue that coat protein VP1 of a picornavirus may be involved in virus neutralization came from studies of protease sensitivity of FMDV. Treatment of FMDV virions with trypsin led to cleavage of only VP1; virions modified in such a manner lost protective immunogenicity [11]. The conclusion drawn from this experiment, that VP1 of FMDV is the protective immunogenic site, was of course, only suggestive because picornavirions are known to undergo dramatic conformational changes upon chemical alteration or in response to changes in the environment (pH, temperature, etc.; see ref. 2).

The role of VP1 of FMDV in a protective immune response was supported, however, by the report that purified VP1 induces virus-neutralizing antibodies [12]. Chemical labeling of polio- and rhinovirions, meanwhile, showed that VP1 is the most surface-exposed of all capsid proteins, a result fitting VP1's role as the major immunogen of picornaviruses (the small VP4 of 7000 daltons appears to be the only picornavirus capsid protein not exposed to the surface; see ref. 13). This was found to be true for polio- and rhino virions [9,10] but remains to be proven for FMDV virions. In any event, the observation that purified VP2 (and only VP2 of coxsackie B3 virus) elicits neutralizing antibodies in rabbits [14], however, hinted that the antigenic makeup of picornaviruses may be more complex than anticipated.

It should be stressed that the neutralizing immunogenic activity of isolated capsid polypeptides (that is, ability to produce neutralizing antibodies to FMDV) is at best 0.1% to 1% of that of the intact virion. In the case of poliovirus, the production of neutralizing antibodies by purified capsid polypeptides was overlooked for several years because of the extremely low neutralization titers obtained. Careful analysis of immune sera after injection of VP1 into rabbits [15] or into rats [16], however, revealed low neutralizing activity. It was subsequently observed that as the result of specific isolation procedures, low but significant neutralizing immunogenic activity can be obtained with three of the four capsid proteins of poliovirus (VP1, VP2, VP3) [17,18]. A role of VP3 in neutralization was also suggested by Emini et al. [19], who obtained immune sera that could be freed of neutralizing antibodies by preadsorbtion with VP3. Thus, all three surface-exposed capsid proteins of poliovirus can be considered to be carriers of N-Ags. The function of these N-Ags and their N-Eps, however, is exquisitely sensitive to conformational changes. Indeed, mild heating or treatment with weak base converts poliovirus from the D-antigenic state (strongly immunogenic) to the C-antigenic state, the latter being no more immunogenic for neutralizing antibodies than isolated capsid proteins (for review, see ref. 20). Moreover, N-mcAbs elicited to whole virions generally do not bind to isolated capsid proteins or to C-antigenic particles. A notable exception is N-mcAb C3 that was obtained, remarkably, with heat-inactivated poliovirus type 1. This N-mcAb interacts with virions and also precipitates denatured VP1 and, quite appropriately, heat-inactivated virions [21].

The elucidation of the primary structures of the picornavirion and the availability of complete genetic maps of picornavirions; the preparation of cDNA clones, of neutralizing monoclonal antibodies, and of synthetic peptides; and the elucidation of the crystal structure of the picornavirion have made it possible to implicate directly, specific viral polypeptides with neutralization, to map N-Ags within the polypeptide chain, and to determine the spacial arrangement of the N-Ags.

Because VP1 and only VP1 of FMDV produces reproducibly a neutralizing immune response in experimental animals, several research groups proceeded to clone and express in *Escherichia coli* the genome region coding for VP1 of FMDV. Neutralizing immunogenicity of the genetically engineered product, usually a fusion protein, was observed [22–24]. This result generated excitement in that the expressed fusion proteins may prove useful as a vaccine for domestic animals susceptible to foot-and-mouth disease.

A more precise mapping of N-Ags in VP1 of FMDV was carried out by Strohmaier et al. [25], who dissected the capsid polypeptide chemically and proteolytically and analyzed the immunogenicity of various peptide fragments. It was found that the important neutralizing immunogenic N-Ags of FMDV VP1 reside in the regions of amino acid residues 145–154 and 200–213. Strohmaier's conclusions were supported by a comparison of VP1 sequences of various serotypes and subtypes that revealed hypervariable regions clustering around residues 130–160 and 190–213 [23,26,27]. It was argued that because of immune selection these hypervariable regions undergo more rapid genetic changes than others do, and thus are hallmarks of N-Ags. (For a more recent analysis discussing *antigenic drift* in FMDV, see ref. 28.)

Finally, peptides have been synthesized that corresponded to the above-mentioned regions of FMDV VP1, and neutralizing immunogenic activity was observed, whereas peptides from other regions of VP1 were inactive [29,30]. These experiments strongly support the map positions of N-Ags. Interestingly, the anti-peptide immune sera showed high specificity in regard to the sequence of VP1 of a given serotype [31], an observation that appears to make the development of peptides as anti-FMDV vaccines more difficult.

N-Ag mapping in poliovirus capsid proteins followed a different path due to the problem that isolated capsid polypeptides and the proteolytic fragments thereof possess extremely low or even undetectable neutralizing immunogenic activity. The first direct evidence that VP1 of poliovirus is recognized by neutralizing antibodies was provided by chemical crosslinking of F(ab) fragments derived from two different N-mcAbs to poliovirus and analyzing the complexes for the modification of the capsid polypeptides [32]. Only VP1 appeared to be crosslinked to the F(ab) fragments [32]. Sequence data of all three types of poliovirus ([33], and references therein) then allowed the identification of hypervariable *and* hydrophilic regions in capsid polypeptides likely to harbor N-Ags (see, for example, ref. 34). With the use of large panels of N-mcAbs [21,35–37], synthetic peptides [34,38], and genetically engineered fusion proteins [39], a N-Ag was mapped to the region of amino acids 93–103 of VP1 of both poliovirus type 1 and type 3. Although straight-forward on first sight, the experiments that led to this conclusion revealed an unexpected complexity of the antigenic makeup of poliovirus.

Minor et al. [37] and Evans et al. [40] analyzed a large number of non-neutralizable polio 3 variants selected by various type 3-specific N-mcAbs. Analysis of the variants by cross-neutralization tests, by fingerprinting, and by sequence analysis, suggested to these authors that (a) all N-Eps of polio 3 cluster into a single N-Ag (that is, no N-Ep that functions independently from all other N-Eps was found); (b) nearly all mutations leading to neutralization resistance map in amino acid sequence 93–103 of VP1; (c) region 93–103 represents the predominant, if not sole, surface structure of polio 3 that elicits neutralizing antibodies; and (d) a limited number of mutations in this region leads to neutralization resistance.

Emini et al. [36], in collaboration with R. Crainic and his colleagues [35], working with polio 1, also selected nonneutralizable variants with N-mcAb and found six N-Eps that can function independently from each other. Peptides, synthesized according to amino acid sequences of regions 70–80 and 93–103 of VP1, were each capable of binding to a distinct panel of N-mcAbs in an enzyme-linked immunosorbent assay (ELISA) [33]. Peptides from other regions of VP1, or derived from sequences of VP2 and VP3, were not recognized by any of the available N-mcAbs. These experiments were interpreted to mean that regions 70–80 and 93–103 represent N-Ags in VP1 of polio 1 (termed N-Ag1 and N-Ag2), a result that appeared to be corroborated by immunization and priming with conjugated peptides. This interpretation, however, was later modified (see below).

A big surprise was the observation by Diamond et al. [6] that none of the point mutations in these nonneutralizable variants of polio 1 was found to map to N-Ag1 or N-Ag2. That is, in contrast to the work on polio 3, mutations leading to neutralization resistance were found downstream on VP1 (at amino acids 221, 222, or 223) or in different capsid polypeptides (VP3 at amino acids 60 or 73; and VP2 at amino acids 72 or 270) (see Figure 19.1). It was thought that these apparently distant mutations (distant in terms of linear sequence) leading to neutralization resistance might neighbor N-Ag1 and 2 in three dimensions and thus alter the local environment or conformation of individual epitopes. As we will see below, the mutations lie nowhere near N-Ag1 or 2, but can be assigned to polypeptide chains emerging at the surface of the virion and are thus likely to indicate neutralization antigenic sites.

Apart from neutralization resistance, the phenotypes of these mutants are interesting also in that some retain their ability to bind the N-mcAb whereas others do not [6]. This observation showed that a structural change of an N-Ep can convert it into an epitope that is capable of binding a neutralizing antibody *without* neutralization ("site A," see above). Of particular interest was the mutant VP3:T>K, selected for by N-mcAb ICJ27 (see Figure 19.1). Here the single point mutation accomplished the interconversion of type 1 (Mahoney)-specific N-Eps to type 1 (Sabin)-specific N-Eps without change of the virulent phenotype [6]. That the Sabin strain-specific epitope(s) and

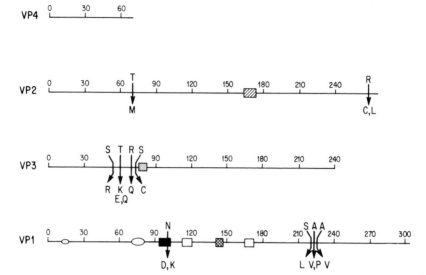

Figure 19.1 Antigenic architecture of poliovirus, type 1 (Mahoney). The position within the four capsid proteins of neutralization antigenic sites (N-Ags) and amino acid substitutions found in diverse N-mcAb-resistant variants are shown. N-Ags that have been identified through peptide immunization and priming are represented as boxes and ovals. The boxes represent peptides that can directly elicit neutralizing antibodies; these sites are exposed on the exterior surface of the virion. The ovals represent sites on the interior surface of the virion and peptides that elicit a neutralizing response only through priming (see text).

Mahoney strain-specific epitopes are mutually exclusive was shown previously by Crainic et al. [35,41], who analyzed field strains of polio 1.

A N-Ep of polio 1 with special properties is that recognized by N-mcAb C3 [21], as this epitope appears to be continuous and entirely insensitive to denaturation. Van der Werf and her colleagues [38] mapped this unique epitope to nucleotides 95–110 of VP1 by site-directed deletions within VP1. The deletion mapping experiment was carried out with poliovirus cDNA inserted into suitable fusion protein expression vectors of *E. coli*. It should be noted that VP1 of poliovirus is rapidly degraded in several strains of *E. coli* [42,43]. Poliovirus fusion proteins are more stable and they are immunogenic, albeit only weakly [43]. A neutralization escape mutant selected for by N-mcAb C3 was found to have an amino acid substitution, N>K or D (see Figure 19.1), at position 100 of VP1 [44]. Thus C3 appears to bind to a region of polio 1 that represents the major N-Ag of polio 3. As mentioned before, whereas heat-inactivated virions of polio 1 have elicited N-mcAb C3, intact polio 1 virions have so far failed to produce antibodies specific for this region.

Synthetic peptides were also tested for their ability to induce neutralizing antibodies. Peptides corresponding to the region of amino acids 93–103 of

VP1 [34,38,45] and corresponding to amino acids 162–173 of VP2 [46] were found to induce a neutralizing immune response in rabbits. Chow et al. [47] studied in rats and rabbits the immune response to a series of peptides derived from sequences spanning nearly the entire length of VP1 of polio 1, and found four immunogenic peptides: amino acids 61–80, 91–109, 182–201, and 222–241. In view of the three-dimensional structure of the picornavirion now available, some of these results are difficult to explain because regions 61–80 and 182–201 are considered to be located internally (see below). Moreover, all studies with peptides have suffered from the rather arbitrary method by which the length of the synthetic "immunogenic" probes are chosen. Clearly, length and/or the borders of a peptide are likely to influence the conformation(s) it can assume and thus are likely to influence a peptide's immunogenic potential [7,34,45,47].

Of particular interest is the observation reported by Emini et al. [34] that conjugated peptides can "prime" an immune response in rabbits. This is true even of peptides that were found unable to directly elicit neutralizing antibodies (for example, amino acids 11–17 of VP1 of polio 1) yet produced a strong priming effect. An FMDV-specific peptide, amino acids, 141–160 of VP1, coupled to keyhole limpet hemocyanin was also shown to "prime" a neutralizing antibody response in guinea pigs [48]. Van der Marel et al. [17] reported strong "priming" with isolated capsid polypeptides VP1, VP2, and VP3 of polio 1, an observation supporting the notion that these peptides carry N-Ags. The phenomenon of *priming* is poorly understood. It is particularly difficult to explain for peptides of those regions (11–17 and 70–80 of type 1 VP1) that are considered "internal" in the virion structure [9,10]. The possibility exists, however, that the picornavirion undergoes structural rearrangement(s) upon interaction with host cellular components such as the cell receptor or at specific ionic conditions that lead to the exposure of "internal" polypeptide segments. For example, the exposure of the N-terminal sequence of the capsid protein of tomato bushy stunt virus at specific ionic conditions has been reported [49].

Recent studies with HAV have shown that at least two N-Ags defined by two independent monoclonal antibodies reside on the virion's VP1 protein [50]. These results were obtained by chemical crosslinking of the antibodies' F(ab) fragments to purified virions (see above). Comparisons of the VP1 amino acid sequences of poliovirus and HAV have led to the identification of several putative HAV N-Ags which appear to correspond to those of polio 1. A synthetic peptide containing the HAV amino acid sequence of one of these sites (residues 13–24) has been found to elicit directly, HAV-specific neutralizing antibody [51]. This finding and the ability of this peptide to compete with the virus for antibody binding lends some credence to the idea that the N terminus of VP1 is exposed under some conditions.

Analysis of neutralization-resistant mutants of human rhinovirus type 14 (HRV14) initially led to the conclusion that two independent N-Ags map one each on VP1 and VP3 [52]. These mutations leading to resistance were assigned by isoelectric focusing of the virion proteins. More recently, sequence analysis of genome RNAs of these and other neutralization-resistant mutants in conjunction with the soultuion of the HRV14 crystal structure [9,53] has yielded the specific location of four independent N-Ags, with two on VP1 and one each on VP3 and VP2. Whereas in HRV14 these four N-Ags are of comparable neutralizing immunogenicity, this appears not to be the case in the other picornaviruses that have been examined.

Correlation of Antigenic Sites with the Surface Structure of the Picornavirion

The elucidation of the crystal structures of two picornavirions published very recently by Rossmann et al. [9] (human rhinovirus 14) and by Hogle et al. [10] (poliovirus type 1, Mahoney) provided the information now needed to place the primary sequences implicated in neutralization antigenic sites on the surface of the virion. The correlation of sequence and structure has led to the characterization of domains consisting either of continuous or discontinuous amino acid sequences that are likely to induce and bind to neutralizing antibodies.

Briefly, the x-ray cyrstallographic studies confirmed that VP4 is an internal virion component whereas VP1, VP2, and VP3 are exposed to the outside (see Figure 19.2) and are to some extent intertwined. Remarkably, it was found that the tertiary folds and quaternary organization of VP1, VP2, and VP3 are very similar to one another, whereas the chemical structures of these polypeptides, as we have known for some time, and hence their antigenic properties, are different. On the basis of the mapping of the neutralization escape mutations and, to some extent, the immunogenicity of synthetic peptides, the major neutralization antigenic sites of rhino-, polio-, and foot-and-mouth disease virus could be correlated with surface protrusions. With no exception, all mutations occurred in residues that face outward to the viral exterior and are thus most likely indicators of antibody binding sites. This important result made unnecessary our own speculations of general epitope modification caused by conformational distortions through amino acid substitutions, although this may be true in isolated instances [6].

Rossmann et al. [9] compared the large capsid proteins of poliovirus, HRV14, and FMDV and indicated the relation between primary sequence, higher-order structure, and known antigenic sites, conclusions also reached by Hogle et al. [10] for poliovirus. Poliovirus and HRV14 are very similar, but not identical in their antigenic makeup. Both viruses appear to possess a continuous site in region 93–103 of VP1 (N-AgI* for polio, NIm-IA for HRV14) that corresponds

---

*We have changed our nomenclature of the N-Ag to facilitate discussion of the results of virion structure but have maintained the difference from HRV14 because of apparent differences.

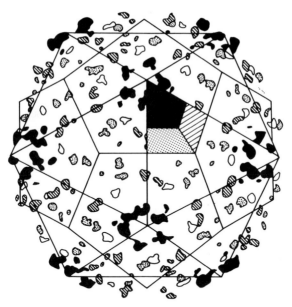

Figure 19.2 Location of the major antigenic sites of poliovirus. The solid (N-AgI), hatched (N-AgII), and stippled (N-AgIII) regions generally correspond to the similarly shaded regions described in Figure 19.1. Solid: VP1 residues 89–103, 168, and 154. Hatched: VP1 residues 222–224 and VP2 residues 166, 169, 170, and 270. Stippled: VP3 residues 58–60 and 71–73, and VP2 residue 72 (this last residue has some influence on the previous site). Open: VP1 residues 284–287. The relative positions of the capsid proteins are shown in one protomer (solid = VP1, hatched = VP2, and stippled = VP3). (Compiled from computer graphics prepared by A.J. Olson, Research Institute of Scripps Clinic.)

to a loop at the apexes of the icosahedral structure (filled shapes in Figure 19.2) This sequence is deleted in VP1 of FMDV, and hence, FMDV does not possess a N-Ag in this position. The other N-Ags of poliovirus and HRV14 are probably discontinuous in structure. For poliovirus they are shown in Figure 19.2: N-AgII (hatched areas) consists of regions 220–222 of VPI and regions 168–178 and 270 of VP2 (it corresponds roughly to NImII of HRV14), and N-AgIII (stippled areas) consists of region 55–85 of VP3 and possibly region 72 in VP2. Because of the properties of N-mcAb used for the selection of neutralization escape mutants, we favor the possibility that the T>M mutation in VP2 (see Figure 19.1) belongs to N-AgII. Hogle et al. [10] suggest a fourth N-Ag in poliovirus (type 3) in region 271–295 of VP1 but allow the possibility that this site may be part of either N-AgII or N-AgIII. A fourth, independent N-Ag (NImIB) has been described for HRV14 in region 83–85 and 138–139 of VP1 [9]. N-mcAbs to such a site have not been found in poliovirus, although mutations mapping in region 138–148 have been identified in type 1 strains isolated from vaccinated children [54] and may thus be an indication of a N-Ag (see cross-hatched box in Figure 19.1). The region corresponding to 138–148 of poliovirus

VP1 is also deleted in VP1 of FMDV. However, the major N-Ag of FMDV (region 145–154 of VP1) maps to a large loop predicted to be at the outside of the FMD virion because it corresponds to the large 220–280 loop of poliovirus VP1 which constitutes part of N-AgII [9].

The significance of the immunogenic activity of certain synthetic peptides that correspond to internal portions of the virion structure remains to be elucidated. It is possible that in the process of adsorption and penetration, structural changes of the virion may occur [55] that lead not only to the loss of VP4 but expose amino acid sequences for the attachment of neutralizing antibodies [2]. Another most intriguing puzzle is the dominance of one N-Ag over others in polioviruses. In type I, N-AgII and N-AgIII are the predominantly active sites, whereas in type 3 (and also type 2, P. Minor, personal communication) these sites are "recessive" and N-mcAb are formed primarily against N-AgI. Most recently it has been found that the preponderance of one polio virus N-Ag over others is seen only in inbred mouse strains and not in outbred animals such as the horse or in man (J. Hogle, personal communication).

Conclusion

A variety of methods have been employed to identify N-Ags and their epitopes of picornaviruses. The emerging patterns are complex and suggest, not unexpectedly, that various surface components of the picornavirion can participate in the process of neutralization, and that the immune response is distinct for different genera, different species, or even different types of picornaviruses. Some N-Ags appear to be more readily recognized by the immune system than others; those N-Ags hae been termed *immunodominant* (see discussion in ref. 3). Immunodominant N-Ags reside in VP1 of aphthoviruses (FMDV), of poliovirus type 3 and of HAV; in VP1, VP2, and VP3 of human rhinovirus 14 and of poliovirus type 1; and very likely in VP2 of coxsackie B3 virus (precise mapping of N-Ags on coxsackie B viruses is currently in progress).

The assignment of a dominant N-Ag is based upon studies with isolated capsid proteins or its fragments, on the one hand, or by N-mcAb, on the other hand. If the assignment is based solely upon the production of N-mcAbs and analyses using N-mcAbs as probes, one must be cautious to remember that this strategy may lead to only an imcomplete description of the antigenicity of a virion of the following: First, for reasons entirely unknown, there is considerable variation in the specificity of hybridoma-produced immunoglobulins, depending on which laboratory prepares N-mcAb. For example, Emini et al, [36] obtained no N-mcAbs directed against N-AgIII (see Figure 19.1), whereas Crainic et al. [35] did, even though both groups used polio 1 (Mahoney) as immunogen and also used nearly identical hybridoma technologies. Second, the immune response to N-Ags may differ from animal to animal even within species. Thus, a specific N-Ag may be recognized in one but not in another, a phenomenon not necessarily borne out by a panel of mouse-specific N-mcAb. Considering the extent of genetic variation RNA viruses can undergo even in a few rounds of replication, it seems mandatory that an immune system can recognize several distinct N-Ags and their N-Eps, even though the response may be weak to some N-Eps structures and strong to others. It appears, though, that some viruses, such as aphthovirus, express only a small number of immunodominant N-Ags. Moreover, genetic variation of these FMDV N-Ags appears not to interfere with virion stability or with biological function. This may be the explanation for the *antigenic drift* observed with aphthovirus that leads to new virulent field strains [28]. Rhinoviruses, that occur in more than 100 subtypes, may also have evolved to greatly vary their N-Ags recognized by the human immune system. Poliovirus, on the other hand, exists only in three serotypes, which indicates constraints to variability of virus structure, including the virus attachment site to the host cell receptor and N-Ags. Nevertheless, all picornaviruses are considered to be descendants of a single progenitor virus (see, for example, ref. 56) that evolved presumably by immune selection and host-cell (receptor) adaptation, a process that is likely to lead to new virulent strains or even new disease syndromes (see for example, refs. 57, 58).

Finally, the fascinating phenomenon of neutralization itself remains to be explained. How do N-Ab bind to the virion and inactivate infectivity? As with the complexity of the antigenic makeup of picornavirions, antibody-mediated host defense against picornavirus disease can be anticipated to be complex. Molecular studies with monoclonal antibodies (for example, refs. 59, 60) and the solution of the crystal structure of several picornaviruses will help to solve this important problem.

Acknowledgments

We are grateful to J.M. Hogle and A.J. Olson of the Research Institute of Scripps Clinic for sending us their computer graphics of poliovirus antigenic sites prior to publication. We thank S. Burns and S. Studier for help in preparing this manuscript. Our work is supported by U.S. Public Health Services grants AI-15122 and CA-28146.

References

1. White J, Kielian M, Helenius A (1983) Membrane fusion proteins of enveloped animal viruses. Rev Biophys 16:151–195
2. Mandel B (1982) Interaction of viruses with neutralizing antibodies. *In* Fraenkel-Conrat, H, Wagner, RR (eds) Comprehensive Virology. Plenum Press, New York, pp 37–121
3. Wimmer E, Jameson BA, Emini EA (1984) Poliovirus antigenic sites and vaccines. Nature 308:19

4. Emini EA, Jameson BA, Wimmer E (1984) The identification of multiple neutralization antigenic sites on poliovirus type 1 and the priming of immune response with synthetic peptides. *In* Chanock R, Lerner R (eds) Modern Approaches to Vaccines. Cold Spring Harbor Laboratory, Cold Spring Harbor, New York, pp 65–75

5. Fields BN (1984) Viral genes and tissue tropism. In Notkins AL, Oldstone MBA (eds) Concepts in Viral Pathogenesis. Springer-Verlag, New York, pp 102–108

6. Diamond DC, Jameson BA, Bonin J, Kohara M, Abe S, Itoh H, Komatsu T, Arita M, Kuge S, Osterhaus ADME, Crainic R, Nomoto A, Wimmer E (1985) Antigenic variation and resistance to neutralization in poliovirus type 1. Science 229:1090–1093

7. Lerner RA (1982) Tapping the immunological repertoire to produce antibodies at predetermined specificity. Nature 299:592–596.

8. Kitamura N, Semler BL, Rothberg PG, Larsen GR, Adler CJ, Dorner AJ, Emini EA, Hanecak R, Lee JJ, van der Werf S, Anderson CW, Wimmer E (1981) Primary structure, gene organization and polypeptide expression of poliovirus RNA. Nature 291:547–553

9. Rossman MG, Arnold E, Erichson JW, Frankenberger EA, Griffith JP, Hecht J-J, Johnson J, Kamer G, Luo M, Mosser AG, Rueckert RR, Sherry B, Vriend G (1985) Structure of a human common cold virus and functional relationship to other picornaviruses. Nature 317:145–153

10. Hogle JM, Chow M, Filman DJ (1985) Three-dimensional structure of poliovirus at 2.9 angstrom resolution. Science 229:1358–1363

11. Wild TF, Burroughs JN, Brown F (1969) Surface structure of foot-and-mouth disease virus. J Gen Virol 4:313–320

12. Laporte J, Grosclaude J, Wantighem J, Bernard S, Rouze P (1973) Neutralisation en culture cellulaire du pouvoir infectieux du virus de la fièvre aphteuse pardés serums provenant de porcs immunisés à paidé d'une proteine virale purifiée. C R Séances Acad Sci [III] 276:3399–3401

13. Rueckert RR (1976) On the structure and morphogenesis of picornaviruses. *In* Fraenkel-Conrat H, Wagner RR (eds) Comprehensive Virology, vol 6. Plenum Press, New York, pp 131–213

14. Beatrice ST, Katze MG, Zajac BA, Crowell RL (1980) Induction of neutralizing antibodies by the coxsackie B3 virion polypeptide, VP2. Virology 104:426–438

15. Blondel B, Crainic R, Horodniceanu F (1982) Le polypeptide structural VP1 du poliovirus type 1 induit des anticorps neutralisants. C R Séances Acad Sci 294:91–94

16. Chow M, Baltimore D (1982) Isolated poliovirus capsid protein VP1 induces a neutralizing response in rats. Proc Natl Acad Sci USA 79:7518–7521

17. van der Marel P, Hazendank TG, Henneeke MAC, van Wezel AL (1983) Induction of neutralizing antibodies by poliovirus capsid polypeptides VP1, VP2 and VP3. Vaccine 1:17–22

18. Dernick R, Heukeshoven Y, Hilbrig M (1983) Induction of neutralizing antibodies by the three structural poliovirus polypeptides. Virology 130:243–246

19. Emini EA, Dorner AJ, Dorner LF, Jameson BA, Wimmer E (1983) Identification of a poliovirus neutralization epitope through use of neutralizing anti-serum raised against a purified viral structural protein. Virology 124:144–151

20. Emini EA, Jameson BA, Wimmer E (1985) Antigenic structure of poliovirus. *In* Neurath AR, van Regenmortel MHV (eds) Immunochemistry of Viruses—the Basis of Serodiagnosis and Vaccines. Elsevier Biomedical Press, Holland, pp. 281–294

21. Blondel B, Akacem O, Crainic R, Corillin P, Horodniceanu F (1983) Detection by monoclonal antibodies of an antigenic determinant critical for poliovirus neutralization present on VP1 and on heat-inactivated virions. Virology 126:707–710

22. Kupper H, Keller W, Kurz C, Forss S, Schaller H, Franze R, Strohmaier K, Marquardt O, Zaslasky VG, Hofschneider PH (1981) Cloning of cDNA of major antigen of foot and mouth disease virus and expression in *E. coli*. Nature 289:555–559

23. Kleid DG, Yansura D, Small B, Doubenko D, Moore DM, Grubman MJ, McKercher PD, Morgan DO, Robertson BH, Bachrach HL (1981) Cloned viral protein vaccine for foot-and-mouth disease: Responses in cattle and swine. Science 214:1125–1129

24. Boothroyd JC, Highfield PE, Cross GAM, Rowlands DJ, Lowe PA, Brown F, Harris TJR (1981) Molecular cloning of foot and mouth disease virus genome and nucleotide sequences in the structural protein genes. Nature 290:800–802

25. Strohmaier K, Franze R, Adam KH (1982) Location and characterization of the antigenic portion of the FMDV immunizing protein. J Gen Virol 59:295–306

26. Kurz C, Forss S, Kupper H, Strohmaier K, Schaller H (1981) Nucleotide sequence and corresponding amino acid sequence of the gene for the major antigen of foot and mouth disease virus. Nucleic Acids Res 9:1919–1930

27. Boothroyd JC, Harris TJR, Rowland DJ, Lowe PA (1982) The nucleotide sequence of cDNA coding for the structural proteins of foot-and-mouth disease virus. Gene 17:153–161

28. Weddel GN, Yansura DG, Doubenko DJ, Hartlin ME, Grubman MJ, Moore DM, Kleid DG (1985) Sequence variation in the gene for the immunogenic capsid protein VP1 of foot-and-mouth disease virus type A. Proc Natl Acad Sci USA 82:2618–2622

29. Bittle JL, Houghten RA, Alexander H, Shinnick TM, Sutcliffe YG, Lerner RA, Rowlands DJ, Brown F (1982) Protection against foot-and-mouth disease by immunization with a chemically synthesized peptide predicted from the viral nucleotide sequence. Nature 298:30–33

30. Pfaff E, Mussgay M, Bohm HO, Schulz GE, Schaller H (1982) Antibodies against a preselected peptide recognize and neutralize foot-and-mouth disease virus. EMBO J 1:869–874

31. Rowlands DJ, Clarke BE, Carroll AR, Brown F, Nicholson BH, Bittle JL, Houghten RA, Lerner RA (1983) The chemical basis for variation in the major antigenic site eliciting neutralizing antibodies in foot-and-mouth disease. Nature 306:694–697

32. Emini EA, Jameson BA, Lewis AJ, Larsen GR, Wimmer E (1982) Poliovirus neutralization epitopes: Analysis and localization with neutralizing monoclonal antibodies. J Virol 43:997–1005

33. Toyoda H, Kohara M, Kataoka Y, Suganuma T, Omata T, Imura N, Nomoto A (1984) The complete nucleotide sequence of all three poliovirus serotype genomes (Sabin): Implication for the genetic relationship, gene function and antigenic determinants. J Mol Biol 174:561–585

34. Emini EA, Jameson BA, Wimmer E (1983) Priming for and induction of anti-poliovirus neutralizing antibodies by synthetic peptides. Nature 304:699–703

35. Crainic R, Coullin P, Blondel B, Caban N, Bone A, Horodniceanu F (1983) Natural variation of poliovirus neutralization epitopes. Infect Immun 41:1217–1225

36. Emini EA, Kao S-Y, Lewis AJ, Crainic R, Wimmer E (1983) The functional basis of poliovirus neutralization determined with monospecific neutralizing antibodies. J Virol 46:466–474
37. Minor PD, Schild GC, Bootman J, Evans DMA, Ferguson M, Reeve P, Spitz M, Stanway G, Cann AJ, Hauptmann R, Clarke LD, Mountford RC, Almond JW (1983) Location and primary structure of the antigenic site for poliovirus neutralization. Nature 301:674–679
38. Wychowski C, van der Werf S, Siffert O, Crainic R, Bruneau P, Girard M (1983) A poliovirus type 1 neutralization epitope is located within amino acid residues 93 to 104 of viral capsid polypeptide VP1. EMBO J 2:2019–2024
39. van der Werf S, Wychowski C, Bruneau P, Blondel B, Crainic R, Horadniceanu F, Girard M (1983) Localization of a poliovirus type 1 neutralization epitope in a viral capsid polypeptide VP1. Proc Natl Acad Sci USA 80:5080–5084
40. Evans DMA, Minor PD, Schild, GS Almond JW (1983) Critical rate of an eight-amino acid sequence of VP1 in neutralization of poliovirus type 3. Nature 304:452–462
41. Crainic R, Blondel B, Horaud F (1984) Antigenic variation of poliovirus studied by means of monoclonal antibodies. Rev Infect Dis 6:S535–S539
42. van der Werf S (1984) Clonage moleculaire du poliovirus type 1 et expression de ses proteines de capside chez *Escherichia coli:* Identification d'un epitode de neutralization. Thèse De Doctorat, D'Etat Universitat, Paris 7
43. Enger-Valk BE, Jore J, Pouwels PH, van der Marel P, van Wezel TL (1984) Expression in *Escherichi coli* of capsid protein VP1 of poliovirus type 1. In Chanock RM, Lerner RA (eds) Modern Approaches to Vaccines. Cold Spring Harbor Laboratory, Cold Spring Harbor, New York, pp 173–178
44. Blondel B, Crainic R, Fichot O, Onfraise G, Candrea A, Diamond D, Girard M, Horaud F (1986) Mutations conferring resistance of neutralization with monoclonal antibodies in type 1 poliovirus can be located outside or inside the antibody-binding site. J Virol 57:81–90
45. Ferguson M, Evans DMA, Magrath DI, Minor PD, Almond JW, Schild GC (1985) Introduction by synthetic peptides of broadly reactive, type-specific neutralizing antibody to poliovirus type 3. Virology 143:505–515
46. Emini EA, Jameson BA, Wimmer E (1984) Peptide induction of poliovirus neutralizing antibodies: Identification of a new antigenic site on coat protein VP2. J Virol 52:719–721
47. Chow M, Yabrov R, Bittle J, Hogle J, Baltimore D (1985) Synthetic peptides from four separate regions of the poliovirus type 1 capsid protein VP1 induce neutralizing antibodies. Proc Natl Acad Sci USA 82:910–914
48. Francis MJ, Fry CM, Rowlands DJ, Brown F, Bittle JL, Houghten RA, Lerner RA (1985) Priming with peptides of foot-and-mouth disease virus. In Lerner RA, Chanock RM, Brown F (eds) Vaccines 85. Cold Spring Harbor Press, New York, pp 203–210
49. Robinson IK, Harrison SC (1982) Structure of the expanded state of tomato bushy stunt virus. Nature 297:563–568
50. Hughes JV, Stanton LW, Tomassini JE, Long WJ, Scolnick EM (1984) Neutralizing monoclonal antibodies to hepatitis A virus: Partial localization of a neutralizing antigenic site. J Virol 52:465–473
51. Emini EA, Hughes JV, Perlow DS, Boger J (1985) Induction of hepatitis A virus-neutralizing antibody by a virus-specific synthetic peptide. J Virol 55:836–839

52. Sherry B, Rueckert RR (1985) Evidence for at least two dominant neutralization antigens on human rhinovirus 14. J Virol 53:137–143

53. Sherry B, Mosser AG, Colonno RJ, Rueckert RR (1986) Use of monoclonal antibodies to identify four neutralization immunogens on a common cold picornavirus, human rhinovirus 14. J Virol 57:246–257

54. Jameson BA, Bonin J, Wimmer E, Kew OM (1986) Natural variants of the Sabin type 1 vaccine strain of poliovirus and correlation with a poliovirus neutralization site. Virology 143:337–341

55. Crowell RL, Landau BJ (1983) Receptors in the initiation of picornavirus infections. *In* Fraenkel-Conrat H, Wagner RR (eds) Comprehensive Virology. Plenum Press, New York, pp 1–42

56. Argos P, Kamer G, Nicklin MJH, Wimmer E (1984) Similarity in gene organization and homology between proteins at animal picornaviruses and a plant comovirus suggest common ancestry of these virus families. Nuceic Acids Res 12:7251–7276

57. Cherry JD, Donald MD, Nelson DB (1966) Enterovirus infections: Their epidemiology and pathogenesis. Clin Pediatr 5:659–664

58. Brown F, Wild F (1974) Variation in the coxsackievirus type B5 and its possible role in the etiology of swine vesicular disease. Intervirol 3:125–128

59. Icenogle J, Shiwen H, Duke G, Gilbert S, Rueckert R, Anderegg J (1983) Neutralization of poliovirus by a monoclonal antibody: Kinetics and stoichiometry. Virology 127:412–425

60. Emini EA, Ostapchuk P, Wimmer E (1983) Bivalent attachment of antibody onto poliovirus leads to conformational alteration and neutralization. J Virol 48:547–550

CHAPTER 20
CTL Recognition of Transfected *H-2* Gene and Viral Gene Products

THOMAS J. BRACIALE AND VIVIAN L. BRACIALE

The host response to viral infection consists of both virus-specific and non-specific elements. Among the specific (immune mediated) responses to viral infection, several lines of evidence point toward the cytolytic T lymphocyte (CTL) as an important effector in antiviral immunity [1]. Notably, specific antiviral cell-mediated cytolytic activity is readily demonstrable during the course of many experimental viral infections [2]. More importantly, both heterogeneous and cloned populations of antiviral CTL have been shown, upon adoptive in vivo transfer into infected recipients, to specifically inhibit virus replication and to alter the outcome of lethal infection [2–5]. Because the antiviral activity of this T lymphocyte subset is monitored in vitro by the capacity of sensitized CTL to directly destroy virally infected target cells by cell-to-cell contact [1], it is reasonable to assume that CTL would function in a similar manner in vivo to eliminate virus, i.e., by direct destruction of infected cells. However, antiviral CTL have also been shown to release lymphokines, including interferon-γ (IFNγ), after contact with antigen [6]. Thus, CTL may inhibit, through a variety of mechanisms (both antigen-specific and nonspecific), virus replication and spread in the body [7].

Central to the recognition of virus-infected cells by CTL and to T lymphocyte function in general is the phenomenon of major histocompatibility gene complex (MHC) restriction. This phenomenon has been discussed in detail elsewhere in this series [8,9]. Briefly stated, T lymphocytes can recognize foreign (viral) antigens on cell surfaces only in the context of self MHC structures [1,8]. This requirement for the antigen receptor on CTL to recognize simultaneously a given foreign antigenic epitope along with MHC products on cell surfaces may insure that these effector T lymphocytes will focus their activity on sites of viral replication, i.e., virus-infected cells, rather than free viral particles in tissues and body fluids.

T Cell Receptor

The demonstration of MHC restriction in foreign antigen recognition by T lymphocytes raised fundamental issues concerning the molecular nature of the antigen receptor on T lymphocytes and the topological relationship of MHC gene products and foreign (e.g., viral) antigens on the target cell surface [10]. Until recently, little information has been available on the structure of the T lymphocyte receptor molecule and the structure and organization of the gene(s) coding for this receptor. Recent studies employing biochemical and molecular genetic techniques have revealed that the antigen receptor on cytolytic and helper T lymphocytes is a heterodimeric structure of 85–95 kilodaltons, each chain of which is the product of a gene somatically rearranged in a fashion similar to the immunoglobulin genes [11,12]. In the germ line, the genes for these two chains (α chain and β chain) are organized as discrete genetic elements or families (V_α, D_α, J_α, C_α; V_β, D_β, J_β, C_β) which upon rearrangement generate the diverse repertoire of the T lymphocyte antigen receptor. On the basis of available data, this heterodimeric antigen receptor is likely to be similar in overall structure to that of other members of the immunoglobulin supergene family.

In spite of information on the molecular structure and genetic organization of the T lymphocyte receptor, important questions remain unanswered: Notably, what is the precise contribution of each receptor chain to the recognition of self MHC products and foreign antigens? Although this question will be definitively answered only by analysis of the specificity of individual α- and β-chain combinations expressed by gene transfer techniques, available evidence suggests that both chains contribute to the recognition of both the MHC and foreign antigenic moieties [11]. It seems likely, therefore, that information on the antigenic epitopes displayed by MHC products and foreign target antigens as well as their topological relationships will be required in order for us to appreciate fully the interaction between the T cell receptor and the target cell antigen complex.

H-2 Genes

As initailly described by Zinkernagel and Doherty [1,8], antiviral CTL are restricted in their recognition of virus-infected cells by genes of the MHC that encode class I MHC gene products [1]. The class I molecule is comprised of a 45-kilodalton integral membrane glycoprotein which is noncovalently associated with the 12-kilodalton β_2-microglobulin polypeptide. The class I molecule is divided into five domains or regions, including three external domains of approximately 90 amino acids each along with a transmembrane and cytoplasmic domain. Early studies mapped the genetic elements restricting the CTL-target interaction to the MHC loci encoding the *K, D,* and

L class I gene products in the mouse (*HLA-A* and *-B* loci in the human). Although antibodies directed to the protein products of these class I genetic loci were found to specifically inhibit cytolytic activity of antiviral CTL, these and many other studies only indirectly linked the restricting element for CTL recognition to the 45-kilodalton products of these loci. Recently, the genes corresponding to a large portion of the murine MHC have been cloned [13], and several distinct class I MHC genes have been expressed by DNA-mediated gene transfer. Several groups have now unequivocally identified the 45-kilodalton products of the murine *K, D,* and *L* loci as the restricting elements for antiviral CTL directed to a variety of different viruses, including influenza, lymphocytic choriomeningitis virus, and vesicular stomatitis virus ([13–16], T.J. Braciale, unpublished observation).

Murine class I MHC genes have a characteristic exon–intron organization that is remarkably similar from gene to gene. Each gene is divided into eight exons that correspond to the domains of the class I polypeptides. Exon 1 codes for the leader sequence, whereas exons 2, 3, and 4 encode the three external domains (α_1, α_2, and α_3) of the polypeptide. Exon 5 codes for the transmembrane hydrophobic domain and the cytoplasmic domain, whereas three prime untranslated portions of the gene are encoded by exons 6, 7, and 8 [13]. Because of this exon–intron organization, it has been possible to alter the structure of these class I MHC genes and assess the impact of these alterations on CTL recognition. One experimental approach has been to examine the possible role of the cytoplasmic domain of these class I MHC products in transmembrane signaling events leading to target cell destruction. Exons corresponding to the cytoplasmic domains have been replaced by truncated gene segments or by exons corresponding to the cytoplasmic domains of unrelated proteins. To date, these studies suggest that the cytoplasmic domain may have a quantitative effect on the efficiency of CTL recognition and subsequent lysis. The impact of this domain alteration appears, however, to vary from virus to virus. It is not certain whether the effect of the mutated cytoplasmic domain is on the level of class I product expression on the target cell, the stability of the molecule, etc. It is clear from these studies that there is not a unique class I product cytoplasmic domain that is necessary for the signaling of intercellular events leading to CTL-mediated lysis.

Another type of in vitro mutagenesis of class I MHC genes that has been used in this analysis of antiviral CTL recognition is the shuffling of exons between class I genes. This approach relies on the presence of relatively conserved restriction endonuclease cleavage sites within introns of class I genes and on the fact that the progeny of a given antiviral CTL precursor are restricted in viral antigen recognition by only one of the three major class I MHC products: K, D, or L. In these studies exons corresponding to the α_1, α_2, or α_3 domain of two distinct class I polypeptides are exchanged. The capacity of CTL restricted by either of the parental class I molecules to recognize transfected cells expressing these hybrid class I molecules is then determined. All studies to date have consistently shown that the sites restricting CTL recognition map to the outer two domains, i.e., the α_1 and α_2 of the class I polypeptides [17–19]. This finding is consistent with available sequence data, which has mapped the major amino acid sequence variability among this highly polymorphic family of proteins to the outer two domains of these molecules [13].

Attempts to map CTL restriction sites to the α_1 or α_2 domain of a given class I molecule by exon shuffling have yielded less clear-cut results. On one hand, exchange of α_1 and α_2 gene segments between the *L* gene and *D* gene of the *H-2^d* haplotype has yielded chimeric molecules, e.g., α_1^L, α_2^D, α_3^D, or α_1^D, α_2^L, α_3^L, which can serve as functional restriction elements, thereby allowing for the mapping of antiviral CTL restriction sites to either the α_1 or α_2 domains [19]. In contrast, shuffling of the α_1 and α_2 exons between the K^d and D^d genes ([17], I. Stroynowska and T. Braciale, unpublished observations) or between the K^b and D^b genes [18] yields a chimeric molecule which lacks the appropriate recognition sites for CTL recognition. Thus, the choice of gene pairs used in this exon exchange appears to affect profoundly the outcome of the analysis.

The simplest hypothesis consistent with the above observations is that the site (or sites) on class I MHC molecules that restricts CTL recognition resides within the outer two external domains of the polypeptide. This CTL recognition site on the molecule is generated by the interaction of the α_1 domain with the α_2 domain of the polypeptide and is highly conformationally sensitive. In some instances, such as the aforementioned exon shuffling between L^d and D^d genes, the interactions of α_1 and α_2 domains in the hybrid molecule result in the generation of the appropriate conformation for CTL recognition. In the case of the K^d/D^d and K^b/D^b exon shuffling, hybrid molecules are formed which are expressed at the cell surface in an antigenically stable configuration, but the interaction of the α_1 and α_2 domains in these hybrid molecules does not yield a suitable CTL recognition site. In vitro mutagenesis experiments at the level of individual nucleotides are currently underway in several laboratories to assess the impact of single amino acid substitution in α_1 or α_2 on CTL recognition. These experiments will better define specific CTL recognition sites on class I molecules. Ultimately, detailed x-ray crystallographic information will be required to define the conformation of the α_1 and α_2 domains in the molecule and possible sites of domain–domain interaction.

Viral Genes

Considerable progress has recently been made in defining the specific viral polypeptides that serve as target antigens for MHC-restricted CTL. An important impetus for this work has been the need to get detailed information on the nature of viral antigenic epitopes that stimulate CTL response. Such information is of obvious importance for future vaccine design. In addition, studies on the fine specificity of antiviral CTL have revealed unusual reactivity patterns among viruses which are unrelated to their conventional serologic reactivity patterns [1,20]. These findings raise the possibility that CTL recognition of viruses and virus-infected cells differs fundamentally from antibody recognition.

One ot the first and most striking examples of unusual cross-reactivity at the level of CTL recognition was observed in the influenza system [21]. Here, CTL directed to a given type A influenza strain were found to exhibit a high level of cytolytic activity on target cells infected with any type A influenza strain, irrespective of subtype. Early studies indicated that this cross-reactivity could not be totally accounted for by CTL recognition of conserved epitopes on the serologically distinct viral glycoproteins (hemagglutinin and neuraminidase) expressed on the infected cell surface [22]. Rather, these early studies raised the possibility that internal virion polypeptides with type-common antigenic determinants are expressed on the surface of influenza-infected cells and serve as target antigens for type A influenza-specific, cross-reactive CTL.

Recent studies utilizing target cells displaying the influenza hemagglutinin (HA) and nucleocapsid (NP) gene products expressed by DNA-mediated gene transfer and a cloned population of anti-influenza CTL have begun to clarify the nature of these CTL target antigens on the influenza-infected cells [23–25]. It appears that the HA is the major, if not primary, target antigen for CTL, with specificity for either the immunizing virus strain alone or viruses of the same subtype. There is also at least one epitope common to HAs of the major human type A influenza subtypes that stimulates a cross-reactive CTL response. This response represents a very small component of this cross-reactive CTL response. Current evidence indicates that the internal virion NP antigen is a major cell-surface target for cross-reactive CTL.

The finding in the influenza system that both an integral membrane glycoprotein, i.e., the HA, and an internal virion polypeptide, i.e., the NP, can serve as CTL target antigens is in keeping with recent observations on the CTL response to SV40 virus where the SV40 large T antigen appears to be the primary CTL target antigen on SV40-transformed cells [26,27]. The fact that ''soluble'' molecules like the NP and T antigen that accumulate within the cell and lack conventional transmembrane hydrophobic domains are recognized at the cell surface by CTL raises new and important questions concerning the mechanism of CTL recognition of viral antigens. Preliminary data from several laboratories indicate that recognition of internal viral polypeptides as well as nonstructural gene products expressed in infected cells may be a general feature of antiviral CTL recognition. It is possible that these unconventional CTL target antigens are stably associated with the target cell membrane by a posttranslational modification, e.g., fatty acylation. Alternatively, CTL may recognize processed forms of viral polypeptides displayed on the infected cell surface in some as yet undefined manner, rather than intact molecules stably associated with the plasma membrane. Resolution of this issue will have profound implications for understanding CTL-target interactions at the molecular level. Obviously, target cells displaying altered or truncated forms of specific viral polypeptides expressed by transfection of mutagenized genes will provide powerful tools in the analysis of the mechanism of CTL recognition.

In summary, considerable strides have now been made in defining both the molecular structure of the antigen receptor on CTL and the structural basis for MHC restriction in CTL recognition of viral antigens. DNA-mediated gene transfer has been a valuable tool in this analysis. The imaginative application of in vitro mutagenesis and transfection techniques to the study of CTL recognition is likely to provide definitive answers to some old questions and in the process raise many important new questions for cellular immunologists and virologists.

Acknowledgments

The authors thank Mrs. Cathy Hamby and Ms. Betsy Klein for assistance in the preparation of this manuscript. This work was supported by NIH Grants AI-15608, HL-33391, AI-15353, and by research support from the following companies: Brown and Williamson Tobacco Corp., Philip Morris, Inc., and the United States Tobacco Company.

References

1. Zinkernagel RM, Doherty PC (1979) MHC-restricted cytotoxic T cells: Studies on the biological role of polymorphic major transplantation antigens determining T cell restriction-specificity, function, and responsiveness. Adv Immunol 27:51–177
2. Ada GL, Leung KN, Ertl H (1981) An analysis of effector T cell generation and function in mice exposed to influenza A or Sendai viruses. Immunol Rev 58:5–24
3. Lin YL, Askonas BA (1981) Biological properties of an influenza A virus-specific killer T cell clone: Inhibition of virus replication in vivo and induction of delayed-type hypersensitivity reactions. J Exp Med 54:225–234
4. Byrne JA, Oldstone MBA (1984) Biology of cloned cytolytic T lymphocytes specific for lymphocytic choriomeningitis virus: Clearance of virus in vivo. J Virol 51:682–686
5. Lukacher AE, Braciale VL, Braciale TJ (1984) In vivo effector function of influenza virus-specific cytotoxic T lymphocyte clones is highly specific. J Exp Med 160:814–826
6. Morris AG, Lin YL, Askonas BA (1982) Immune interferon release when a cloned cytotoxic T-cell line meets its correct influenza-infected target cell. Nature 295:150–152
7. Zinkernagel RM, Althage A (1977) Antiviral protection by virus-immune cytotoxic T lymphocytes: Infected target cells are lysed before infectious virus progeny is assembled. J Exp Med 145:644–651

8. Doherty PC, Zinkernagel RM (1984) MHC restriction and cytotoxic T lymphocytes. *In* Notkins AL, Oldstone MBA (eds) Concepts in Viral Pathogenesis, vol I. Springer Verlag, New York, pp 53–57

9. McDevitt HO (1984) Host genes controlling the immune response. *In* Notkins AL, Oldstone MBA (eds) Concepts in Viral Pathogenesis, vol I. Springer Verlag, New York, pp 79–85

10. Doherty PC, Blanden RV, Zinkernagel RM (1976) Specificity of virus-immune effector T cells for H-2K or H-2D compatible interactions: Implications for H-antigen diversity. Transplant Rev 29:89–107

11. Haskins K, Kappler J, Marrack P (1984) The major histocompatibility complex-restricted antigen receptor on T cells. Ann Rev Immunol 2:51–66

12. Davis M (1985) Molecular genetics of T-cell receptor β chain. Ann Rev Immunol 3:537–560

13. Hood L, Steinmetz M, Malissen B (1983) Genes of the major histocompatibility complex of the mouse. Annu Rev Immunol 1:529–568

14. Orn A, Goodenow, RS, Hood L, Braylon PR, Woodward JG, Harmon RC, Frelinger JA (1982) Product of a transferred H-2L^d gene acts as restriction element for LCMV-specific killer T cells. Nature 297:415–417

15. Reiss CS, Evans G, Margulus DH, Seidman JG (1983) Allospecific and virus-specific cytolytic T lymphocytes are restricted to the N or C1 domain of H-2 antigens expressed on L cells after DNA-mediated gene transfer. Proc Natl Acad Sci USA 80:2709–2712

16. Forman J, Goodenow RS, Hood L, Ciavarra R (1983) Use of DNA mediated gene transfer to analyze the role of H-2L^d in controlling the specificity of anti-VSV cytotoxic T cells. J Exp Med 157:1261–1269

17. Arnold B, Burgert HG, Hamman U, Hammerling G, Kees U, Kvist S (1984) Cytolytic T cells recognize the two amino-terminal domains of H-2 antigens in tandem in influenza A-infected cells. Cell 38:79–87

18. Allen H, Wraith D, Pala P, Askonas BA, Flavell RA (1984) Domain interactions of H-2 class I antigens alters cytotoxic T cell recognition sites. Nature 309:279–281

19. Murre C, Reiss CS, Bernabeu C, Chen LB, Burakoff S, Seidman JC (1984) Construction, expression, and recognition of an H-2 molecule lacking its carboxyl terminus. Nature 307:432–436

20. Zinkernagel RM, Rosenthal KL (1981) Experiments and speculation on the specificity of T cells and B cells. Immunol Rev 58:131–155

21. Effros RB, Doherty PC, Gerhard W, Bennink J (1977) Generation of both cross-reactive and virus-specific T-cell populations after immunization with serologically distinct influenza A viruses. J Exp Med 145:557–568

22. Braciale TJ (1977) Immunolgic recognition of influenza virus-infected cells. II. Expression of influenza A matrix protein on the infected cell surface and its role in recognition by cross-reactive cytotoxic T cells. J Exp Med 146:673–689

23. Braciale TJ, Braciale VL, Henkel TJ, Sambrook J, Gething MJ (1984) Cytotoxic T lymphocyte recognition of the influenza hemagglutinin gene product expressed by DNA-mediated gene transfer. J Exp Med 159:341–352

24. Townsend ARM, McMichael, AJ, Carter NP, Huddeston JA, Brownlee GG (1984) Cytotoxic T cell recognition of the influenza nucleoprotein and hemagglutinin expressed in transfected mouse L cells. Cell 39:13

25. Yewdell JW, Bennink JR, Smith GL, Moss B (1985) Influenza A virus nucleo-protein is a major target antigen for cross-reactive anti influenza A virus cytotoxic T lymphocytes. Proc Natl Acad Sci USA 82:1785–1789
26. Campbell AE, Foley FL, Tevethia SS (1983) Demonstration of multiple antigenic sites of the SV40 transplantation rejection antigen by using cytotoxic T lymphocyte clones. J Immunol 130:490–496
27. Gooding LR, O'Connell KA (1983) Recognition by cytotoxic T lymphocytes of cells expressing fragments of the SV40 tumor antigen. J Immunol 131:2580–2586

CHAPTER 21
Heterogeneity of CTL Reactive Antigenic Sites on SV40 Tumor Antigen

SATVIR S. TEVETHIA

Simian virus 40 (SV40), a DNA virus, is capable of transforming almost any mammalian cell in culture. The tumorigenic potential of these transformed cells can be demonstrated rather easily in athymic nude mice but not in immunocompetent mice [1]. Adult mice, depending upon the haplotype, respond to syngeneic SV40-transformed cells by generating a vigorous cytotoxic T lymphoctye (CTL) response that is H-2 restricted. In the $H-2^b$ haplotype both $H-2K^b$ and $H-2D^b$ act as restriction elements, whereas in the $H-2^k$ haplotype only the $H-2K^k$ acts as a restriction element [2,3]. Direct evidence indicates that the CTL are involved in restricting the progressive growth of tumors in vivo, as a line of SV40-transformed C3H cells that does not express $H-2K^k$ glycoprotein neither is susceptible to lysis by the CTL in vitro, nor can be rejected by the SV40-immunized mice in vivo [4].

The CTL response of mice to SV40 is highly specific, and the antigen to which the CTL responds is cross-reactive on all SV40-transformed cells. This antigen specificity is provided by the SV40-encoded tumor (or T) antigen of 94,000 MW and is one of the two SV40 nonvirion proteins synthesized by the SV40-transformed cells [5,6]. The smaller of the two proteins, the t antigen, which is located in the cytoplasm as well as in the nucleus of the transformed cells, appears not to be involved in the generation of CTL nor to provide a target for CTL. The large T antigen is a multifunctional protein that possesses all of the viral functions required for the transformation of cells in culture and for tumorigenicity. In addition, T antigen plays an essential role in virus replication. The large T antigen is located primarily in the nucleus of virus-infected and -tranformed cells [7]. However, a small fraction can also be localized in cell membranes, as has been demonstrated by the binding of anti-T antibodies to the cell surface and by immunoprecipitation of T antigen in the cell membranes [8–10]. The fact that T antigen in the cell membranes is acylated provides strong biochemical evidence that surface T antigen is not a contaminant from the nuclear T antigen [11].

Overwhelming evidence supports the conclusion that surface T antigen provides a target for the CTL. First, T antigen purified to homogeneity can immunize mice against SV40 tumor transplantation and can prime mice in vivo for the generation of pre-CTL, which, upon in vitro stimulation, are able to lyse syngeneic SV40-transformed cells [12,13]. Additional evidence comes from the experiments of Gooding and Edwards [14], who demonstrated that it is possible to prime mice for the generation of pre-CTL by immunizing them with either allogeneic or xenogeneic SV40-transformed cells. This indicates that the T antigen from these allogeneic or xenogeneic SV40-transformed cells is processed by Ia$^+$ cells. Another line of evidence suggests that the target antigen that reacts with SV40 specific CTL is present only on cells synthesizing T antigen. Flyer et al. [15] showed that polyoma virus tumor cells, when retransformed by SV40, express SV40 T antigen and become sensitive to lysis by the SV40 specific CTL. However, these cells lose the susceptibility to be lysed by SV40-specific CTL concurrently with the loss of T antigen. Supportive evidence that CTL recognize T antigen has been provided by Pan and Knowles [16], who demonstrated that a monoclonal antibody to T antigen can block the ability of an SV40-specific CTL clone to lyse SV40 transformed cells. We realize that most of the evidence gathered so far leads to the conclusion that CTL recognize the surface T antigen. Nonetheless, the possibility, although remote, that the CTL target may be a protein other than T antigen, remains.

The demonstration that surface antigens on SV40-transformed cells were mediated in tumor rejection, coupled with the observation that all cells transformed by SV40, regardless of species of origin, immunize animals against a SV40 tumor cell challenge, suggested that a single antigenic specificity was involved. That argument was further strengthened by the demonstration that a DNA fragment encoding the carboxy-terminal 220 amino acids of T antigen was sufficient to immunize mice against a SV40 tumor cell challenge [17]. Large-size T antigen fragments encoded by SV40 DNA integrated into adenovirus 2 DNA yielded similar results.

The first indication that there is more than one distinct antigenic site at the surface of SV40 transformed cells was provided from the studies of Campbell et al. [18]. These investigators isolated two independent clones of CTL, K11 and K19, both of which lysed SV40 transformed cells of H-2^bhaploytpe. However, only the K19 clone lysed H-2^b cells transformed by human papovavirus BK which synthesize cross-reactive T antigen. These results indicated that more than one antigenic site, presumably determined by the T antigen, exists at the surface of SV40-transformed cells, and these sites can be distinguished by the use of CTL clones and the appropriate target cells. Later studies [19] have localized the antigenic sites identified by CTL clones K11 and K19 in the amino-terminal half of T antigen. Cells (B6/pSV3T3-20-GV) expressing a T antigen composed of the first 368 amino acids were susceptible to lysis by both these clones. The existence of distinct antigenic sites has also been reported by others who isolated CTL clones that either mapped in the amino-terminal half or the carboxy-terminal half of T antigen [20]. It now appears that there are more than two antigenic sites in the carboxy-terminal half of T antigen (Gooding, personal communication) and at least two (and probably more) in the amino-terminal half of T antigen ([18,19], Anderson and Tevethia, unpublished observations].

The question arises whether the expression of epitopes on T antigen that are defined by their reactivity with CTL clones depends upon the integrity of T antigen domains in which the antigenic sites reside. The use of mutants carrying extensive deletions in the carboxy- as well as the amino-terminal half of T antigen has led to the conclusion that either the domain structure, in which the antigenic site resides, is not affected by these extensive deletions, or that the individual antigenic site within a given domain is independent of the interdomain interactions. For example, CTL clones K11 and K19 efficiently lyse B6/WT-19 and B6/pSV3T3GV cells which produce 94K and 48K T antigens, respectively. Similarly, CTL clones that map in the carboxy-terminal half of T antigen lyse SV40 WT-transformed cells and Ad2-ND1-infected cells equally well [20]. Ad2-ND1 produces a T antigen of 220 amino acids in length extending from the carboxy-terminal end. It is possible that the antigenic sites in the carboxy- and amino-terminal region of T antigen are tightly clustered. The use of point mutants and deletion mutants of smaller size will be needed to narrow the antigenic sites defined by CTL clones.

Because T antigen is present in minute quantities at the cell surface and because serological tests have not provided unequivocal evidence for its localization at the cell surface, its reactivity with CTL remains the most sensitive way to demonstrate its presence. The evidence that CTL are indeed reacting with the T antigen is impressive. However, additional genetic evidence that the loss of an antigenic site closely parallels the deletion of sequences from the T antigen coding regions that define that particular antigenic site without affecting other antigenic sites would strengthen the conclusion. We have recently identified a short stretch of amino acids in the amino-terminal half of T antigen, which when deleted, abolishes the reactivity of surface T antigen to one CTL clone but not to another, which also maps in the amino-terminal region of T antigen (Anderson and Tevethia, unpublished).

The presence of multiple and distinct antigenic sites on surface T antigen that react with CTL provides a very powerful immunosurveillance weapon in the host reacting to a developing neoplasm. We have recently shown that a CTL clone is capable of abrogating the transformation of primary mouse embryo cells by SV40 DNA in vitro [21]. It is interesting that a virus-coded protein that mediates transformation and the development of neoplasia in vivo induces the host to reject the cell it is trying to convert to a neoplastic cell. The result of these diametrically opposing functions of T antigen is obviously to the benefit of the host as SV40 is rarely oncogenic in hosts that have vigorous T cell responses. As a matter of fact, SV40 neoplasia may be directly related to the capacity of the host class I antigen to associate with SV40 T antigen [22].

The physiological function performed by the surface T in the virus lytic cycle or during transformation, and whether its expression at the cell surface is even necessary in these virus–cell interactions, is not known. The use of CTL clones to isolate mutants in T antigen that do not home to the cell surface may provide a clue to the question as to whether surface T expression plays an essential role in the biology of SV40.

Acknowledgments

The work reported in this study was supported by research grant CA25000 from the National Cancer Institute, Bethesda, MD. The author thanks Dr. M. Judith Tevethia for critical reading of the manuscript, and Ms. Patricia Thompson for assistance in its preparation.

References

1. Tevethia SS (1980) Immunology of simian virus 40. *In* Klein G (ed) Viral Oncology. Raven Press, New York, pp 581–601
2. Knowles BB, Koncar M, Pfizenmaier K, Solter D, Aden DP, Trinchieri, G (1979) Genetic control of the cytotoxic T cell response to SV40 tumor-associated specific antigen. J Immunol 122:1798–1806
3. Gooding LR (1979) Specificities of killing by T lymphocytes generated against syngeneic SV40 transformants: Studies employing recombinants within the H-2 complex. J Immunol 122:1002–1008
4. Gooding LR (1982) Characterization of a progressive tumor from C3H fibroblasts transformed in vitro with SV40 virus. Immunoresistance in vivo correlates with phenotypic loss of H-2K^k. J Immunol 129:1306–1312
5. Tooze J (ed) (1980) Molecular Biology of Tumor Viruses: DNA Tumor Viruses. 2nd ed, Cold Spring Harbor Laboratory, Cold Spring Harbor, New York
6. Rigby PWL, Lane DP (1983) The structure and function of the simian virus 40 large T antigen. Adv Viral Oncol 3:31–57
7. Rapp F, Butel JS, Melnick JL (1964) Virus induced intranuclear antigen in cells transformed by papovavirus SV40. Proc Soc Exp Biol Med 116:1131–1135
8. Deppert W, Pates R (1979) Simian virus 40 specific proteins on surface of HeLa cells infected with adenovirus 2–SV40 hybrid virus Ad2$^+$ ND2. Nature 277:322–324
9. Soule HR, Lanford RE, Butel JS (1980) Antigenic and immunogenic characteristics of nuclear and membrane-associated simian virus 40 tumor antigen. J Virol 33:887–901
10. Gooding LR, Geib RW, O'Connell KA, Harlow E (1984) Antibody and cellular detection of SV40 T antigenic determinants on the surfaces of transformed cells. *In* Levine AJ, Vande Woude GF, Topp WC, Watson JD (eds) Cancer Cells, the Transformed Phenotype, Cold Spring Harbor Laboratory, Cold Spring Harbor, New York, pp 263–269

11. Klockman, H Deppert W (1983) Acylated simian virus 40-specific proteins in the plasma membrane of HeLa cells infected with adenovirus 2–simian virus 40 hybrid virus Ad2⁺ND2. Virology 126:717–720

12. Tevethia SS, Flyer DC, Tjian R (1980) Biology of simian virus 40 (SV40) transplantation antigen (TrAg). VI. Mechanism of induction of SV40 transplantation immunity in mice by purified SV40 T antigen (D2 protein). Virology 107:13–23

13. Chang C, Martin RG, Livingston DM, Luborsky SW, Hu CP, Mora PT (1979) Relationship betwen T-antigen and tumor specific transplantation antigen in simian virus 40-transformed cells. J Virol 29:69–75

14. Gooding L, Edwards CB (1980) H-2 antigen requirements in the in vitro induction of SV40 specific cytotoxic T lymphocytes. J Immunol 124:1258–1262

15. Flyer DC, Pretell J, Campbell AE, Liao WSL, Tevethia MJ, Taylor JM, Tevethia SS (1983) Biology of simian virus 40 (SV40) transplantation in antigen (TrAg). X. Tumorigenic potential of mouse cells transformed by SV40 in high responder C57B1/6 mice and correlation with the persistence of SV40 TrAg, early proteins and viral sequences. Virology 131:207–220

16. Pan S, Knowles BB (1983) Monoclonal antibody to SV40 T antigen blocks lysis of cloned cytotoxic T cell line specific for SV40 TASA. Virology 125:1–71

17. Jay G, Jay FT, Chang C, Friedman RM, Levine AS (1978) Tumor specific transplantation antigen: Use of the Ad⁺ ND1, Hybrid virus to identify the protein responsible for simian virus 40 tumor rejection and its genetic origin. Proc Natl Acad Sci USA 75:3055–3059

18. Campbell AE, Foley LF, Tevethia SS (1983) Demonstration of multiple antigenic sites of the SV40 transplantation rejection antigen by using cytotoxic lymphocyte clones. J Immunol 130:490–492

19. Tevethia SS, Tevethia MJ, Lewis AJ, Reddy VB, Weissman SM (1983) Biology of simian virus 40 (SV40) transplantation antigen (TrAg). IX. Analysis of TrAg in mouse cells synthesizing truncated SV40 large T antigen. Virology 128:319–330

20. O'Connell KA, Gooding LR (1984) Cloned cytotoxic T cells recognize cells expressing discrete fragments of SV40 tumor antigen. J Immunol 132:953–958

21. Karjalainen HE, Tevethia MJ, Tevethia SS (1985) Abrogation of simian virus 40 DNA mediated transformation of primary C57B1/6 mouse embryo fibroblasts by exposure to an SV40 specific cytotoxic T lymphocyte clone. J Virol 56:373–377

22. Abramczuk J, Pan S, Maul G, Knowles B (1967) Tumor induction by simian virus 40 in mice is controlled by long term persistence of the viral genome and the immune response of the host. J Virol 49:540–548

CHAPTER 22
Virus-specific HLA Class II-restricted Cytotoxic T Cells

STEVEN JACOBSON AND WILLIAM E. BIDDISON

A hallmark of the immune response is that all T cell recognition functions are associated with components of the major histocompatability complex (MHC) [1,2]. Foreign antigens are recognized by T cells, only when these antigens are presented in association with self MHC molecules. The human MHC (HLA) is located on the short arm of chromosome six and encodes three general classes of molecules [3,4]. The HLA class I molecules, designated HLA-A, -B, and -C, consist of two subunits; a 44,000-dalton polymorphic heavy chain which is noncovalently associated with a nonpolymorphic 11,500-dalton light chain (β_2-microgloubulin). HLA class I molecules are expressed on the surface of virtually every cell type [3,4]. The HLA class II molecules, designated HLA-DR, -DP, and -DQ and collectively termed HLA-D, are composed of two noncovalently associated glycoproteins consisting of a 34,000-dalton heavy or α chain and a 29,000-dalton light or β chain. HLA class II molecules are present on a limited set of cell types, most notably B cells, macrophages, Langerhans cells, and activated T cells. The HLA class III molecules consist of several components of the complement system and are beyond the scope of this chapter.

Until recently, the HLA class I molecules were thought to be the primary, if not the only, HLA recognition structures for virus-specific cytotoxic T cells (CTL) [5,6]. CTL are effector T cells that bind to target cells by antigen-specific receptors, and through a still poorly defined process deliver a "lethal hit" to the target cell, causing it to lyse [7]. The CTL is then free to engage and lyse other target cells. Virus-specific CTL recognize a target cell infected with the appropriate virus only in association with self HLA antigens [1,2,5,6]. The exact nature of the antigen-HLA association that is recognized by the CTL antigen-specific receptor is not known.

Virus-specific CTL have been demonstrated for a wide variety of viruses both in vitro and in vivo [1,5,6]. In each case, there is considerable evidence to indicate that virus-immune CTL recognize their appropriate virus-infected target in association with self HLA class I molecules. The purpose of this review is to (a) introduce the concept that in addition to HLA class I virus-specific CTL, HLA class II-restricted virus-specific CTL also exist; and (b) discuss the possible implications that class II-restricted CTL responses might have on the overall immune response to viruses.

A Historical Perspective

The ability to demonstrate virus-specific HLA class II-restricted CTL has been facilitated by the use of Epstein–Barr virus (EBV)-transformed lymphoblastoid lines as the target cell [8–11,13,14,16]. These B cell lines have been the target cell of choice in all these studies because they express both HLA class I and II determinants, are susceptible to a wide variety of viruses, can hold radioactive Na_2CrO_4 with low levels of spontaneous release, and, being immortalized cells, can be continually grown. Historically, the target cells that had been used in assays to demonstrate virus-specific CTL activity were mitogen-stimulated T cells or human skin fibroblasts infected with an appropriate virus [5,6]. Both cell types express low levels of HLA class II antigens and would therefore be inadequate targets for HLA class II-restricted CTL. It is important to keep this historical perspective in mind when discussing the relative importance of a particular virus-specific HLA class I-versus an HLA class II-restricted CTL population. In order to demonstrate class II-restricted CTL, one must use the appropriate target cell.

T Cell Clones

The bulk of information on HLA class II-restricted, virus-specific CTL has come from recent studies that have used T cell clones as the effector population. In a panel of EBV-specific T cell clones that were generated from an EBV-seropositive donor, 13 of 59 (22%) expressed the T4$^+$ phenotype, and three of these were shown to be cytotoxic for EBV-transformed B cell lines that expressed the DR7 HLA specificity [8]. This cytotoxicity could be blocked with anti-class II and anti-T4 antibodies, but not by anti-class I or anti-T8 antibodies. This initial report served to demonstrate the existence of an HLA class II-restricted CTL population specific for EBV and supported the hypothesis that the T4 molecule is involved in T cell recognition of class II HLA antigens [8].

HSV-1-specific, HLA class II-restricted T cell clones have also been reported [9]. All of these clones expressed the T4$^+$ phenotype and could specifically lyse HSV-1-infected B cell targets in association with the appropriate HLA class II specificity (either DR1 or MB1). Monoclonal antibodies that react with either DR1 or MB1 could block the cytolytic effect of the appropriate T cell clone. The exact percentage of such T4$^+$, class II-restricted, HSV-1 specific T cell clones generated from HSV-1-seropositive individuals was not given. Subsequent studies identified both HSV type-specific and HSV type-common specific T cell clones that had both HLA class II-restricted cytolytic activity and helper cell activity[10].

Lastly, experiments from our laboratory have also demonstrated the existence of measles virus-specific CTL clones that are HLA class II-restricted [11]. T cell clones were derived from a patient with multiple sclerosis (MS) who had been described previously as a strong proliferative responder to measles virus [12]. Of the 14 clones tested, all had the T4$^+$ phenotype, and a majority (78%) were cytotoxic for autologous measles virus-infected B cell targets. Selected clones were shown to be restricted by either DR2 or DR4, could be blocked by an anti-class II (L243) but not an anti-class I (W6/32) monoclonal antibody, and were specific for measles virus-infected targets. Subsequent T cell cloning experiments have demonstrated that measles virus-specific class II-restricted CTL could also be generated from a normal individual; both DR1- and SB3-restricted CTL clones have been isolated (Jacobson and Richert, unpublished observations). Measles virus-specific, HLA DR2-restricted CTL clones have been useful in defining which HLA-DR molecules could serve as genetic restriction elements for HLA class II-restricted CTL. It was demonstrated that molecules that are apparently recognized by T cells in the allogeneic mixed lymphocyte reaction (the DRβ2 polypeptide) could also serve as the restriction element for these class II-restricted T cell clones [13].

Bulk Culture Studies

Although T cell clones have been useful in initially defining virus-specific HLA class II-restricted CTL, they may not be representative of the HLA class specificity of CTL in vivo. Stimulation of peripheral blood lymphocytes (PBL) with virus or viral antigens and the analysis of their subsequent cytolytic effect on appropriate virus-infected targets (bulk culture studies) are more informative with regard to the "normal" CTL response to viruses. As discussed previously, prior reports using bulk culture techniques had shown that virus-specific CTL were HLA class I-restricted [1,5,6]. However, the target cells used in these studies were clearly inadequate to demonstrate HLA class II-restricted CTL in that they expressed low levels of HLA class II determinants. Therefore, the proportion of HLA class I- relative to HLA class II-restricted CTL for a given virus is not known.

Experiments in our laboratory have attempted to address this issue by comparing the normal CTL response to influenza and measles virus by generating CTL in bulk culture to both viruses and comparing the ability of these CTL to lyse influenza and measles virus-infected B lymphoblastoid cell targets [14]. These studies have demonstrated that the majority of normal individuals possess measles virus-specific CTL activity, that these CTL could be blocked by an anti-class II (L243) monoclonal antibody but not by an anti-class I antibody (W6/32), and that the cytolytic activity could be blocked by anti-T4 but not by anti-T8 antibodies. In contrast, influenza virus-specific CTL could be partially blocked by both anti-T8 and anti-T4 antibodies. An anti-class I monoclonal antibody produced a 70% reduction in influenza virus-specific cytolytic activity, whereas an anti-class II antibody produced a 40% reduction. These studies indicate that (a) measles virus-specific CTL generated in bulk culture are predominantly HLA class II-restricted; and (b) although the major portion of influenza virus-specific CTL are HLA class I-restricted, there may also be a component of the CTL response to influenza virus that is HLA class II-restricted. The CTL response to more viruses must be examined to determine the proportion of HLA class I- versus HLA class II-restricted CTL in the overall T cell population.

Significance of Virus-specific HLA Class II-restricted CTL

The existence of virus-specific HLA class II-restricted CTL may be of benefit to the host. If the entire repertoire of HLA-D specificities (class II antigens) is now added to the HLA-A, -B,and -C determinants that could potentially serve as genetic restriction elements for virus-specific CTL, then the immune response to viruses would be significantly increased by the generation of both HLA class I- *and* HLA class II-restricted CTL. This would serve to decrease the probability that an individual would be unable to generate a CTL response to any virus. It is known that some cell types that normally lack class II HLA antigens can be induced to express HLA class II determinants (e.g., by induction with γ-interferon) [4]. The generation of HLA class II-restricted CTL could serve to lyse these cells that express newly synthesized HLA antigens.

The exact mechanism by which CTL recognize the viral antigen–HLA complex on target cell membranes is not known. It is possible that different viruses or different viral antigens of the same virus are preferentially expressed in the context of one or another HLA class specificity. Therefore, the generation of HLA class II-restricted CTL could expand the overall response of CTL that recognize foreign antigens in association with either HLA class I or class II determinants and would serve to increase the cellular immune response to a wider variety of antigens.

Alternatively the generation of virus-specific HLA class II-restricted CTL could be pathologic. Class II HLA determinants have been demonstrated on a wide variety of cell types [3,4]. If these cell types are within selected sites of specialized organs (e.g., Ia-positive astrocytes in the central nervous system) and upon virus infection become targets for virus-specific HLA class II-restricted CTL, then destruction of these cells could be detrimental to the host. This may be one mechanism involved in autoimmune disorders, particularly in a disease associated with a viral etiology. Thus, it becomes essential to define the total CTL response to viruses with regard to HLA class specificity of virus-specific CTL.

Whereas the previous discussion has dealt with a generalized view of HLA class II-restricted CTL, studies in our laboratory on measles virus-specific CTL responses may serve to highlight the potential significance of virus-specific HLA class II-restricted CTL. Measles virus has been long considered as a possible etiolgic factor in the pathogenesis of multiple sclerosis (MS) (reviewed in ref. 15). Using T cell clones and bulk culture populations, we have demonstrated that measles virus-specific CTL are predominantly HLA class II-restricted [11,13,14]. To assess whether a virus-specific immune defect may be involved in MS, we have examined the ability to generate measles virus-specific and influenza virus-specific CTL (predominantly HLA class II- and HLA class I-restricted, respectively) in patients with MS, normal individuals, and other neurological and inflammatory disease controls [16]. The vast majority of MS patients failed to generate measles virus-specific CTL or had significantly lower CTL responses than normal individuals or other disease controls. Moreover, this decreased CTL response in MS patients was specific for measles virus because all three groups generated indistinguishable CTL responses to influenza virus [16]. These results suggest that a measles virus-specific immune abnormality exists in MS and supports the hypothesis that measles virus is involved in the etiology of this disease. An examination of the HLA class I- versus HLA class II-restricted CTL response to other viruses will enable a more precise analysis of the "normal" CTL response to a given virus and possible aberrations of these responses in disease states.

References

1. Zinkernagel RM, Doherty PC (1979) MHC-restricted cytotoxic T cells: Studies on the biological role of polymorphic major transplantation antigens determining T-cell restriction-specificity, function, and responsiveness. Adv Immunol 27:52–180
2. Benacerraf, B (1981) Role of MHC gene products in immune regulation. Science 212:1229–1234
3. Bodmer WF (1978) The HLA system. Br Med Bull 34:213–246
4. Kaufman JF, Auffray C, Korman AK, Shackelford DA, Strominger JL (1984) The class II molecules of the human and murine major histocompatability complex. Cell 36:1–13
5. McMichael AJ (1980) HLA restriction of human cytotoxic T cells. Springer Semin Immunopathol 3:3–22
6. Biddison WE (1982) The role of the human major histocompatability complex in cytotoxic T-cell responses to virus-infected cells. J Clin Immunol 2:1–9
7. Berke G (1980) Interaction of cytotoxic T lymphocytes and target cells. Prog Allergy 27:69–133

8. Meuer SC, Hodgdon JC, Cooper DA, Hussey RE, Fitzgerald KA, Schlossman SF, Reinherz EL (1983) Human cytotoxic T cell clones directed at autologous virus transformed targets: Further evidence for linkage of genetic restriction to T4 and T8 surface glycoproteins. J Immunol 131:186–192

9. Yasukawa M, Zarling JM (1984) Human cytotoxic T cell clones directed against herpes simplex virus-infected cells. I. Lysis restricted by HLA class II MB and DR antigens. J Immunol 133:422–427

10. Yasukawa M, Zarling JM (1984) Human cytotoxic T cell clones directed against herpes simplex virus-infected cells. II. Bifunctional clones with cytotoxic and virus-induced proliferative activities exhibit herpes simplex virus type 1 and 2 specific or common reactivities. J Immunol 133:2736–2742

11. Jacobson S, Richert JR, Biddison WE, Satinsky A, Hartzman RJ, McFarland HF (1984) Measles virus-specific T4 + human cytotoxic T cell clones are restricted by class II HLA antigens. J Immunol 133:754–757

12. Greenstein JI, McFarland HF (1982) Immunological responses in multiple sclerosis: Cell mediated immunity. Clin Immunol Allergy 2:371–383

13. Jacobson S, Nepom GT, Richert JR, Biddison WE, McFarland HF (1985) Identification of a specific HLA-DR2 Ia molecule as a restriction element for measles virus-specific HLA class II-restricted cytotoxic T cell clones. J Exp Med 161:263–268

14. Jacobson S, McFarland HF (1984) Studies of measles virus-specific HLA class II-restricted cytotoxic T lymphocytes generated in bulk culture. *In* Symposium on the Pathobiology and Immunopathology of Virus Infection, Sendai, Japan (abstract)

15. McFarlin DE, McFarland HF (1982) Multiple sclerosis. N Engl J Med 307:1183–1188, 1246–1251

16. Jacobson S, Flerlage ML, McFarland HF (1985) Impaired measles virus-specific cytotoxic T cell responses in multiple sclerosis. J Exp Med 162:839–885

New Principles in Viral Immunology and Immunopathology

CHAPTER 23
Molecular Mimicry

MICHAEL B.A. OLDSTONE AND ABNER L. NOTKINS

Prologue

Molecular mimicry defines similar structures shared by molecules from dissimilar genes or by their protein products. Either the molecules' linear amino acid sequences or their conformational fit may be shared, even though their origins are as separate as, for example, a virus and a normal host self determinant. Because guanine–cytosine (GC) sequences and introns designed to be spliced away may provide, respectively, *false hybridization signals* and nonsense *homologies,* we focus here on molecular mimicry at the protein level. Such homologies between proteins have been detected either by use of immunologic reagents, humoral or cellular, that cross-react with two presumably unrelated protein structures, or by computer searches to match proteins described in storage banks. Regardless of the methods used for identification, it is now clear that molecular mimicry between proteins encoded by numerous DNA and RNA viruses and host "self" proteins is a relatively common event [1–3]. Among the broad implications of these data are leads for understanding virally induced autoimmunity and disease [2–8] as well as mechanisms by which viral proteins are processed inside cells [9]. Further, the unexpected cross-reactivities attendant to mimicry warrant cautious use or reagents in diagnostic virology and microbiology, even though these materials originated from hybridomas or from animals immunized with predetermined (peptide) amino acid sequences. This chapter is presented to define molecular mimicry and emphasize its effect on viral pathogenesis.

Molecular Mimicry Between Viruses and Host Cell Proteins

Examples of molecular mimicry were first described as such in the early 1980s by investigators who found that monoclonal antibodies against SV 40 T antigens cross-reacted with host cell proteins [10]. However, the importance of this observation became apparent only when others realized that the monoclonal antibodies against a battery of viruses were cross-reacting with host determinants [1,3,9,11]. For example, Fujinami et al. [3] showed cross-reactivity between measles virus phosphoprotein (72K molecular weight) and the cytoskeleton component keratin (54K molecular weight), and between a herpes simplex virus glycoprotein of 140K and a separate epitope on keratin from that recognized by the measles virus phosphoprotein. Dales et al. noted shared homology between vaccinia virus hemagglutinin and the cytoskeleton protein vimentin [9], whereas Sheshberadaran and Norrby found homolgy between the fusion protein of measles virus and a heat shock protein [11]. These and other observations of immunologic reactivity between viral proteins and cytoskeletal determinants suggested the hypothesis of shared determinants on cell linker proteins that might help to guide viral proteins along highways and stop points traveled inside cells (see Figure 7 of ref. 9). Others found that monoclonal antibody directed against a specific viral determinant to which it was raised, reacted with determinants of another totally unrelated virus [12].

Molecular Mimicry and Autoimmune Manifestations

The potential of molecular mimicry as an important event in pathogenesis leading to disease became evident from two groups of observations. The first was cross-reactivity at the monoclonal antibody level between viral protein and host self proteins. In such a system, antibodies to hormones, lymphocyte subsets, or cells of the nervous system developed as a consequence of virus infection, with all the inherent capacity to participate in disease. The cross-reactivity between viruses and particular tissues offers some insight into the association of viral infection and specific diseases. For example, coxsackievirusB4 has been found in individuals with myocarditis or inflammatory disease of the heart muscle. Of related interest, a monoclonal antibody directed against the neutralizing domain of coxsackievirus also interacted with the heart muscle (see figure 23.1). Equally intriguing was a link between Theiler's virus and the demyelinating disease it causes. A monoclonal antibody directed against the major neutralizing domain of Theiler's virus also reacted with galactocerebroside, the main component on the surface of oligodendrocytes [13]. Because oligodendrocytes are cells that make the myelin lamellae wrapped around axons, their destruction leads to demyelination [14]. Interestingly, inoculation of this monoclonal antibody induced demyelination [13]. Figure 23.1 shows several examples of molecular mimicry detected by using normal host tissues to screen monoclonal antibodies against various viral proteins (see ref. 1 for details).

The second sequela associated with molecular mimicry is the formation and trapping of immune complexes [15]. In this instance, cytoskeletal or other self proteins of an infected host are released into fluids either as a result of normal cell turnover or enhanced turnover and lysis occurring during viral infection. Antibodies induced against proteins of the infecting virus,

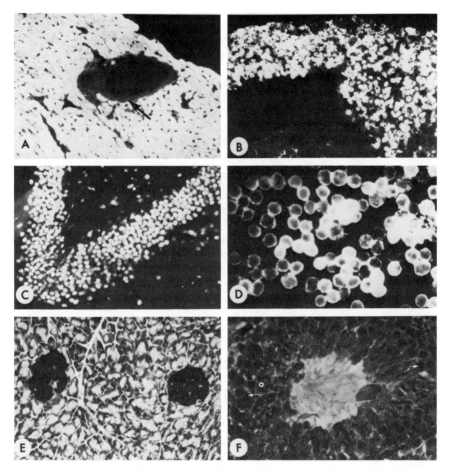

Figure 23.1. Reactivity of monoclonal antiviral antibodies with normal tissues detected by indirect immunofluorescence (**A,B,C,D**) or immunoperoxidase (**E,F**). Monoclonal antibody to coxsackievirus B_4 reacts with mouse myocardium; smooth muscle in coronary artery shows no reaction (arrow) (**A**). Monoclonal antibody to Japanese encephalitis virus reacts with mouse anterior pituitary cells (**B**). Two monoclonal antibodies to measles virus are shown, one reacting with nuclei of cells in mouse hippocampus (**C**), and the other with human T lymphocytes (**D**). A monoclonal antibody to herpes simplex virus that reacts with both hamster pancreatic B cells in the islets of Langerhans (**E**) and nuclei of spermatogonia (arrows) in the testis (**F**).

but cross-reactive with host proteins, can form antigen–antibody complexes in the circulation. These complexes may become trapped in vessels with fenestrated endothelial linings such as the renal glomeruli, small arteries, and capillaries in the choroid plexus. Here, they can accumulate to set in motion the events of immune complex disease. After viral infection of even antiviral immuniztion [9], the host antigen–antibody complexes may form

an important component of total immune-complex deposits. The principles
of immune-complex disease have been detailed earlier in this series [15].

How Common Is Molecular Mimicry?

To determine the frequency of molecular mimicry, Srinivasappa and his colleagues at the NIH acquired from many laboratories over 600 monoclonal
antibodies raised against viral polypeptides. These investigators then charted
the incidence of the monoclonals' cross reactivity with host proteins expressed in a large panel of normal tissues [1] (Figure 23.1). In the analysis
were antibodies against 11 different viruses, including such commonly found
representatives of DNA and RNA viruses as the herpesvirus group, vaccinia
virus, myxoviruses, paramyxoviruses, arenaviruses, flaviviruses, alphaviruses, rhabdoviruses, and coronaviruses. The results were that approximately
4% of such monoclonals cross-reacted with host-cell determinants expressed
on uninfected tissues. Moreover, some of these monoclonal antiviral antibodies reacted with antigens in more than on organ [1,16]. From these data,
it is clear that molecular mimicry is common and not restricted to any specific
class or group of virus.

Using Peptides to Induce Monoclonal Antibodies

We now know that a minimum of six to seven peptides are required for the
induction of monoclonal antibodies [17]. Thus, the probability that the requisite 20 amino acids occur in six identical sequences between two proteins
is 20^6 or 1 to 128,000,000, assuming all amino acids are represented equally
and at random. After a search through the 2511 amino acid sequences in the
Dayhoff protein data base (which includes 470,158 residues) to discover
overlapping peptides, Wilson and colleagues [18] found 2469 hexamers, 186
septamers, and 17 octamers with homologies. This data provides a ballpark
estimate of the expected frequency for mimicry by use of linear sequences.

Amino Acid Homologies and Immune Responses Between Important Host Protein and Virus as Mechanisms for Autoimmunity

Because, on the basis of antibody cross-reactivity, many viruses clearly share
antigenic sites with normal host-cell components, the next step was to look
for cross-reactivity capable of eliciting autoimmunity and related disease.
Myelin basic protein was chosen as the host component to study because
its entire amino acid sequence is known, and its encephalitogenic site of 8–
10 amino acids has been mapped in several animal species. With the use of

computer assisted analysis, several viral proteins listed in the Dayhoff files showed significant homology with the encephalitogenic site of myelin basic protein. Included were similarities and/or fits between myelin basic protein and the nucleoprotein and hemagglutinin of influenza virus, coat protein of polyoma virus, core protein of the adenovirus, polyprotein of poliomyelitis virus, EC-LF2 protein of Epstein–Barr virus, hepatitis B virus polymerase, and others. However, the best fit occurred between the myelin basic protein encephalitogenic site in the rabbit and hepatitis B virus polymerase (HBVP):

66	75	Encephalitogenic site, rabbit myelin basic protein
THR-THR-HIS-TYR-GLY-SER-LEU-PRO-GLN-LYS,		

589	598	
ILE-GLY-CYS-TYR-GLY-SER-LEU-PRO-GLN-GLU,		HBVP

Interestingly, products of the immune responses, both humoral and cellular, generated in rabbits inoculated with the octomer or dexomer viral peptide reacted with whole myelin basic protein. Further, inoculation of the HBVP peptide into rabbits caused perivascular infiltration localized to the central nervous system reminiscent of the disease induced by inoculation of either whole myelin basic protein or the encephalitogenic site of myelin basic protein [4]. This outcome clearly exemplifies the potential of molecular mimicry to cause both autoimmune responses and autoimmune disease.

Mechanisms by Which Molecular Mimicry Occurs and Causes Disease

The most likely explanation for how molecular mimicry causes disease is that an immune response against the determinant shared by host and virus can bring forth a tissue-specific immune response, presumably capable of destroying cells and eventually the tissue. The probable mechanism is the generation of cytotoxic cross-reactive effector lymphocytes or antibodies that recognize specific determinants of "self proteins" located on target cells. Interestingly, the induction of cross-reactivity would not require a replicating agent, and the immunologically mediated injury could occur after removal of the immunogen—a hit-and-run event. Clearly, the virus infection that initiates an autoimmune phenomenon need not be present at the time overt disease develops. A likely scenario would be that the virus responsible for inducing a cross-reacting immune response is cleared initially, but the components of that immunity continue to assault host elements. The cycle continues as the autoimmune response itself leads to tissue injury that, in turn, releases more self antigen, thereby inducing more antibodies, and so on. Such a sequence might account for the virus encephalopathies occurring in humans after measles, mumps, vaccinia, or herpes–zoster virus infectious;

in these postinfectious diseases, recovery of the inducing agent has been rare [19]. This theory is reinforced by studies showing that, after several types of acute viral infection, mononuclear cells from peripheral blood or cerebral spinal fluid proliferate in response to host antigens, one of which is myelin basic protein. Interestingly, several clonal populations of lymphocytes have been harvested from central nervous system fluid of humans with encephalitis, that proliferate to the infecting virus as well as to nervous system antigens (B. Waksman, personal communication). Several relevant human diseases that may be associated with a molecular mimicry pathogenesis are recorded elsewhere [2,20,21]. Viruses also play by other game plans. For example, a virus with the capacity to persist in its host may continuously or cyclically express its antigens. Although expression of a viral genome may be restricted so that no infectious virus replicates, production of a viral determinant in common with that of the host might continue. This would allow initiation of an immune response and/or autoimmunity, either one leading to cyclic, chronic, or progressive disease.

In any case, molecular mimicry would occur only when the virus and host determinants are sufficiently similar to induce a cross-reactive response yet different enough to break immunologic tolerance. Weigle and his colleagues [22] have mapped the induction and breaking of tolerance at both the B cell and T cell levels by using heterologous serum proteins, and the same principles undoubtedly govern microbially induced molecular mimicry.

With the revolution in current technology allowing cloning and sequencing of genes and their proteins, more data on viral polypeptides will soon be available. Then, just as homologies are being found between the acetylcholine receptor and/or the insulin receptor with several viral proteins [2] and between important proteins of both the central and peripheral nervous systems with several viruses [2,20], other similarities will surely emerge. However, unless such homology and the subsequent immunologic cross-reactivity involve a host protein that precipitates disease, e.g., the restricted encephalitogenic site of myelin rather than multiple sites on myelin basic protein, disease is unlikely to follow, despite an autoimmune response.

Acknowledgments

This is Publication No. 4112-IMM from the Department of Immunology, Scripps Clinic and Research Foundation, La Jolla, California 92037.

This research was supported by U.S.P.H.S. grants NS-12428, AI-07007, and AG-04342.

References

1. Srinivasappa J, Saegusa J, Prabhakar BS, Gentry MK, Buchmeier MJ, Wilktor TJ, Koprowski H, Oldstone MBA, Notkins AL. (1986) Molecular minicry: Frequency of reactivity of monoclonal antiviral antibodies with normal tissues. J Virol 57:397–401

2. Dyrberg T, Oldstone MBA (1986) Peptides as probes to study molecular mimicry and virus induced autoimmunity. Curr Top Microbiol Immunol (in press)
3. Fujinami RS, Oldstone MBA, Wroblewska Z, Frankel ME, Koprowski H (1983) Molecular mimicry in virus infection: Cross-reaction of measles phosphoprotein or of herpes simplex virus protein with human intermediate filaments. Proc Natl Acad Sci USA 80:2346–2350
4. Fujinami RS, Oldstone MBA (1985) Amino acid homology and immune responses between the encephalitogenic site of myelin basic protein and virus: A mechanism for autoimmunity. Science 230:1043–1045
5. Onodera T, Toniolo A, Ray UR, Jenson AB, Knazek RA, Notkins AL (1981) Virus-induced diabetes mellitus. XX. Polyendocrinopathy and autoimmunity. J Exp Med 153:1457–1473
6. Onodera T, Ray UR, Melez KA, Suzuki H, Toniolo A, Notkins AL (1982) Virus-induced diabetes mellitus: Autoimmunity and polyendocrine disease prevented by immunosuppression. Nature 297:66–68
7. Notkins AL, Onodera T, Prabhakar B (1984) Virus-induced autoimmunity. In Notkins AL, Oldstone MBA (eds) Concepts in Viral Pathogenesis, vol 1. Springer-Verlag, New York, pp 210–215
8. Garzelli C, Taub FE, Scharff JE, Prabhakar BS, Ginsberg-Fellner F, Notkins AL (1984) Epstein–Barr virus-transformed lymphocytes produce monoclonal autoantibodies that react with antigens in multiple organs. J Virol 52:722–725
9. Dales S, Fujinami RS, Oldstone MBA (1983) Serologic relatedness between Thy-1.2 and actin revealed by monoclonal antibody. J Immunol 131:1332–1338
10. Lane DP, Hoeffler WK (1980) SV40 large T shares an antigenic determinant with a cellular protein of molecular weight 68,000. Nature 288:167–170
11. Sheshberadaran H, Norrby E (1984) Three monoclonal antibodies against measles virus F protein cross-react with the cellular stress proteins. J Virol 52:995–999
12. Norrby E, Sheshberadaran, Rafner B (1985) Antigen mimicry involving the measles virus hemagglutinin and the human respiratory syncytial virus nucleoprotein. J Virol 53:456–460
13. Fujinami R, Powell H (1986) A monoclonal antibody that neutralizes Theiler's virus, reacts with galactocerebroside and causes demyelination. (Manuscript submitted)
14. Lampert P, Rodriguez M (1984) Virus-induced demyelination. In Notkins AL, Oldstone MBA (eds) Concepts in Viral Pathogenesis. Springer-Verlag, New York pp 260–268
15. Oldstone MBA (1984) Virus-induced immune complex formation and disease: Definition, regulation, importance. In Notkins AL, Oldstone MBA (eds) Concepts in Viral Pathogenesis. Springer-Verlag, New York pp 201–209
16. Notkins AL, Prabhakar BS (1986) Monoclonal autoantibodies that react with multiple organs: Basis for reactivity. Ann NY Acad Sci (in press)
17. Wilson IA, Haft DH, Getzoff ED, Tainer JA, Lerner RA, Brenner S (1985) Identical short peptide sequences in unrelated proteins can have different conformations: A testing ground for theories of immune recognition. Proc Natl Acad Sci USA 82:5255–5259
18. Wilson I, Niman H, Houghten R, Cherenson A, Connolly M, Lerner RA (1984) The structure on an antigenic determinant in protein. Cell 37:767–778
19. Paterson P (1971) Immunologic Disease. Little, Brown, Boston, p 1400

20. Alvord EC (1985) Disseminated encephalomyelitis: Its variation in form and their relationships to other diseases of the nervous system. *In* Handbook of Clinical Neurology. Elsevier Science Publishing Company, Inc., New York, pp 467–502
21. Jahnke U, Fischer EH, Alvord EC (1985) Sequence homology between certain viral proteins and proteins related to encephalomyelitis and neuritis. Science 229:282–284
22. Weigle WO (1980) Adv Immunol 30:159

CHAPTER 24
Virus-induced Autoimmune Demyelinating Disease of the Central Nervous System

RICHARD T. JOHNSON AND DIANE E. GRIFFIN

Acute perivenular demyelinating disease of the brain and spinal cord can complicate a number of human viral infections. It has been most frequent late in the course of the exanthematous viral infections, particularly measles and vaccinia, and, to a lesser extent, varicella and rubella. The discontinuation of vaccination against smallpox and the successful immunization against measles and rubella in the United States have decreased the incidences of these infections and their parainfectious complications. Currently, postinfectious encephalomyelitis in the United States is most frequent after varicella, where mortality and morbidity rates are low, and after upper respiratory infections, where the etiologic agent is usually undetermined [1].

Nevertheless, measles is not controlled in most of the world; 1.5 million children still die annually from measles. Although most measles-associated deaths result from virus dissemination and secondary infections, encephalomyelitis occurs in about 1:1000 children [2] and remains a significant cause of mortality and the major cause of permanent neurological morbidity. Encephalomyelitis after vaccinia virus inoculation was highly variable; incidence rates ranged from 1 to 63 in one Dutch experience to 1 in 300,000 in a national survey in the United States. The reason for this variability has never been explained. Although postvaccinal encephalomyelitis appeared of only historical interest after the worldwide eradication of smallpox, this complication has assumed new importance. One of the most promising strategies for future vaccines is the use of recombinant vaccinia virus carrying sequences coding for multiple virus antigens [3]. A better understanding of the determinants of the postvaccinal encephalomyelitis will be critical before clinical trials of these vaccines are initiated.

Unfortunately, there is no animal model for acute postinfectious encephalomyelitis. The experimental disease that best simulates the acute perivenular demyelination is experimental allergic encephalomyelitis. Subacute or chronic coronavirus infections of rat nervous system do lead to probable autoimmune-mediated demyelination [4]; however, studies of measles encephalomyelitis in humans indicate that neurotropism or direct infection of the nervous system may not be prerequisites to the development of acute demyelinating disease [5].

Similarities of Postinfectious Encephalomyelitis to Experimental Allergic Encephalitis

Postinfectious encephalomyelitis is remarkably similar clinically and pathologically to the neuroparalytic complications observed after multiple injections of rabies virus vaccine prepared in adult animal brains. This similarity led Rivers and Schwentker [6] to inoculate monkeys with homogenates of normal brain in an attempt to reproduce this disease. This resulted in the first induction of experimental allergic encephalomyelitis, the prototype autoimmune disease. Extensive studies of experimental allergic encephalomyelitis have shown that the encephalitogenic antigen is the basic protein of central nervous system myelin, and that the disease can be passively transferred by sensitized lymphocytes of the T-helper subset [7].

The parallels between the three diseases are remarkable (Table 24.1). In experimental allergic encephalomyelitis and post-rabies-vaccine encephalomyelitis, lymphocytes show lymphoproliferative responses to myelin basic protein. Similar responses have been observed in lymphocytes from patients with postinfectious encephalomyelitis following measles, varicella, and upper respiratory infections [5,8].

The inexplicable factor in these comparisons is how viruses can induce sensitization to neural proteins. Certainly, normal individuals have immune cells capable of reacting against myelin proteins. Expresssion of this reactivity is normally actively suppressed, but this suppression can be overcome by injection of excess antigen, as in experimental allergic encephalomyelitis and postrabies encephalomyelitis. Possibly, viral infections overcome this suppression by modification of myelin, by release of myelin products into the general circulation, or by disruption of immune regulation.

Coronavirus-induced Autoimmune Demyelination in Mice

One potential mechanism for autoimmune demyelinating disease induced by viruses comes from studies of a coronavirus, the JHM strain of mouse hepatitis virus. This virus was originally isolated from a spontaneous paralytic disease of mice. The virus causes acute demyelinating encephalitis by selective infection of oligodendrocytes [9]. The virus persists in the nervous system of mice, and late subclinical inflammatory demyelinating foci develop.

Table 24.1. Comparisons of experimental allergic encephalomyelitis with encephalomyelitis after rabies vaccine or viral infections.

	Experimental allergic encephalomyelitis	Post-rabies-vaccine encephalomyelitis	Postinfectious encephalomyelitis
Inducing event	Inoculation with CNS tissue or myelin basic protein	Inoculation with CNS tissue	Infection with enveloped viruses
Latency	10–21 days	10–41 days	10–40 days[a]
Clinical features			
Acute onset	+	+	+
Monophasic	+	+	+
Occasional			
Chronic or relapsing	+	+	+
Pathologic findings			
Perivenular lymphocytic infiltrates	+	+	+
Perivenular demyelination	+	+	+
Immunologic studies			
Lymphocytes stimulated in vitro by myelin basic protein	+	+	+
In vitro demyelination by lymphocytes	+		+

[a] From beginning of incubation period
Source: Modified from ref. 20.

Wantanabe and co-workers [4] inoculated rats with this murine virus and described a subacute demyelinating encephalomyelitis that developed several weeks to two months later. Viral antigen was found primarily in glial cells in the neighborhood of demyelinating plaques. Most rats subsequently recovered despite the persistence of virus. Lymphocytes from sick rats were cultivated in vitro in the presence of myelin basic protein, and these lymphocytes were transferred passively to syngeneic rats. In four to five days, mild clinical disease was seen in many recipients, and the white matter lesions resembled experimental allergic encephalomyelitis. Thus, viral infection of the oligodendrocytes appears capable of inducing an autoimmune response against myelin proteins produced by the oligodendrocytes. This might occur by incorporation of the oligodendrocyte membranes into the virus envelope, by release of myelin from damaged cells, or by modification of myelin proteins.

Postmeasles Encephalomyelitis in Humans

Alternate mechanisms have been suggested in human studies of postmeasles encephalomyelitis. Measles is seldom isolated from the nervous system, and the cerebrospinal fluid shows no evidence of intrathecal immunoglobulin synthesis, as seen in most viral encephalitides associated with direct virus invasion of the nervous system [5]. Finally, immunocytochemical studies have failed to demonstrate viral antigen in the brains of patients with acute measles encephalomyelitis [10]. Measles virus infections are known for their striking immunosuppressive effects [11]. Thus, there is a paradox in that the secondary infections causing pneumonia and gastrointestinal disease are thought to be secondary to the immunosuppressive effect, whereas the encephalomyelitis is presumed to be a hypersensitivity response.

Studies of a number of immune parameters during measles and measles encephalomyelitis show a variety of abnormal immune responses with suppression of some responses on one hand and release of normal inhibition on the other (Table 24.2). Most patients with measles show prolonged suppression of lymphoproliferative responses to mitogens, which is similar in uncomplicated measles, measles pneumonia, and encephalomyelitis. Many patients who are exposed to measles nevertheless show normal lymphocyte responses to mitogens. Despite the normal ratio of lymphoctyes bearing T helper–inducer and T suppressor–cytotoxic surface markers, functional studies show an increase in spontaneous suppressor cell activity [12]. C-reactive proteins are elevated during acute measles and show an apparent secondary elevation during postmeasles encephalomyelitis [13]. In patients with measles, and particularly with encephalomyelitis, serum IgE is elevated; this immunoglobulin is normally down-regulated by suppressor cells [14]. Significantly, lymphocytes cultivated from patients with measles may show proliferative responses to normal human myelin basic protein, and this response is more frequent in lymphocytes from patients with encephalomyelitis [5]. A variety of abnormalities in immune regulation occur, rather than simply "immunosuppression."

Because there is no evidence that virus regularly invades the nervous system and because there is abnormal immune regulation, the hypothesis arises that a deregulation of autoreactive cells may occur secondary to viral infection of lymphoid cells.

Unanswered Issues

Attempts to devise an animal model for postinfectious encephalomyelitis to facilitate studies of pathogenesis have remained unsuccessful. Therefore, studies require human investigation.

Two interrelated questions remain unanswered: Why are only a small number of individuals affected, and why is the central nervous system the

Table 24.2. Immunologic studies in patients with acute measles and measles encephalomyelitis.

Findings in uncomplicated measles	Comparison of encephalomyelitis	Reference
Antibody responses to nucleoprotein, fusion, hemagglutinin proteins of virus in all patients; low responses in one-half of patients to matrix protein	same response	21
Prolonged suppression of lymphoproliferative responses to mitogens	same suppression of similar duration	12
Decreased T cells but no change in helper/suppressor ratios	ND[a]	12
Spontaneous suppressor activity in cultured lymphocytes	ND	12
Elevated serum C-reactive protein levels	apparent second elevation at onset	13
Elevated serum IgE levels	prolonged	14
Lymphoproliferative responses to myelin basic protein	more frequent (47%)	5

[a] ND: not done.
Source: Modified from ref. 1.

sole target organ? If the disease is mediated by viral invasion of the central nervous system, then the answer to both may be that virus infects the brain only infrequently and then triggers the disease. To date, in measles it has been difficult to find virus in the brain, but more sensitive techniques or examinations of tissue taken at earlier times in the disease may be necessary. If this disease is autoimmune, then genetic susceptibility of the host to autoimmune disease may be important. Multiple sclerosis, for instance, is associated with the presence of the DR2 or DW2 allotypes [15]. In a limited study, we found no evidence for an increase in these or other HLA types in postmeasles encephalitis. Such an association could be revealed with the study of larger numbers of patients or it may be that other unstudied background genes are important for susceptibility, as they are in determining the susceptibility of mice to experimental allergic encephalomyelitis [16] or demyelination following Theiler's virus infection ([17]; see also Lipton et al., Chapter 29, this volume). The central nervous system could be targeted because of antigenic similarities between myelin proteins and viral proteins. Sequence data suggest regions of homology between myelin basic protein and numerous polypeptides from viruses that are or are not associated with postinfectious encephalomyelitis, so the biological significance of this finding is not yet clear ([18,19]; see Oldstone and Notkins, Chapter 23, this volume). Because the CNS is relatively isolated, alterations in access to the systemic immune response may occur during infection. Virus may increase permeability by infecting vascular endothelium or the infectious process with attendant immune responses may indirectly damage the normal blood–brain barrier by release of lymphokines, vasoactive amines, or other factors affecting permeability.

Acknowledgement

These studies were supported in part by grants from the Rockefeller Foundation and NIH (POl N21916-01).

References

1. Johnson RT, Griffin DE, Gendelman HE (1985) Postinfectious encephalomyelitis. Semin Neurol 5:180–190
2. Miller HG, Stanton JB, Gibbons JL (1956) Parainfectious encephalomyelitis and related syndromes. Quart J Med 25:427–505
3. Perkus ME, Piccini A, Lipinskas BR, Paoletti E (1985) Recombinant vaccinia virus: Immunization against multiple pathogens. Science 229:981–984
4. Watanabe R, Wege H, ter Meulen V (1983) Adoptive transfer of EAE-like lesions from rats with coronavirus-induced demyelinating encephalomyelitis. Nature 305: 150–152

5. Johnson RT, Griffin DE, Hirsch RL, Wolinsky JS, Roedenbeck S, Lindo de Soriano I, Vaisberg A (1984) Measles encephalomyelitis: Clinical and immunological studies. New Engl J Med 310:137–141
6. Rivers TM, Schwentker FF (1935) Encephalomyelitis accompanied by myelin destruction experimentally produced in monkeys. J Exp Med 61:689–702
7. Hashim GA (1985) Neural antigens and the development of autoimmunity. J Immunol 135:838s–842s
8. Lisak RP, Zweiman B (1977) In vitro cell-mediated immunity of cerebrospinal fluid lymphocytes to myelin basic protein in primary demyelinating diseases. N Engl J Med 297:850–853
9. Weiner LP (1973) Pathogenesis of demyelination induced by mouse hepatitis virus (JHM virus) Arch Neurol 28:298–303
10. Gendelman HE, Wolinsky JS, Johnson RT, Pressman NJ, Pezeshkpour GH, Boisset GF (1984) Measles encephalomyelitis: Lack of evidence of viral invasion of the central nervous system and quantitative study on the nature of demyelination. Ann Neurol 15:353–360
11. Casali P, Rice GPA, Oldstone MBA (1984) Viruses disrupt functions of human lymphocytes: Effects of measles virus and influenza virus on lymphoctye-mediated killing and antibody production. J Exp Med 159:1322–1337
12. Hirsch RL, Griffin DE, Johnson RT, Cooper SJ, Lindo de Soriano I, Roedenbeck S, Vaisberg A (1984) Cellular immune responses during complicated and uncomplicated measles virus infections of man. Clin Immunol Immunopathol 31:1–12
13. Griffin DE, Hirsh RL, Johnson RT, Lindo de Soriano I, Roedenbeck S, Vaisburg A (1983) Changes in serum C-reactive protein during complicated and uncomplicated measles virus infections. Infect Immun 41:861–864
14. Griffin DE, Cooper SJ, Hirsch RL, Johnson RT, Lindo de Sorinao I, Roedenbeck S, Vaisberg A (1985) Changes in plasma IgE levels during complicated and uncomplicated measles virus infections. J Allergy Clin Immunol 76:206–213
15. Spielman RS, Nathanson N (1982) The genetics of susceptibility to multiple sclerosis. Epidemiol Rev 4:45–65
16. Linthicum DS, Frelinger JA (1982) Acute autoimmune encephalomyelitis in mice. II. Susceptibility is controlled by the combination of H2 and histamine sensitization genes. J Exp Med 155:31–39
17. Melvold RM, Knobler R, Lipton H (1986) Linkage between Theiler's virus demyelinating disease and T-cell receptor beta chain (constant region) gene. J Immunol (in press)
18. Fujinami RS, Oldstone MBA (1985) Amino acid homology and immune responses between the encephalitogenic site of myelin basic protein and virus: A mechanism for autoimmunity. Science 230:1043–1045
19. Jahnke U, Fischer EH, Alvord EC (1985) Sequence homology between certain viral proteins and proteins related to encephalomyelitis and neuritis. Science 229:282–284
20. Johnson RT (1982) Viral Infections of the Nervous System. Raven Press, New York
21. Graves M, Griffin DE, Johnson RT, Hirsch RL, Lindo de Soriano I, Roedenbeck S, Vaisberg A (1984) Development of antibody to measles virus polypeptides during complicated and uncomplicated measles virus infections. J Virol 49:409–412

Evolving Concepts in Viral Pathogenesis Illustrated by Selected Plant and Animal Models

CHAPTER 25
The Enigma of Viroid Pathogenesis

T.O. DIENER

Aside from their obvious agricultural interest [1], viroids are fascinating molecules in their own right. They are the smallest known agents of infectious disease, consisting solely of short strands of covalently closed circular, single-stranded RNA (240–380 nucleotides) [22]. Because of extensive intramolecular complementarity, viroids can assume rodlike, quasi double-stranded configurations, in which short, base-paired regions alternate with mismatched, single-stranded loops [3] (Figure 25.1). Despite their small size and severely limited genetic information content, viroids replicate (or, more accurately, are replicated) in susceptible cells without the assistance of recognizable helper viruses [4]. Their autonomous replication clearly distinguishes viroids from plant viral satellite RNAs, which may be of similar size but which can replicate only in the presence of a specific helper virus in the same cell [2].

Here, then, is the first challenging problem posed by viroids: How can an RNA of only about 300 nucleotides induce its own replication in susceptible cells? A priori, one might suppose that although most of the biochemical machinery for viroid replication must be supplied by the host cell, the viroid could supply the genetic information for some essential component, such as a small polypeptide. All evidence indicates, however, that viroids do not code for proteins at all.

It has long been known that in various cell-free, protein-synthesizing systems, viroids are inactive and do not interfere with the translation of genuine mRNAs (reviewed in refs. 1 and 2). No viroid-specific proteins have been identified in infected cells [5]. Furthermore, there is no consistency among viroids regarding structural prerequisites of translation. Some do and some do not contain AUG initiation codons: with the potato spindle tuber viroid (PSTV), for example, none is present in the viroid itself or in its complement.

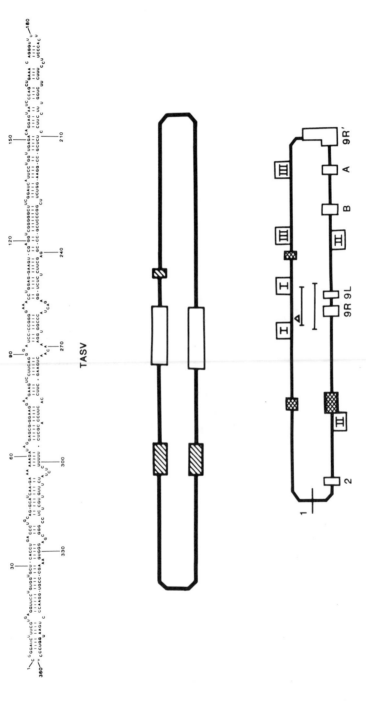

Figure 25.1. (**Top**) Primary and most probable secondary structure of a typical viroid (tomato apical stunt viroid, which causes disease problems in tropical Africa). Note the arrangement of short helical regions alternating with mismatched single-stranded loops. (**Bottom**) Putative functional regions within the "native" structure of PSTV. The positions of the upper and lower portions of the central conserved region are indicated by the bars within the rodlike structure; the positions of sequences forming the stems of secondary hairpins I–III are shown as boxes containing the appropriate roman numerals, whereas the shaded boxes indicate the presumed "pathogenicity-modulating" regions. The open boxes marked 9R', A, B, 9L, 9R, and 2 (right to left) indicate sequences homologous to the "box" sequences of class I introns; Δ, the unique cleavage site in PSTV cDNA for Bam HI; 1, nucleotide No. 1 in conventional representation.

Some viroids have potential open reading frames, but others do not. Because of their close structural and biological similarities, it seems unlikely that some viroids would be translated and others would not. Hence, their inconsistent translational features further weaken the case for viroid-specified proteins.

Hence, two important conclusions follow by necessity: (a) Viroids must be replicated entirely by preexisting (although possibly activated) host biochemical systems; and (b) all effects that viroids exert on their hosts must be a consequence of direct interaction of the viroid (or its complement) with host constituents. Both properties sharply distinguish viroids from viruses: (a) All viruses code for one or more proteins, some of which, at one stage or other of the viral replicative cycle, are synthesized and required for replication; and (b) most viruses are believed to exert their effects on host cells indirectly, via virus-specified proteins.

Here, then, is the second challenging problem posed by viroids: How does the viroid (or its complement) interact with host constituents to bring about the pathological aberrations in its hosts, by which viroids were discovered in the first place? Conceivably, these aberrations may be caused by viroid replication per se, that is, by diversion of host metabolites toward viroid synthesis or by the monopolization of host enzyme systems, resulting in deficiencies of certain host constituents. Alternatively, viroids may exert their effects in more subtle ways, by interfering with gene expression, for example.

Elucidation of the mechanisms of viroid pathogenesis is important because, as has been pointed out before [2], most types of symptoms induced in higher plants by virus infection have their counterpart in viroid-incited diseases and vice versa, suggesting that, in many cases, viroids and viruses may affect the same or similar metabolic pathways. Because of the relative simplicity of viroids, our complete knowledge of their structure, and lack of complicating factors, such as pathogen-specific proteins, it appears that viroids constitute more favorable systems than viruses with which to investigate the molecular basis of pathogenesis.

Present knowledge does not permit a detailed description of the molecular events that eventually result in the appearance of particular macroscopic symptoms. Lack of knowledge, however, did not stop investigators from proposing speculative schemes, and, by necessity, the following is mostly a description and critical analysis of some hypothetical schemes that have been advanced.

Mechanism of Viroid Replication

To examine the possibility that viroid diseases are a consequence of viroid replication per se, it is necessary to summarize what is known of the mechanism of viroid replication.

The finding, in infected plants, of viroid-complementary RNA strands

[6,7,8], and the absence of viroid-related DNA sequences [8,9] indicate that viroid replication occurs via RNA intermediates and does not involve host DNA or DNA transcribed from the viroid. Although, in vitro, several enzymes (DNA-dependent RNA polymerases II and III, RNA-dependent RNA polymerase) accept purified viroids as templates and transcribe full-length complementary copies (reviewed in ref. 10), the enzyme(s) responsible for in vivo viroid replication have not been established definitely. Viroids and their replication intermediates are associated primarily with cell nuclei [11,12], and considerable evidence suggests that their replication takes place there. Involvement of nuclear enzymes (DNA-dependent RNA polymerases) is therefore more likely than that of cytoplasmic enzymes (RNA-dependent RNA polymerase). The presence in infected cells of multimeric viroid forms of both polarities has been interpreted as evidence for a rolling-circle type of replication mechanism [7,10].

If normally DNA-dependent RNA polymerases are, indeed, responsible for viroid replication, the detrimental effects of viroid infection could be explained by successful competition of the viroid for the respective enzyme(s), resulting in a deficiency of host RNA synthesis—either of mRNAs, in the case of polymerase II, or of rRNAs, in the case of polymerase I. This explanation of viroid pathogenesis, as well as the related one that posits diversion of metabolites from host constituent to viroid synthesis, suffers from two shortcomings. First, plant metabolism is remarkably adaptable, and normally, only a fraction of its synthetic potential is utilized. Second, any theory of viroid pathogenesis must be able to account not only for the observed pathological consequences of viroid infections, but also for the fact that in many host species, viroids are replicated efficiently without detectable damage to the host [1,2]. It is difficult to understand how the commandeering of enzymes by the viroid or the viroid-induced diversion of metabolites could be harmless in one host species, yet cause serious damage in another, closely related, species or even in different cultivars of one species. More likely, the pathogenic consequences of viroid infection are due to more subtle and far more specific causes than the ones considered so far.

Direct Interaction of the Viroid with Host Constituents

Although viroids, of course, could interact with any number of host constituents, their nuclear location suggests that viroid-genomic DNA or viroid-nuclear protein complexes may be involved. The concept that viroids constitute abnormal regulatory molecules that bind to specific regulatory sites on the host's DNA, thereby affecting gene expression, has been advanced repeatedly [4,10].

Such a hypothesis can readily account for specific subtle differences in pathogenicity of viroids in closely related species, but is difficult to test experimentally and suffers from the fact that a regulatory role of certain RNA species in eukaryotic cells has not been demonstrated unequivocally.

It has long been known that, in situ, viroids are associated with host constituents and that this association is abolished by high ionic strength [4,11], suggesting that viroid-host protein complexes are involved. Indeed, recent evidence indicates that viroids are bound in the nucleolus in a complex of 12–15 Svedberg units(S) (which is the size of nucleosomes and nucleosome oligomers), and in vitro reconstitution of viroids with nuclear proteins identified (aside from histones) one particular protein with strong affinity for the viroid[13]. Further work is required, however, before the involvement, if any, of these or other specific viroid-host protein complexes in viroid pathogenesis will be known.

The Intron Connection

Possible connections between viroids and introns have been postulated by several authors (reviewed in refs. 2, 10), and the presence of a nucleotide sequence in the PSTV complement that exhibits complementarity with the 5'-end of mammalian small nuclear RNA U1 (as do nuclear encoded mRNA introns) [14] has added plausibility to such speculations. Evidently because of these sequence similarities, viroids could interfere with the splicing process of mRNAs, thereby causing metabolic disturbances.

The presence, in infected cells, of multimeric forms of viroids (presumably replication intermediates) suggests that monomers are produced by specific cleavage of multimers, followed by ligation of monomers to form covalently closed circular molecules. Conceptually, such a process resembles the cleavage–ligation reaction by which introns are spliced out of precursor RNAs and exons joined to form functional RNA. Indeed, analysis of viroid nucleotide sequences disclosed the presence of features characteristic of class I introns (encompassing nuclear rRNA and certain mitochondrial mRNA and rRNA introns ([15]; Hadidi and Diener, unpublished). These similarities suggest evolutionary and/or functional connections between viroids and class I introns and may, in turn, explain the pathogenic effects of viroid infection. Although, in vitro, the splicing of certain class I introns occurs in the absence of proteins [16], the corresponding in vivo process may require the participation of specific proteins [17]. A similar situation may obtain with viroids, and conceivably, disease processes may be triggered by the greater affinity of viroids than of introns, for these proteins, leading to inadequate or faulty splicing of precursor RNAs.

The A and O of Symptom Formation

Disease induction can be regarded as a causal chain of events that starts with the introduction of the pathogen into a susceptible cell, passes through intermediary stages, and terminates with the appearance of gross symptoms. Experimentally, it is most convenient to concentrate on the beginning and end of this causal chain, that is, on studying the effects of viroid structure on pathogenicity on the one hand, and on a search for qualitative and/or quantitative host constituent abnormalities in infected tissue on the other.

Structure–Function Correlation

The elucidation of the primary and most likely secondary structures of a number of viroids and viroid isolates has permitted comparisons of their nucleotide sequences with biological properties. PSTV isolates differing in symptom severity in tomato (from mild, almost symptomless, isolates to severe, almost lethal, ones) have been sequenced. Despite their drastically different biological effects, these isolates differ from one another by only two to seven nucleotide exchanges, and these are clustered in two regions of the viroid molecule [10]. Although these exchanges result in only minor stability differences of the viroid's secondary structure, it has been noted that in one of these regions, the so-called "pathogenicity-modulating" region (located, in the conventional depiction of viroid secondary structure, in the left half of the molecule; see Figure 25.1), base pairing between the upper and lower strands is strongest in mild strains of PSTV and becomes increasingly weaker from intermediate to severe to (almost) lethal strains [10]. It has been reasoned that with decreasing stability of this region, the molecule becomes more sutiable for interaction with cellular constituents, which, in this model, is considered a prerequisite for pathogenicity [10].

With other viroids, however, no such regularities are apparent. Thus, a number of citrus exocortis viroid (CEV) isolates that vary in the severity of symptom expression in tomato form two sequence classes which differ from one another by at least 26 changes in a total of 370–375 residues [18]. The stability of the CEV region corresponding to the "pathogenicity-modulating" region of PSTV, however, does not decrease with increasing severity of symptom expression. On the contrary, the stability of this region in severe isolates is considerably greater than that of mild isolates [18], indicating that more than one region is probably involved in determining symptom severity. Furthermore, the degree of symptom severity of CEV isolates in tomato may not reflect a similar response of citrus trees to infection with the same isolates [18]. In light of present knowledge, it appears premature to extrapolate limited sequence data and to propose general structural rules for viroid pathogenicity.

The Biochemistry of Symptom Formation

Many efforts are in progress to apply recombinant DNA technology to the elucidation of viroid structure–function correlations. The demonstration that complete DNA copies of viroids, when introduced into viroid hosts, are transcribed into viroid RNA [19] permits in vitro mutagenesis of viroids by the introduction into cDNA of nucleotide exchanges, insertions, or deletions at precisely defined positions in the viroid molecule. Almost certainly, these experiments will lead to a great increase in our understanding of the initial stage of viroid pathogenesis, but one must realize that the identification of the exact structural requirements for different levels of pathogenicity is unlikely to yield clues as to the biochemical mechanisms involved. To obtain this information, classical biochemical approaches will still be required.

One such approach consists in searching for qualitative and/or quantitative differences of host constituents in viroid-infected, as compared to healthy plants. Although one might suspect abnormalities in one or the other specific host nucleic acids, no such abnormalities have been reported. Quantitative changes in host proteins, however, are a common consequence of viroid infection ([20] and others). In CEV-infected plants, for example, the concentrations of two host proteins are greatly increased, and, in viroid-infected tomato, an M_r 140,000 protein is present in relatively large amounts, whereas only traces exist in healthy plants [5]. Additional work will be required, however, to learn whether some or all of these protein abnormalities are directly involved in pathogenesis or whether they are simply by-products of symptom formation.

Conclusion

As far as their molecular structure is concerned, viroids are among the best-known biological macromolecules. Yet, despite this detailed knowledge, dynamic aspects of viroid-host systems are still largely enigmatic. This applies to viroid replication, but even more to the biochemical mechanisms by which these low-molecular-weight RNAs bring about pathological conditions in some of their hosts.

References

1. Diener TO (1979) Viroids and Viroid Diseases. John Wiley, New York
2. Diener TO (1983) Viroids. Adv Virus Res 28:241–283
3. Gross HJ, Domdey H, Lossow C, Jank P, Raba M, Alberty H, Sänger HL (1978) Nucleotide sequence and secondary structure of potato spindle tuber viroid. Nature 272:203–208
4. Diener TO (1971) Potato spindle tuber "virus." IV. A replicating, low molecular weight RNA. Virology 45:411–428
5. Galindo JA, Smith DR, Diener TO (1984) A disease-associated host protein in viroid-infected tomato. Physiol Plant Pathol 24:257–275
6. Grill LK, Semancik JS (1978) RNA sequences complementary to citrus exocortis viroid in nucleic acid preparations from infected *Gynura aurantiaca*. Proc Natl Acad Sci USA 75:896–900
7. Owens RA, Diener TO (1982) RNA intermediates in potato spindle tuber viroid replication. Proc Natl Acad Sci USA 79:113–117
8. Rohde W, Sänger HL (1981) Detection of complementary RNA intermediates of viroid replication by Northern blot hybridization. Biosci Rep 1:327–336

9. Hadidi A, Cress DE, Diener TO (1981) Nuclear DNA from uninfected or potato spindle tuber viroid-infected tomato plants contains no detectable sequences complementary to cloned double-stranded viroid cDNA. Proc Natl Acad Sci USA 78:6932–6935

10. Sänger HL (1984) Minimal infectious agents: The viroids. In Mahy BWJ, Pattison JR (eds) The Microbe. Cambridge University Press, pp 281–334

11. Diener TO (1971) Potato spindle tuber virus: A plant virus with properties of a free nucleic acid. III. Subcellular location of PSTV-RNA and the question of whether virions exist in extracts or in situ. Virology 43:75–89

12. Spiesmacher E, Mühlbach HP, Schnölzer M, Haas B, Sänger HL (1983) Oligomeric forms of potato spindle tuber viroid (PSTV) and of its complementary RNA are present in nuclei isolated from viroid-infected potato cells. Biosci Rep 3:767–774

13. Wolff P, Gilz R, Schumacher J, Riesner D (1985) Complexes of viroids with histones and other proteins. Nucleic Acids Res 13:355–367

14. Diener TO (1981) Are viroids escaped introns? Proc Natl Acad Sci USA 78:5014–5015

15. Dinter-Gottlieb G, Cech TR (1984) Viroids contain homologies with the Tetrahymena IVS and may resemble group I introns in structure. In Abstracts of the Second Cold Spring Harbor Meeting on RNA Processing. Cold Spring Harbor Laboratory, Cold Spring Harbor, New York, p 37

16. Cech TR (1983) RNA splicing: Three themes with variations. Cell 34:713–716

17. Van der Horst G, Tabak HF (1985) Self-splicing of yeast mitochondrial ribosomal and messenger RNA precursors. Cell 40:759–766

18. Visvader JE, Symons RH (1985) Eleven new sequence variants of citrus exocortis viroid and the correlation of sequence with pathogenicity Nucleic Acids Res 13:2907–2920

19. Cress DE, Kiefer MC, Owens RA (1983) Construction of infectious potato spindle tuber viroid cDNA clones. Nucleic Acids Res 11:6821–6835

20. Conejero V, Picazo I, Segado P (1979) Citrus exocortis viroid (CEV): Protein alterations in different hosts following viroid infection. Virology 97:454–456

CHAPTER 26
Prions Causing Scrapie and Creutzfeldt–Jakob Disease

STANLEY B. PRUSINER, RONALD A. BARRY, AND MICHAEL P. MCKINLEY

Scrapie and Creutzfeldt–Jakob disease (CJD) are caused by *prions*, which appear to be different from both viruses and viroids. Prions contain protein that is required for infectivity, but no nucleic acid has been found within them. Prion proteins are encoded by a cellular gene and not by a nucleic acid within the infectious prion particle. A cellular homologue of the prion protein has been identified. The role of this homologue in metabolism is unknown. Prion proteins, but not the cellular homologue, aggregate into rod-shaped particles that are histochemically and ultrastructurally identical to amyloid. Extracellular collections of prion proteins form amyloid plaques in scrapie- and CJD-infected rodent brains as well as CJD-infected human brains. Within the plaques, prion proteins assemble to form amyloid filaments. Elucidating the molecular differences between the prion protein and its cellular homologue may be important in understanding the chemical structure and replication of prions.

Degenerative Neurologic Diseases

In developing models for the study of degenerative neurologic diseases, we have focused our attention on the study of scrapie, a degenerative neurologic disorder of sheep and goats. Scrapie occurs after a prolonged incubation period and is readily transmissible to laboratory rodents. The disease is characterized by progressive neurologic dysfunction caused by degeneration of the CNS. Neuropathologic changes include proliferation of glial cells, vacuolation of neurons, and deposition of extracellular amyloid in the form of plaques.

Three degenerative human diseases similar to scrapie have been described: kuru, Creutzfeldt–Jakob disease (CJD), and Gerstmann–Sträussler syndrome

(GSS). All three of these disorders are transmissible to experimental animals. Kuru presents as a cerebellar ataxia [1]. Cases with incubation periods as long as 30 years have been described. Kuru appears to have been transmitted during ritualistic cannibalism among New Guinea natives. Creutzfeldt–Jakob disease occurs at a rate of one per million population throughout the world. It is a dementing disorder much like Alzheimer's disease, but in addition, myoclonus is frequently seen. Gerstmann-Sträussler syndrome presents, much like kuru, as a cerebellar ataxia [2]. Later, dementia becomes a prominent symptom of GSS prior to death. Gerstmann–Sträussler syndrome is very rare and generally is confined to families; it is apparently inherited as an autosomal dominant trait. In all three of these human diseases, as well as scrapie, the host remains afebrile, and there is no inflammatory response despite an overwhelming and fatal CNS "slow infection."

Purification Studies

The development of effective purification protocols of the infectious particles causing scrapie is leading to an understanding of their chemical structure [3–6]. Numerous attempts have been made to purify the scrapie agent over the past three decades [7–12]. Few advances in this area of investigation were made until a relatively rapid and economical bioassay was developed [13,14]. Over a period spanning nearly a decade, our investigations of the molecular properties of the scrapie agent have been oriented toward developing effective procedures for purification. We began our studies by determining the sedimentation properties of the scrapie agent in fixed-angle rotors and sucrose gradients [15–17]. Subsequent work extended those findings and demonstrated the efficacy of nuclease and protease digestions as well as sodium dodecyl sarcosinate gel electrophoresis in the development of purification protocols [18,19]. Once a 100-fold purification was achieved, convincing evidence was obtained demonstrating that a protein is required for infectivity [3,20].

Prions

Progress in purification of the scrapie agent allowed us to establish that a protein molecule is required for infectivity [3,13,20]. On the other hand, attempts to demonstrate a similar dependence of infectivity upon a nucleic acid continue to be unsuccessful. Based upon these observations, the term *prion* was suggested in order to distinguish this class of novel infectious pathogens from viruses and viroids [21]. *Prions* are defined as "small proteinaceous infectious particles that resist inactivation by procedures which modify nucleic acids." This is an operational definition; it does not prejudge the chemical composition of the scrapie prion except to state that it must contain a protein. The question of whether or not prions contain nucleic acids remains to be answered.

Some investigators have misinterpreted the term prion. They have used it to signify infectious proteins [22] or even as a synonym for scrapie-associated fibrils [23]. This misuse of the term has led to confusion and should be avoided.

Scrapie Prions Contain a Sialoglycoprotein

In our search for a scrapie-specific protein, it became necessary to substitute discontinuous sucrose gradients in vertical rotors for gel electrophoresis [4]. The resulting purification scheme led to the first identification of a macromolecule that appears to be a component of the scrapie prion [5,6,24–26]. This molecule is a sialoglycoprotein designated PrP 27–30 [27]. Many lines of evidence indicate that PrP 27–30 is both required for and inseparable from scrapie infectivity [5,6,25,26]. The development of a large-scale purification protocol has allowed us to determine the N-terminal sequence of PrP 27–30 and to raise antibodies against the protein [6,28,29].

Search for a Prion Genome Continues

The size of the smallest infectious unit remains controversial, largely because of the extreme heterogeneity and apparent hydrophobicity of the scrapie prion [21,30,31]. Early studies suggested a molecular weight of 60,000–150,000 [32]. Although an alternate interpretation of that data has been proposed [31], there is no experimental evidence to suggest that these molecular weight calculations are incorrect. In fact, sucrose gradient sedimentation, molecular sieve chromatography, and membrane filtration studies all suggest that a significant portion of the infectious particles may be considerably smaller than the smallest known viruses [21,30]. However, the propensity of the scrapie agent to aggregate makes molecular weight determinations by each of these methods subject to possible artifact.

With one possible exception, no experimental data has to date been accumulated that indicates that scrapie infectivity depends upon a nucleic acid within the particle [33]. Inactivation of scrapie prions with chlorpromazine after ultraviolet irradiation has been cited as evidence for a scrapie-specific RNA; however, modification of proteins by chlorpromazine is well documented [34]. Attempts to inactivate scrapie prions with nucleases, ultraviolet irradiation at 254 nm, Zn^{2+}-catalyzed hydrolysis, psoralen photoinactivation, and chemical modification by hydroxylamine have all been negative [21,35,36] even when preparations containing one major protein as determined by amino acid sequencing are used [37]. Although these negative results do not establish the absence of a nucleic acid genome within the prion, they make this a possibility worthy of consideration. Attempts to identify a nucleic acid in purified prion preparations by silver staining and ^{32}P-end-labeling have been unsuccessful to date [38].

Some investigators have identified a 4.3S RNA in density gradient fractions from scrapie-infected brain which appears to be absent in uninfected brains according to fingerprinting studies [39]. We have failed to identify a similar molecule in our purified preparations of scrapie prions (K. Gillis, D. Riesner, and S. B. Prusiner, unpublished observations). Our studies show that the molecular properties of scrapie prions with respect to nucleic acids are antithetical to those of the filamentous bacteriophage M13 and the rod-shaped plant viruses, tobacco mosaic virus and potato X virus [40].

Recently, a cDNA encoding PrP 27–30 has been cloned [41]. The cloned PrP cDNA has been used to search for a complementary nucleic acid within purified preparations of prions. To date, we have failed to identify a PrP-related DNA or RNA molecule in these preparations. Assuming PrP is a component of the infectious scrapie particle, our observations then demonstrate that scrapie prions are not typical viruses. As noted, we still cannot eliminate the possibility of a small, nongenomic nucleic acid that lies highly protected within the prion core.

Cellular Gene Encodes Prion Proteins

A cDNA encoding PrP 27–30 has been used as a probe to show that the gene for PrP is found in healthy uninfected hamsters [41]. In contrast to most viruses where their major proteins are encoded within the viral genome, prions contain no nucleic acid encoding PrP 27–30, as described above. Subsequent studies led to the discovery of a cellular homologue of the prion protein designated PrP 33–35^C. This protein can be distinguished from PrP 33–35^{Sc} by its sensitivity to proteases and its inability to polymerize into amyloid filaments [42]. The role of PrP 33–35^C in cellular metabolism is unknown. In contrast, the scrapie prion protein, PrP 33–35^{Sc} accumulates to high levels in infected brains and forms amyloid rods and filaments. Proteolytic digestion converts PrP 33–35^{Sc} to PrP 27–30, which is resistant to further degradation [41]. Both PrP 33–35^C and PrP 33–35^{Sc} appear to be integral membrane proteins [42]. How the accumulation of PrP 33–35^{Sc} disrupts cellular metabolism and nervous system function is unknown. PrP mRNA levels were found to remain constant during scrapie infection. Only a single mRNA of 2.1 kb has been detected in hamsters by Northern blotting [41]. Similar results using a cloned cDNA constructed from scrapie-infected mouse brain poly(A)$^+$ have been reported [43]. In situ hybridization has shown that PrP mRNA within brain is largely confined to neurons [44]. The same mRNA is found at lower levels in organs outside the CNS [41].

Whether or not prions contain macromolecules other than PrP 33–35^{Sc}

remains to be established. The apparent biologic properties of prions argue in favor of a small nucleic acid, but there is no physical or biochemical evidence for such a molecule [21,30]. How prions replicate is unknown. Understanding the chemical differences between PrP 33–35^C and PrP 33–35^{Sc} will be important in unraveling the mechanism of prion biosynthesis.

The discovery of PrP 33–35^C explains one of the most interesting, yet preplexing, features of scrapie. This slow infection progresses in the absence of any detectable immune response [45]. Because PrP 33–35^C and PrP 33–35^{Sc} share epitopes, PrP 33–35^C probably renders the host tolerant to PrP 33–35^{Sc} [41].

Ultrastructural Identification of Prion Aggregates

Many investigators have used the electron microscope to search for a scrapie-specific particle. Spheres, rods, fibrils, and tubules have been described in scrapie, kuru, and CJD-infected brain tissue [46–52]. Notable among the early studies are reports of filamentous, virus-like particles in human CJD brain, measuring 15nm in diameter [50]; and rod-shaped particles in sheep, rat, and mouse scrapie brain, measuring 15–26 nm in diameter and 60–75 nm in length [51,52]. Studies with ruthenium red and lanthanum nitrate suggested that the rod-shaped particles possessed polysaccharides on their surface; these findings are of special interest because PrP 27–30 has been shown to be a sialoglycoprotein [27].

In purified fractions prepared from scrapie-infected brains, rod-shaped particles were found, measuring 10–20 nm in diameter and 100–200 nm in length [4,5]. Although no unit morphologic structure could be identified, most of the rods exhibited a relatively uniform diameter and appeared as flattened cylinders. Some of the rods had a twisted structure, suggesting that they might be composed of protofilaments. In the fractions containing rods, one major protein (PrP 27–30) and $\sim 10^{9.5}$ ID$_{50}$ (median infective dose) units of prions per milliliter were also found. The high degree of purity of our preparations demonstrated by radiolabeling and sodium dodecyl sulfate polyacrylamide gel electrophoresis indicated that the rods are composed of PrP 27–30 molecules. Because PrP 27–30 had already been shown to be required for, and inseparable from, infectivity [25], we concluded that the rods must be a form of the prion [5]. In earlier studies with less purified fractions, we could not determine whether the rods were a pathologic product of infection or an aggregate of prions [4]. Subsequently, others faced the same dilemma because of protein contaminants in their preparation [53]. Recent immunoelectron microscopic studies using antibodies raised against PrP 27–30 have established that the rods are composed of PrP 27–30 molecules [54]. Sonication of the prion rods reduced their mean length to 60 nm and generated many spherical particles without altering infectivity titers (M. P. McKinley, M. B. Braunfeld, and S. B. Prusiner, unpublished results). In contrast, fragmentation of M13 filamentous bacteriophage by brief sonication reduced infectivity significantly [55].

Spherical particles have been found within postsynaptic evaginations of the brains of scrapie-infected sheep and mice as well as CJD-infected humans and chimpanzees [46–48]; these particles measured 23–35 nm in diameter. Because sonication fragmented prion rods and generated spheres measuring 10–30 nm in diameter, the question arises as to whether or not the spherical particles observed in brain tissue are related to the spheres generated by sonication of the prion rods.

Prion Rods and Filaments Are Amyloid

The ultrastructure of the prion rods is indistinguishable from many purified amyloids [5]. Histochemical studies with Congo red dye have extended this analogy in purified preparations of prions [5] as well as in scrapie-infected brain where amyloid plaques have been shown to stain with antibodies to PrP 27–30 [28,56]. In addition, PrP 27–30[Sc] has been found to stain with periodic acid Schiff reagent [27]; amyloid plaques in tissue sections readily bind this reagent. Amyloid plaques have also been found in three transmissible disorders similar to scrapie and CJD; kuru and GSS of humans as well as a chronic wasting disease of mule deer and elk [7,57,58]. These findings raise the possibility that prion-like molecules might play a causative role in the pathogenesis of nontransmissible disorders such as Alzheimer's disease [59]. Amyloid proteins are prevalent in Alzheimer's disease, but for many decades these proteins have been considered a consequence rather than a possible cause of the disease.

Immunocytochemical studies with antibodies to PrP 27–30 have shown that filaments measuring approximately 16 nm in diameter and up to 1500 nm in length within amyloid plaques of scrapie-infected hamster brain are composed of prion proteins [56]. The antibodies to PrP 27–30 did not react with neurofilaments, glial filaments, microtubules, and microfilaments in brain tissue. The prion filaments have a relatively uniform diameter, rarely show narrowings and possess all the morphologic features of amyloid. Except for their length, the prion filaments appear to be identical ultrastructurally to the rods that are found in purified fractions of prions.

Abnormal structures, labeled scrapie-associated fibrils, were distinguished from other filamentous structures in extracts of scrapie-infected rodent brains by their characteristics and well-defined morphology [60]. Published electron micrographs show two or four subfilaments wound into helical fibrils of 300–800 nm in length exhibiting a regular periodicity. Based on these ultrastructural characteristics, the fibrils have been repeatedly reported to be different from amyloid [60,61]. Attempts to stain the fibrils with Congo red dye have yielded negative results [60–62]; however, even a positive result would have been uninterpretable because of impurities in the extracts.

No particles with the ultrastructural morphology of scrapie-associated fibrils have been found either in thin sections of scrapie-infected brain or in purified preparations of scrapie prions. If these fibrils in brain extracts are eventually found to be composed of PrP 27–30 molecules, then the possibility that they are an artifact of the preparative extraction procedure must be entertained. Some investigators have used the term *scrapie-associated fibrils* [53] to describe the rod-shaped particles found in purified preparations of prions [4,5]. Renaming the prion rods is both confusing and misleading [23]. It suggests that scrapie-associated fibrils as originally described are amyloid [63] and that they are found in purified fractions of prions [53]; neither assertion is true. Furthermore, it distorts the sequence of discoveries leading to current knowledge of prions. Some investigators have suggested that scrapie-associated fibrils are filamentous animal viruses [62], whereas others claim that they are pathologic products of infections [23]. We believe that prion rods are neither filamentous viruses nor pathologic products of infection.

Creutzfeldt–Jakob Disease Prions

Investigations of scrapie prions have recently been extended to studies on CJD. The CJD agent has been partially purified via procedures developed for scrapie prions [29,64]. The CJD agents from humans, mice, and guinea pigs contain protease-resistant proteins which exhibit cross-reactivity with PrP 27–30 antisera. Electron microscopy reveals that the CJD preparations contain rod-shaped particles of dimensions similar to those found in scrapie prion preparations. Futhermore, the CJD prion rods stain with Congo red dye and exhibit green-gold birefringence. It is noteworthy that long, helically twisted fibrils have been reported in extracts from human, mouse, and guinea pig CJD brains and are called scrapie-associated fibrils [65]; however, our results with purified preparations of CJD prions show that structures with the morphology of these fibrils are not required for infectivity.

Recent studies have shown that PrP antisera stain amyloid plaques in human CJD and GSS brains as well as rodent brains with experimental CJD [66].

The Prion Hypothesis

New knowledge about the molecular structure of scrapie prions is beginning to accumulate rapidly. If prions were typical viruses, they should then contain a genomic nucleic acid that encodes PrP 27–30. They do not! To explain the apparent biological diversity of prions, a small, nongenomic nucleic acid that does not encode PrP 27–30 has been proposed; however, there continues to be no chemical or physical evidence to indicate the existence of such a nucleic acid. Alternatively, prions may be devoid of nucleic acid. Information for the synthesis of new prion proteins is encoded within the host genome. A cloned PrP cDNA as well as antibodies to the protein provide new tools with which to extend our investigation of the chemical structure of prions.

Once it is determined whether or not prions contain other macromolecules besides glycoproteins, chemical studies to determine the molecular mechanisms by which prions reproduce and cause disease should become possible. Indeed, efforts to purify and characterize the infectious particles causing scrapie and CJD have yielded important new information about the structure and biology of prions.

Acknowledgments

Important contributions from Drs. D. Westaway, C. Bellinger, J. Bockman, D. Bolton, R. Meyer, S. DeArmond, D. Kingsbury, D. Stites, and M. Scott are gratefully acknowledged. Collaborative studies with Drs. L. Hood, C. Weissmann, S. Kent, T. Diener, J. Cleaver, R. Aebersold, B. Oesch, and W. Hadlow have been important to the progress of these studies. The authors thank D. Groth, K. Bowman, M. Braunfeld, P. Cochran, and L. Pierce for technical assistance, as well as L. Gallagher, J. Sleath, and F. Elvin for editorial and administrative assistance. This work was supported by research grants from the National Institutes of Health (AG02132 and NS14069) as well as by gifts from R. J. Reynolds Industries, Inc. and Sherman Fairchild Foundation.

References

1. Gajdusek DC (1977) Unconventional viruses and the origin and disappearance of kuru. Science 197:943–960
2. Masters CL, Gajdusek DC, Gibbs CJ Jr (1981) Creutzfeldt–Jakob disease virus isolations from the Gerstmann–Sträussler syndrome. Brain 104:559–588
3. Prusiner SB, McKinley MP, Groth DF, Bowman KA, Mock NI, Cochran SP, Masiarz FR (1981) Scrapie agent contains a hydrophobic protein. Proc Natl Acad Sci USA 78:6675–6679
4. Prusiner SB, Bolton DC, Groth DF, Bowman KA, Cochran SP, McKinley MP (1982) Further purification and characterization of scrapie prions. Biochemistry 21:6942–6950
5. Prusiner SB, McKinley MP, Bowman KA, Bolton DC, Bendheim PE, Groth DF, Glenner GG (1983) Scrapie prions aggregate to form amyloid-like birefringent rods. Cell 35:349–358
6. Prusiner SB, Groth DF, Bolton DC, Kent SB, Hood LE (1984) Purification and structural studies of a major scrapie prion protein. Cell 38:127–134
7. Millson GC, Hunter GD, Kimberlin RH (1976) The physico-chemical nature of the scrapie agent. In Kimberlin RH (ed) Slow Virus Diseases of Animals and Man. American Elsevier, New York, pp 243–266
8. Siakotos AN, Gajdusek DC, Gibbs CJ Jr, Traub RD, Bucana C (1976) Partial purification of the scrapie agent from mouse brain by pressure disruption and zonal centrifugation in sucrose–sodium chloride gradients. Virology 70:230–237

9. Mould DL, Smith W, Dawson AM (1965) Centrifugation studies on the infectivities of cellular fractions derived from mouse brain infected with scrapie ('Suffolk strain') J Gen Microbiol 40:71–79

10. Diringer H, Hilmert H, Simon D, Werner E, Ehlers B (1983) Towards purification of the scrapie agent. Eur J Biochem 134:555–560

11. Marsh RF, Dees C, Castle BE, Wade WF, German TL (1984) Purification of the scrapie agent by density gradient centrifugation. J Gen Virol 65:415–421

12. Brown P, Green EM, Gajdusek DC (1978) Effect of different gradient solutions on the buoyant density of scrapie infectivity. Proc Soc Exp Biol Med 158:513–516

13. Prusiner SB, Groth DF, Cochran SP, Masiarz FR, McKinley MP, Martinez HM (1980) Molecular properties, partial purification and assay by incubation period measurements of the hamster scrapie agent. Biochemistry 19:4883–4891

14. Prusiner SB, Cochran SP, Groth DF, Downey DE, Bowman KA, Martinez HM (1982) Measurement of the scrapie agent using an incubation time interval assay. Ann Neurol 11:353–358

15. Prusiner SB, Hadlow WJ, Eklund CM, Race RE (1977) Sedimentation properties of the scrapie agent. Proc Natl Acad Sci USA 74:4656–4660

16. Prusiner SB, Hadlow WJ, Eklund CM, Race RE, Cochran SP (1978) Sedimentation characteristics of the scrapie agent from murine spleen and brain. Biochemistry 17:4987–4992

17. Prusiner SB, Hadlow WJ, Garfin DE, Cochran SP, Baringer JR, Race RE, Eklund CM (1978) Partial purification and evidence for multiple molecular forms of the scrapie agent. Biochemistry 17:4993–4997

18. Prusiner SB, Groth DF, Bildstein C, Masiarz FR, McKinley MP, Cochran SP (1980) Electrophoretic properties of the scrapie agent in agarose gels. Proc Natl Acad Sci USA 77:2984–2988

19. Prusiner SB, Groth DF, Cochran SP, McKinley MP, Masiarz FR (1980) Gel electrophoresis and glass permeation chromatography of the hamster scrapie agent after enzymic digestion and detergent extraction. Biochemistry 19:4892–4898

20. McKinley MP, Masiarz FR, Prusiner SB (1981) Reversible chemical modification of the scrapie agent. Science 214:1259–1261

21. Prusiner SB (1982) Novel proteinaceous infections particles cause scrapie. Science 216:136–144

22. Manuelidis L, Valley S, Manuelidis EE (1985) Specific proteins associated with Creutzfeldt–Jakob disease and scrapie share antigenic and carbohydrate determinants. Proc Natl Acad Sci USA 82:4263–4267

23. Multhaup G, Diringer H, Hilmert H, Prinz H, Heukeshoven J, Beyreuther K (1985) The protein component of scrapie-associated fibrils is a glycosylated low molecular weight protein. EMBO J 4:1495–1501

24. Bolton DC, McKinley MP, Prusiner SB (1982) Identification of a protein that purifies with the scrapie prion. Science 218:1309–1311

25. McKinley MP, Bolton DC, Prusiner SB (1983) A protease-resistant protein is a structural component of the scrapie prion. Cell 35:57–62

26. Bolton DC, McKinley MP, Prusiner SB (1984) Molecular characteristics of the major scrapie prion protein. Biochemistry 23:5898–5905

27. Bolton DC, Meyer RK, Prusiner SB (1985) Scrapie PrP 27–30 is a sialoglycoprotein. J Virol 53:596–606

28. Bendheim PE, Barry RA, DeArmond SJ, Stites DP, Prusiner SB (1984) Antibodies to a scrapie prion protein. Nature 310:418–421

29. Bendheim PE, Bockman JM, McKinley MP, Kingsbury DT, Prusiner SB (1985) Scrapie and Creutzfeldt–Jakob disease prion proteins share physical properties and antigenic determinants. Proc Natl Acad Sci USA 82:997–1001

30. Prusiner SB (1984) Prions - novel infectious pathogens. In Lauffer MA, Maramorosch K (eds) Advances in Virus Research, vol 29. Academic Press, New York, pp 1–56

31. Rohwer RG (1984) Scrapie infectious agent is virus-like in size and susceptibility to inactivation. Nature 308:658–662

32. Alper T, Haig DA Clarke MC (1966) The exceptionally small size of the scrapie agent. Biochem Biophys Res Commun 22:278–284

33. Dees C, Wade WF, German TL, Marsh RF (1985) Inactivation of the scrapie agent by ultraviolet irradiation in the presence of chlorpromazine. J Gen Virol 66:845–849

34. Rosenthal I, Ben-Hur E, Prager A, Riklis E (1978) Photochemical reactions of chlorpromazine; chemical and biochemical implications. Photochem Photobiol 28:591–594

35. Diener TO, McKinley MP, Prusiner SB (1982) Viroids and prions. Proc Natl Acad Sci USA 79:5220–5224

36. McKinley MP, Masiarz FR, Isaacs ST, Hearst JE, Prusiner SB (1983) Resistance of the scrapie agent to inactivation by psoralens. Photochem Photobiol 37:539–545

37. Bellinger-Kawahara CG, Cleaver JE, Diener TO, Prusiner SB (1986) (submitted for publication)

38. Bellinger-Kawahara CG, Gillis K, McKinley MP, Riesner D, Prusiner SB (1986) (in preparation)

39. Dees C, McMillan BC, Wade WF, German TL, Marsh RF (1985) Characterization of nucleic acids in membrane vesicles from scrapie-infected hamster brain. J Virol 55:126–132

40. Bellinger-Kawahara CG, Diener TO, Prusiner SB (1986) (in preparation)

41. Oesch B, Westaway D, Wälchi M, McKinley MP, Kent SBH, Aebersold R, Barry RA, Tempst P, Teplow DB, Hood LE, Prusiner SB, Weissmann C (1985) A cellular gene encodes scrapie PrP 27–30 protein. Cell 40:735–746

42. Meyer RK, McKinley MP, Bowman KA, Barry RA, Prusiner SB (1986) Separation and properties of cellular and scrapie prion proteins. Proc Natl Acad Sci USA 83:2310–2314

43. Chesebro B, Race R, Wehrly K, Nishio J, Bloom M, Lechner D, Berstrom S, Robbins K, Mayer L, Keith JM, Garon C, Haase A (1985) Identification of scrapie prion protein-specific mRNA in scrapie-infected and uninfected brain. Nature 315:331–333

44. Kretzschmar H, Prusiner SB, Stowring LE, DeArmond SJ (1986) Scrapie prion proteins are synthesized in neurons. Amer J Pathol 122:1–5

45. Kasper KC, Bowman K, Stites DP, Prusiner SB (1981) Toward development of assays for scrapie-specific antibodies. In Streilein JW, Hart DA, Stein-Streilein J, Duncan WR, Billingham RE (eds) Hamster Immune Responses in Infectious and Oncologic Diseases. Plenum Press, New York, pp 401–413

46. David-Ferreira JF, David-Ferreira KL, Gibbs CJ Jr, Morris JA (1968) Scrapie in mice: Ultrastructural observations in the cerebral cortex. Proc Soc Exp Biol Med 127:313–320

47. Lampert PW, Gajdusek DC, Gibbs CJ Jr (1971) Experimental spongiform encephalopathy (Creutzfeldt–Jakob disease) in chimpanzees. J Neuropathol Exp Neurol 30:20–32

48. Baringer JR, Prusiner SB (1978) Experimental scrapie in mice—ultrastructural observations. Ann Neurol 4:205–211
49. Field EJ, Mathews JD, Raine CS (1969) Electron microscopic observations on the cerebellar cortex in kuru. J Neurol Sci 8:209–224
50. Vernon ML, Horta-Barbosa L, Fuccillo DA, Sever JL, Baringer JR, Birnbaum G (1970) Virus-like particles and nucleoprotein-type filaments in brain tissue from two patients with Creutzfeldt–Jakob disease. Lancet 1:964–966
51. Field EJ, Narang HK (1972) An electron-microscopic study of scrapie in the rat: Further observations on "inclusion bodies" and virus-like particles. J Neurol Sci 17:347–364
52. Narang HK (1974) An electron microscopic study of natural scrapie sheep brain: Further observations on virus-like particles and paramyxocirus-like tubules. Acta Neuropathol (Berl) 28:317–329
53. Diringer H, Gelderblom H, Hilmert H, Özel M, Edelbluth C, Kimberlin RH (1983) Scrapie infectivity, fibrils and low molecular weight protein. Nature 306:476–478
54. Barry RA, McKinley MP, Bendheim PE, Lewis GK, DeArmond SJ, Prusiner SB (1985) Antibodies to the scrapie protein decorate prion rods. J Immunol 135:603–613
55. Bellinger-Kawahara CG, McKinley MP, Prusiner SB (1986) (in preparation)
56. DeArmond SJ, McKinley MP, Barry RA, Braunfeld MB, McColloch JR, Prusiner SB (1985) Identification of prion amyloid filaments in scrapie-infected brain. Cell 41:221–235
57. Klatzo I, Gajdusek DC, Zigas V (1959) Pathology of kuru. Lab Invest 8:799–847
58. Bahmanyar W, Williams ES, Johnson FB, Young S, Gajdusek DC (1985) Amyloid plaques in spongiform encephalopathy of mule deer. J Comp Pathol 95:1–5
59. Prusiner SB (1984) Some speculations about prions, amyloid and Alzheimer's disease. N Engl J Med 310:661–663
60. Merz PA, Somerville RA, Wisniewski HM, Iqbal K (1981) Abnormal fibrils from scrapie-infected brain. Acta Neuropathol (Berl) 54:63–74
61. Merz PA, Wisniewski HM, Somerville RA, Bobin SA, Masters CL, Iqbal K (1983) Ultrastructural morphology of amyloid fibrils from neuritic and amyloid plaques. Acta Neuropathol (Berl) 60:113–124
62. Merz PA, Rohwer RG, Kascsak R, Wisniewski HM, Somerville RA, Gibbs CJ Jr, Gajdusek DC (1984) Infection-specific particle from the unconventional slow virus diseases. Science 225:437–440
63. Somerville RA (1985) Ultrastructural links between scrapie and Alzheimer's disease. Lancet i:504–506
64. Bockman JM, Kingsbury DT, McKinley MP, Bendheim PE, Prusiner SB (1985) Creutzfeldt–Jakob disease prion proteins in human brains. N Eng J Med 312:73–78
65. Merz PA, Somerville RA, Wisniewski HM, Manuelidis L, Manuelidis EE (1983) Scrapie-associated fibrils in Creutzfeldt–Jakob disease. Nature 306:474–476
66. Kitamoto T, Tateishi J, Tashima T, Takeshita I, Barry RA, DeArmond SJ, Prusiner SB (1986) Amyloid plaques in Creutzfeldt–Jakob disease stain with prion protein antibodies. Ann Neurol (in press)

CHAPTER 27
Interferon-induced Disease

ION GRESSER

Interferons constitute a group of closely related cellular proteins that exhibit potent antiviral activity. There is abundant experimental data indicating that the production of interferons in the course of viral infections is an integral part of the host response, and interferon treatment of animals and humans is associated with prophylactic and therapeutic antiviral effects. Although interferon was originally considered to inhibit viral multiplication within the cell without affecting host cell metabolism or function, it is now widely accepted that interferons can also affect cell division and function both in cell culture and in the animal [1]. If this is so, we might expect to find instances in which too much interferon, instead of being beneficial, might even prove inimical to the host. I will summarize here the results of experiments which show that interferon can induce disease in mice and rats, and then try to discuss the relevance of these findings to our understanding of the pathogenesis of some viral diseases.

Mouse Interferon Inhibits the Growth of Suckling Mice, Causes Liver Cell Necrosis, and Results in Death

When newborn mice were subcutaneously inoculated daily with potent mouse interferon-α/β, there was a progressive inhibition of growth in the first week of life and death of all mice in the second week [2]. Suckling mice of different strains including nude and axenic mice were sensitive to this effect of interferon. Two factors were important in the demonstration of the lethal effects of interferon: the timing of injection and the amount of interferon-α/β injected. It was important to start injecting interferon on the day of birth and continue daily injections. Initiation of daily interferon administration on the seventh day of life did not prove lethal [2]. Likewise, there was a requisite minimal dose of interferon. Using highly purified interferon-α/β, we estimated that one 50% lethal dose (LD_{50}) of interferon-α/β ranged between 10 and 100 ng of interferon/mouse/day (administered for seven days) [3,4].

Autopsy of moribund mice showed a pale gray and fatty liver, which on microscopic examination revealed marked steatosis, large areas of cell degeneration, and necrosis without any inflammatory reaction. The necrosis was initially subcapsular, but then extended to involve the whole liver [2,3]. Electron microscopic examination of hepatocytes showed a marked accumulation of fat, a decrease in glycogen content, and characteristic tubular aggregates associated with the endoplasmic reticulum [5,6]. Biochemical investigations showed that interferon treatment resulted in a marked increase in hepatic triglycerides, accompanied by a pronounced decrease in certain phospholipids [7]. Organs other than the liver appeared normal [2,4].

Mouse Interferon Can Also Induce Glomerulonephritis in Mice and Rats

When interferon treatment, begun at birth, was discontinued on the seventh day of life (a time when steatosis and discrete foci of liver cell necrosis were present), most of the mice appeared to recover and gained weight. In the ensuing months, however, a number of these mice died. At autopsy, although the liver and other organs appeared normal, the kidneys were pale, and the surface was granular. Histologic examination revealed a severe glomerulonephritis [8] characterized initially by enlarged glomeruli and thickening of the mesangium and capillary loops, and subsequently by hyalinization of glomerular tufts with epithelial crescents, and voluminous subendothelial deposits. In well-advanced disease, virtually all glomeruli were sclerotic, and there was extensive sclerosis and diffuse atrophy of tubules. By immunofluorescence, there were coarse granular deposits of IgG, IgM, and C3 along the glomerular basement membrane (GBM) [3,4,8]. By sacrificing mice at intervals, it was apparent that the renal lesions were progressive. Whereas all the interferon-treated mice had minimal to moderate glomerular lesions at one month, they all had very severe lesions at two months. Furthermore, the progressive nature of the lesions could also be documented by examining serial renal biopsies of individual mice [8]. Although our findings had at first suggested a latent period between cessation of interferon administration (at seven days of life) and the first appearance of glomerular lesions seen by light microscopy (at 21 days) [8], an ultrastructural study of the development of nephritis showed that at 8 days of life there was already a marked thickening of the GBM [9]. It is important to emphasize that glomerulonephritis developed only in mice injected in the neonatal period with highly purified potent interferon preparations and not in those injected with a variety of control preparations [3,4]. We believe, therefore, that there is no doubt that interferon itself is the agent responsible for inducing these lesions.

When newborn rats were injected daily subcutaneously with potent rat virus-induced interferon preparations, there was also a clear-cut inhibition of growth, a delay in the maturation of several different organs, and the late development of glomerulonephritis [10]. Liver lesions were, however, not observed. Administration of mouse or human interferon to suckling rats did not affect their growth or induce disease.

Interferon Induces Pulmonary Cysts in A2G Mice

Like other strains of mice tested, interferon-treated suckling A2G mice grew poorly and showed a marked liver necrosis. Unlike other strains of mice, they did not develop glomerulonephritis [4]. Furthermore, unlike other strains of mice, interferon treatment in the neonatal period resulted in the subsequent development of multiple, large pulmonary cysts in all mice [11].

As in the previous experiments, it was necessary to inject sufficient amounts of interferon in the neonatal period in order to induce the subsequent development of pulmonary cysts. Initiation of interferon treatment after the eighth day of life did not result in pulmonary cysts. These cysts appeared to result from the rupture of dilated lymphatics into the interstitial tissue of the interlobular septae [11]. To our knowledge, comparable lesions have not been described in experimental animals or humans. One further point may be emphasized: These cysts have been observed to date only in A2G mice, suggesting that certain interferon-induced lesions are influenced (or determined) by the mouse genotype.

Injection of Newborn Mice with Lymphocytic Choriomeningitis Virus Also Causes Liver Cell Necrosis and Induces Glomerulonephritis: Role of Endogenous Interferon in the Induced Syndrome

Why did the suckling mice die with massive liver necrosis? One possibility was that somehow interferon activated a latent virus, such as lymphocytic choriomeningitis (LCM) virus or mouse hepatitis virus. We had been particularly interested in LCM virus, because it was known that inoculation of newborn mice with this virus results in a syndrome characterized by stunted growth, liver cell necrosis, and death. In other words, the syndrome induced in suckling mice infected with LCM virus at birth [12–16] appeared to be identical to the syndrome induced in mice by administration of interferon. To complete the similarity between the two syndromes, LCM virus-infected mice surviving the acute disease developed a glomerulonephritis identical to interferon-induced glomerulonephritis [14,17,18].

How were we to explain the similarity of the two syndromes? Was interferon-induced disease, in reality, LCM virus disease in which interferon merely activated a latent LCM virus? No virus or other infectious agent could be recovered, however, from a pool of the livers or kidneys of interferon-treated mice either in cell culture or by passage in newborn mice [2]. No viral particles had been seen on electron microscopic examination of diseased livers [2,3,5,6].

The other possibility was that LCM virus induced this syndrome by inducing interferon [19–21], and it was the endogenous interferon that was in large part the responsible factor and not the virus itself. To test this hypothesis, we injected newborn mice with LCM virus and then injected them with potent sheep antibody to mouse interferon-α/β or control immunoglobulins. Injection of the anti-mouse interferon globulin neutralized the LCM virus-induced endogenous interferon and resulted in a 100-fold increase in the serum LCM virus titer [22]. Despite this marked increase in circulating virus, mice treated with antibody to interferon grew and developed normally and showed no liver lesions; moreover, the incidence of death was much decreased [22]. As in interferon-treated mice, tubular aggregates were present in the endoplasmic reticulum of liver cells from LCM virus-infected suckling mice, but not in liver cells from virus-infected mice treated with antibody to interferon [4,5].

Furthermore, injection of this anti-interferon antibody also markedly inhibited the subsequent appearance of glomerulonephritis characteristic of late LCM-virus disease [23]. Thus, although all of the control virus-infected mice rapidly developed severe glomerular lesions with heavy coarse deposits of IgG and C3 along the GBM, a marked delay was observed in the appearance of glomerular lesions in mice treated with antibody to interferon. Despite the marked inhibition of the development of glomerulonephritis in LCM virus-infected mice treated with antibody to interferon, LCM virus was present in the blood and kidneys of these mice in amounts comparable to control virus-infected mice developing glomerulonephritis [23].

Because electron microscopic examination of the kidneys of interferon-treated mice revealed very early and marked changes in the GBM, we examined the kidneys of suckling mice injected at birth with LCM virus. Again, the same thickening of the lamina rara interna with areas of rarefaction was observed, and the lesions were identical to those seen in interferon-treated suckling mice. In contrast, the kidneys of LCM virus-infected mice injected with antibody to interferon were normal, and there was no thickening of the GBM in these mice [24].

Another series of experiments added further evidence that the disease observed in suckling mice injected with LCM virus was related to the interferon induced by the virus. The lethality of LCM virus for suckling mice varied markedly with the mouse strain. Virus-infected BALB/c mice exhibited minimal liver lesions and none died, whereas C3H mice had extensive liver lesions and all mice died [25]. An intermediate pattern was observed with Swiss mice (36% mortality). The results of our experiments suggested that the genetic control of susceptibility or resistance to LCM virus was determined not by the extent of viral multiplication, but by the amount of interferon produced. Thus, although there was no difference in the titers of LCM virus in the plasma or liver between these three strains of mice, there was a marked difference in the amount of interferon produced and the duration of interferonemia: BALB/c mice produced only small amounts of interferon for one day, whereas C3H mice produced large amounts of interferon for several days [25]. An intermediate response was observed for Swiss mice. Our results suggested that the amplitude of the interferon response in C3H mice was in large part responsible for the severity of LCM disease. This interpretation was further supported by experiments showing a marked decrease in the incidence of mortality in virus-infected C3H mice when they were injected with antibody to interferon. The relative resistance of BALB/c mice was not caused by insensitivity to interferon because administration of interferon did inhibit their growth and did induce liver lesions. The minimal disease occuring in BALB/c mice would appear to be related, therefore, to their minimal interferon response to LCM virus [25]. *However, several disease manifestations occurring with LCM virus infection via a variety of LCMV strains and mouse strains need not be associated with interferon [26]. In these instances, similar preparations of antibodies to interferon that blocked disease [25] have no effect on resultant tissue injury [26].*

Of the Pathogenesis of Some Viral Diseases

We still do not know how the presence of large amounts of exogenous or endogenous interferon in the immature animal can result in such marked disease. Limitations of space do not permit consideration of the possible mechanisms involved, and the interested reader may find discussion of the various hypotheses in reference 4. Several points germane to concepts in viral pathogenesis may be treated here.

Virus-induced Substances (i.e., Interferon) May Cause Disease

It is usually considered that most manifestations of viral diseases are due to direct destruction of cells by the virus. However, the experimental results summarized herein show that disease can also be caused by virus-induced host substances such as interferon. It may be that other lesions considered heretofore as due to cellular destruction by the virus may also be due in part to interferon or other host substances. For example, one may speculate [22] that the embryotoxic effects ascribed to rubella virus may be related to interferon induced by the virus either in the mother or in the embryo itself. In this regard, Lebon and co-workers have demonstrated an acid-labile interferon-α in sera from fetuses (between the 21st and 29th weeks of gestation) of mothers infected with rubella virus [27].

Is Interferon Toxicity Always Reversible?

It may be argued that under most conditions interferon is not toxic, and that in the experiments described, immature mice or rats were treated with large amounts of interferon or infected experimentally with an unusual virus at birth. The particular interferon-induced diseases we have described were not observed when weanling or adult mice were treated with interferon. In the past few years, however, it has become quite evident that interferon is not as innocuous as once believed. Many of the systemic manifestations of acute viral disease (fever, myalgia, fatigue, leucopenia, etc.) are also seen in patients receiving large amounts of highly purified "natural" or recombinant interferon [28]. It has been suggested that interferon induced by infection of the respiratory mucosa may play a role in "precipitating or potentiating attacks of bronchial asthma" [29].

To date, all the systemic or local effects in patients appear to be reversible with cessation of interferon therapy. It may be, however, that the effects of interferon may be more marked and may even induce irreversible changes in some individuals. This type of genetic predisposition might be analogous to the experimental model in which interferon induced pulmonary cysts only in A2G mice and not in other strains of mice [11].

Effect of Interferon May Not Become Obvious Until Long After Exposure to Interferon

In our experiments we treated suckling mice with interferon for only the first week of life. Nevertheless, despite discontinuation of interferon injections, the mice went on to develop a progressive and lethal disease (i.e., glomerulonephritis), which only became manifest later in life. Likewise, in newborn mice infected with LCM virus, some of the effects that appeared long after viral infection, and therefore long after the short period of the interferon response, may also have been due to interferon. In these instances, had we shown the sick adult mouse (with glomerulonephritis or pulmonary cysts) to observers unfamiliar with the experimental protocol, they could not have supposed that we had treated the mouse with interferon for only the first week of life and that the pathologic process once engaged, had continued to progress inexorably.

Some patients appear to develop a disease suddenly—some forms of glomerulonephritis, for example. Is it possible that the cause of their disease is already far in the past? If such speculations are allowed for glomerulonephritis, may they also be allowed for other diseases such as degenerative diseases of the central nervous system? It is of interest in this regard that Hotchin and Seegal showed that mice injected at birth or as young adults with LCM virus subsequently exhibited long-lasting behavioral abnormalities [30].

What Is the Effect of Persistent Interferonemia?

If we accept the concept that under some conditions interferon may exert harmful effects, we may ask whether chronic interferonemia is associated with harmful or beneficial effects. Virelizier, Dayan and Allison induced a carrier state for mouse hepatitis virus (MHV-3) in C3H and A2G mice that was associated with the development of a progressive neurologic disease [31]. These virus-carrier mice also had a chronic interferonemia [32], and thus the possibility exists that interferon may be exerting untoward effects in some of these mice. Low levels of interferon have been found in the sera of patients with a variety of autoimmune diseases such as systemic lupus erythematosus (SLE), rheumatoid arthritis, scleroderma, and Sjögren's syndrome [33–36], and recently in patients with acquired immune deficiency syndrome (AIDS) [37]. Furthermore, the tubuloreticular structures associated with the endoplasmic reticulum of cells from patients with SLE and AIDS [38–42] are very similar, if not identical, to the aforementioned tubular aggregates found in hepatocytes of suckling mice treated with interferon or injected with LCM virus [5,6].

For the moment, no one can say whether the persistent presence of interferon in the sera or secretions of mice or patients with some subacute or chronic diseases is of biologic significance. Is this interferon part of the host's response to a viral infection, exerting a beneficial effect, or does the interferon contribute to some of the manifestations and pathology of the disease? Is the interferonemia in these patients merely a marker of infection, like an elevated sedimentation rate, and therefore possibly without much relevance to the disease process?

Could Antibody to Interferon Be Useful Therapeutically?

The therapeutic value of exogenous interferon in virus-infected animals is well established. Injection of mice with antibody to interferon markedly enhanced the evolution of diseases induced by several different viruses, attesting to the importance of interferon in the resistance of the host to virus infection [43,44]. Nevertheless, it was quite clear from the experiments with LCM virus that administration of antibody to interferon-α/β was effective in diminishing the severity of this viral disease. Although these results are seemingly paradoxical, we were successfully treating a viral disease with antibody to interferon. It may be that in some specific instances administration of antibody to interferon may prove of therapeutic value in patients [33].

On the Therapeutic Usefulness of Interferon

Can interferon cause disease in humans? No one knows. Interferon has now been administered to thousands of patients. Some patients of Hans Strander in Stockholm were treated for more than two years, 12 years ago, and there is no evidence that interferon exerts permanent harmful effects [45]. Nevertheless, in view of our observations in mice and rats, we should certainly bear in mind the possibility that interferon may either cause or contribute to human pathology. One should be especially careful in the administration of interferon to infants or young children.

In case these speculations might be misunderstood, I should like to emphasize that I do not feel that the results presented herein constitute an argument against the use of interferon in patients. On the contrary, we have been too often witness to interferon's therapeutic efficacy in viral and neoplastic diseases of mice, not to believe that comparable results would be obtained in humans, were interferon properly used. We have also been witness to its extraordinary activity, and to the varied effects interferon exerts on cells [1]. In some instances, interferon inhibits specialized cellular functions, whereas in other instances it enhances cellular functions [1,46]. Factors such as the amount of interferon and time of treatment determine the biologic effects observed. Aside from the theoretical interest, these considerations are of utmost importance in determining how to use interferon in patients. Our results do emphasize that we are dealing with a most potent substance, and, as with hormones, use of interferon must be based on knowledge of its effects and, if possible, on its mode of action.

Acknowledgments

I am grateful to the Richard Lounsbery Foundation, the Direction des Recherches Etudes et Techniques, the Association pour la Recherche sur le Cancer à Villejuif, the C.N.R.S., I.N.S.E.R.M., and the Fondation pour la Recherche Medicale for support.

References

1. Gresser I (1977) On the varied biologic effects of interferon. Cell Immunol 34:406–415
2. Gresser I, Tovey MG, Maury C, Chouroulinkov I (1975) Lethality of interferon preparations for new-born mice. Nature 258:76–78
3. Gresser I, Aguet M, Morel-Maroger L, Woodrow D, Puvion-Dutilleul F, Guillon JC, Maury C (1981) Electrophoretically pure mouse interferon inhibits growth, induces liver and kidney lesions and kills suckling mice. Am J Pathol 102:396–402.
4. Gresser I (1982) Can interferon induce disease? *In* Gresser I (ed) Interferon, vol 4. Academic Press, New York, p 95
5. Moss J, Woodrow D, Sloper JC, Rivière Y, Guillon JC, Gresser I (1982) Interferon

as a cause of endoplasmic reticulum abnormalities within hepatocytes in newborn mice. Br J Exp Pathol 63:43–49

6. Moss J, Woodrow DF, Gresser I (1985) Cytochemistry of the tubular aggregates found in hepatocytes of interferon-treated suckling mice. Histochem J 17:33–42

7. Zwingelstein G, Meister R, Malak NA, Maury C, Gresser I (1985) Interferon alters the composition and metabolism of lipids in the liver of suckling mice. J Interferon Res 5:315–325

8. Gresser I, Morel-Maroger L, Maury C, Tovey MG, Pontillon F (1976) Progressive glomerulonephritis in mice treated with interferon preparations at birth. Nature 263:420–422

9. Morel-Maroger L, Sloper JC, Vinter J, Woodrow D, Gresser I (1978) An ultra-structural study of the development of nephritis in mice treated with interferon in the neonatal period. Lab Invest 39:513–522

10. Gresser I, Morel-Maroger L, Châtelet F, Maury C, Tovey MG, Bandu MT, Buywid J, Delauche M (1979) Delay in growth and the development of nephritis in rats treated with interferon preparations in the neonatal period. Am J Pathol 95:329–346

11. Woodrow D, Moss J, Gresser I (1984) Interferon induces pulmonary cysts in A2G mice. Proc Natl Acad Sci USA 81:7937–7940

12. Traub E (1938) Factors influencing the persistence of choriomeningitis virus in the blood of mice after clinical recovery. J Exp Med 68:229–250

13. Hotchin JE, Cinits M (1958) Lymphocytic choriomeningitis infection of mice as a model for the study of latent virus infection. Can J Microbiol 4:149–163

14. Hotchin J (1962) The biology of lymphocytic choriomeningitis infection: Virus-induced immune disease. Cold Spring Harbor Symp Quant Biol 27:479–499

15. Traub E, Kesting F (1964) Experiments on heterologous and homologous interference in LCM-infected cultures of murine lymph node cells. Arch Gesamte Virusforch 14:55–64

16. Mims CA (1970) Observations on mice infected congenitally or neonatally with lymphocytic choriomeningitis (LCM) virus. Arch Ges Virusforsch 30:67–74

17. Hotchin J, Collins DN (1964) Glomerulonephritis and late onset disease of mice following neonatal virus infection. Nature 203:1357–1359

18. Oldstone MBA, Dixon FJ (1971) Immune complex disease in chronic viral infections. J Exp Med 134:35–40

19. Rivière Y, Bandu MT (1977) Induction d'interféron par le virus de la chorio-méningite lymphocytaire chez la souris. Ann Microbiol (Paris) 128a:323–329

20. Bro-Jorgensen K, Knudtzon S (1977) Changes in hemopoiesis during the course of acute LCM virus infection in mice. Blood 49:47–57

21. Merigan TC, Oldstone MBA, Welsh RM (1977) Interferon production during lymphocytic choriomeningitis virus infection of nude and normal mice. Nature 268:67–68

22. Rivière Y, Gresser I, Guillon JC, Tovey MG (1977) Inhibition by anti-interferon serum of lymphocytic choriomeningitis virus disease in suckling mice. Proc Natl Acad Sci USA 74:2135–2139

23. Gresser I, Morel-Maroger L, Verroust P, Rivière Y, Guillon JC (1978) Anti-interferon globulin inhibits the development of glomerulonephritis in mice infected at birth with lymphocytic choriomeningitis virus. Proc Natl Acad Sci USA 75:3413–3416

24. Ronco P. Woodrow D. Rivière Y, Moss J, Verroust P, Guillon JC, Gresser I, Sloper J, Morel-Maroger L (1980) Further studies on the inhibition of lymphocytic choriomeningitis-induced glomerulonephritis by anti-interferon globulin. Circulating immune complexes and ultrastructural studies. Lab Invest 43:37–46

25. Rivière Y, Gresser I, Guillon JC, Bandu MT, Ronco P, Morel-Maroger L, Verroust P (1980) Severity of LCM virus disease in different strains of suckling mice correlates with increasing amounts of endogenous interferon. J Exp Med 152:633–640

26. Oldstone MBA, Ahmed R, Buchmeier MJ, Blount P, Tishon A (1985) Perturbation of differentiated functions during viral infection in vivo. I. Relationship of lymphocytic choriomeningitis virus and host strains to growth hormone deficiency. Virology 142:158–174

27. Lebon P, Daffos F, Checoury A, Grangeot-Keros L, Forestier F, Toublanc JE (1985) Presence of an acid-labile alpha-interferon in sera from fetuses and children with congenital rubella. 21:775–778

28. Scott GM (1983) The toxic effects of interferon in man. *In* Gresser I (ed) Interferon, vol 5. Academic Press, New York, p 87

29. Hooks JJ, Moutsopoulos HM, Notkins AB (1980) The role of interferon in immediate hypersensitivity and autoimmune diseases. *In* Vilcek I, Gresser I, Merigan TC (eds) Regulatory function of interferons, Ann N Y Acad Sci 350:21–32

30. Hotchin J, Seegal R (1977) Virus-induced behavioral alteration of mice. Science 196:671–674

31. Virelizier JL, Dayan AD, Allison AC (1975) Neuropathological effects of persistent infection of mice by mouse hepatitis virus. Infect Immun 12:1127–1139

32. Virelizier JL, Virelizier AM, Allison AC (1976) The role of circulating interferon in the modification of immune responsiveness by mouse hepatitis virus (MHV-3). J. Immunol 117:748–753

33. Skurkovich SV, Eremkina EI (1975) The probable role of interferon in allergy. Ann Allergy 35:356–360

34. Hooks JJ, Moutsopoulos HM, Geis SA, Stahl NI, Decker JL, Notkins AL (1979) Immune interferon in the circulation of patients with autoimmune disease. N Engl J Med 301:5–8

35. Preble OT, Black RJ, Friedman RM, Klippel JH, Vilcek J (1982) Systemic lupus erythematosus: Presence in human serum of an unusual acid-labile leukocyte interferon. Science 216:429–431

36. Hooks JJ, Jordan GW, Cupps T, Moutsopoulos HM, Fauci AS, Notkins AL (1982) Multiple interferons in the circulation of patients with systemic lupus erythematosus and vasculitis. Arthritis Rheum 25:396–400

37. DeStefano E, Friedman RM, Friedman-Kien AE, Goedert JJ, Henriksen D, Preble OT, Sonnabend JA, Vilcek J (1982) Acid-labile human leukocyte interferon in homosexual men with Kaposi's sarcoma and lymphadenopathy. J Infect Dis 146:451–455

38. Grimley PM, Decker JL, Michelitch HJ, Frantz MM (1973) Abnormal structures in circulating lymphocytes from patients with systemic lupus erythematosus and related diseases. Arthritis Rheum 16:313–323

39. Grimley PM, Schaff Z (1976) Significance of tubuloreticular inclusions in the pathobiology of human diseases. *In* Iochim HL (ed) Pathobiology Annual. Appleton-Century Crofts, New York, vol 6, p 221

40. Grimley PM, Kang YH, Frederick W, Rook AH, Kostianovsky M, Sonnabend JA, Macher AM, Quinnan GV, Friedman RM, Masur H (1984) Interferon-related leucocyte inclusions in acquired immune deficiency syndrome: Localisation in T cells. Am J Clin Pathol 81:147–155

41. Rich SA (1981) Human lupus inclusions and interferon. Science 213:772–775

42. Gyorkey F, Sinkovics JG, Gyorkey P (1982) Tubuloreticular structures in Kaposi's sarcoma. Lancet ii:984–985

43. Gresser I, Tovey MG, Bandu MT, Maury C, Brouty-Boyé, D (1976) Role of interferon in the pathogenesis of virus diseases in mice as demonstrated by the use of anti-interferon serum. I. Rapid evolution of encephalomyocarditis virus infection. J Exp Med 144:1305–1315

44. Gresser I, Tovey MG, Maury C, Bandu MT (1976) Role of interferon in the pathogenesis of virus diseases in mice as demonstrated by the use of anti-interferon serum. II. Studies with herpes simplex, Moloney sarcoma, vesicular stomatitis, Newcastle disease and influenza viruses. J Exp Med 144:1316–1323

45. Strander H (1986) Interferon treatment of human neoplasia. Adv in Cancer Res (in press)

46. Vilcek J, Gresser I, Merigan TC (1980) Regulatory functions of interferons. Ann N Y Acad Sci 350

CHAPTER 28
African Swine Fever Virus

Eladio Viñuela

African swine fever (ASF) was first described in 1921 by Montgomery, who reported several disease outbreaks of domestic pigs in Kenya since 1910 with a mortality close to 100%. Montgomery recognized the viral nature of the disease, its likely transmission by wild swine which probably acted as virus carriers, and the lack of protection by passive immunization [1]. ASF is a menace to the pig population in the world because there is no vaccine, the virus multiplies in ticks and mutates easily, and different virus isolates can produce diseases with different clinical forms or no disease at all. The control and eradication of ASF require rapid diagnosis, drastic slaughter, and quarantine.

In 1957, the virus was found for the first time outside Africa in Portugal, and in 1960, it spread to Spain. In the 1960s and 1970s the virus was found in France, Italy, Sardinia, and Malta and, in the Western hemisphere, in Cuba, Brazil, the Dominican Republic, and Haiti. In early 1985, there was a ASF outbreak in Belgium. Today the disease is present in sub-Saharan Africa, Portugal, Spain, Sardinia, and Haiti.

ASF virus is an icosahedral cytoplasmic deoxyvirus morphologically similar to the members of the family Iridoviridae that infect vertebrates. The virion consist of a nucleoprotein core surrounded by a lipid membrane probably associated with the morphological units of the capsid. The extracellular virions have an external membrane, derived by budding through the plasma membrane, and an external diameter of about 200 nm [2]. The virion is built up by about 35 structural polypeptides, some of which might be host proteins [3,4] or viral proteins that cross-react with cellular antigens [4,5]. In some respects other than morphology, some properties of ASF virus resemble those of poxviruses. The DNA is a double-stranded molecule of about 170 kilobase pairs (kb) with 2.4-kb-long terminal inverted repeats [6] and hairpin loops of 37 nucleotides, composed almost entirely of A and T residues that are incompletely paired. The loops at each DNA end are present in two equimolar forms which, when compared in opposite polarities, are inverted and complementary (flip-flop), as is seen in vaccinia virus DNA [7,8] The virion contains a DNA-dependent RNA polymerase, guanylyl and methyl transferases, a poly (A) polymerase, a topoisomerase, and a protein kinase [9–11]. The RNA synthesized in vitro by the viral RNA polymerase is indistinguishable from that synthesized in ASF virus-infected cells in the presence of cycloheximide or cytosine arabinoside [12]. Like the poxviruses, the virus-specific RNA synthesized in infected cells is independent of host RNA polymerase II [13]. The characteristics of ASF virus altogether exclude it from all the families defined by the International Committee on Taxonomy of Viruses and support the establishment of a new family, of which ASF virus would be the only known representative [14].

ASF virus infects only species of the *Suidae* family and ticks of the genus *Ornithodoros*. It is the only arbovirus that contains DNA. In the pig, virus replication takes place mainly in mononuclear phagocytes. Fluorescent antibody tests carried out in tissues of infected pigs indicate that macrophages and reticular cells are the primary target cells in lymphoid tissues, whereas in blood and bone marrow the main cells involved are monocytes and macrophages and, occasionally, polymorphonuclear leukocytes and megakaryocytes [15]. The sensitivity of lymphocytes and endothelial cells to ASF virus is controversial [16,17].

The key postmortem lesion in ASF is hemorrhage [18]. ASF virus-infected pigs are unable to maintain normal hemostasis, which depends on platelets, several coagulation proteins, endothelial cells, and components of the vascular subendothelium. Both virulent and attenuated ASF viruses cause longer activated partial thromboplastin and thrombin times, an increase of the plasma factor VIII-related antigen, and a decrease of fibrinogen and human platelet agglutination times [19]. Thrombocytopenia develops rapidly over a two-day period several days after the viremia onset. A few days later, platelet count returns to normal values despite the presence of high viremia. The amount of megakaryocytes in the bone marrow increases, but only 2–10% contain viral antigens. There is no evidence of arrest of megakaryocyte maturation and platelet production. Probably, the ASF virus-induced thrombocytopenia is due to platelet consumption [20].

One of the most striking aspects of ASF virus infection is the induction of nonneutralizing, virus-specific antibodies in both pigs and other virus-resistant animal species inoculated with the virus. It has been suggested that the inability to produce neutralizing antibodies might be due to the nature of the virus rather than to the host, because both pigs recovered from virulent infections and those infected chronically by attenuated viruses respond normally to other virus infections by producing neutralizing antibodies [21]. The possibility of a relationship among the induction of nonneutralizing antibodies, the sensitivity of porcine macrophages to the virus, and the role of these cells as antigen-presenting cells is unlikely because virus-resistant animal species whose macrophages are not infected by the virus likewise do not produce neutralizing antibodies [21].

A possibility that accounts for the lack of detection of neutralizing antibodies is antigenic variation of critical antigens. Indications that ASF virus undergoes antigenic changes include the following:

1. Pigs infected with attenuated viruses derived from virulent ones by passage in porcine macrophage cultures, are partially resistant to the original but not to other virulent isolates.
2. An antigen isolated from the spleen of infected pigs seems to be isolate specific [22];
3. The hemadsorption–inhibition reaction [23], complement-mediated cytolysis [24], and a comparison of DNA restriction maps [25] and of the binding properties of a collection of monoclonal antibodies [26] all have allowed different virus isolates to be distinguished.

Altogether, these data indicate the existence of a complex variety of virus serotypes, but the lack of knowledge of the critical antigen(s) (mainly those involved in the attachment of the virus to, or its penetration into the target cells) makes uncertain the immunological significance of the antigenic variation.

A comparison of the restriction maps of 23 ASF virus field isolates (9 African, 11 European, and 3 American) has revealed the existence, in ASF virus DNA, of a central, highly conserved region of about 125 kb and two variable regions close to the DNA ends [25]. Nucleotide sequence analysis of the variable regions has shown the existence of a multigene family with homologous genes at either DNA end, close to the terminal inverted repeat (unpublished observations). A second gene family has been found between 10 and 20 kb from the left DNA end, within the most variable region of ASF virus DNA, with a complex pattern of repeats that seem to account for the deletions or additions found in that region in different virus isolates [27]. The biological significance of the gene families found in ASF DNA—in particular, their possible relation to the virus escape from the host immune system—will be unclear until the function of the putative proteins encoded within those regions is known. Other factors that might possibly account for the escape of ASF virus from the host immune system, such as the existence of blocking antibodies or suppressor antigens, have been considered elsewhere [14].

References

1. Montgomery RE (1921) On a form of swine fever occurring in British East Africa (Kenya Colony) J Comp Pathol 34:159–191, 243–262
2. Carrascosa JL, Carazo JM, Carrascosa AL, García N, Santisteban A, Viñuela E (1984) General morphology and capsid fine structure of African swine fever virus particles. Virology 132:160–172

3. Carrascosa, AL, Santarén JF, Viñuela, E (1982) Production and titration of African swine fever virus in porcine alveolar macrophages. J Virol Methods 3:303–310

4. Sanz A, García-Barreno B, Nogal ML, Viñuela E, Enjuanes L (1985) Monoclonal antibodies specific for African swine fever virus proteins. J Virol 54:199–206

5. Carrascosa JL, González P, Carrascosa, AL, García-Barreno B, Enjuanes L, Viñuela E (1986) Localization of structural proteins in African swine fever virus particles by immunoelectron microscopy. J Virol 58:377–384

6. Sogo JM, Almendral JM, Talavera A, Viñuela E (1984) Terminal and internal inverted repetitions in African swine fever virus DNA. Virology 133:271–275

7. Baroudy BM, Venkatesan S, Moss B (1981) Incompletely base-paired flip-flop terminal loops link the two DNA strands of the vaccinia virus genome into one uninterrupted polynucleotide chain. Cell 28:315–324

8. González A, Almendral JM, Talavera A, Viñuela E. 1986 (to be published)

9. Kuznar J, Salas ML, Vinñuela E (1980) DNA-dependent RNA polymerase in African swine fever virus. Virology 101:169–175.

10. Salas ML, Kuznar J, Viñuela E (1981) Polyadenylation, methylation and capping of the RNA synthesized in vitro by African swine fever virus. Virology 113:484–491

11. Salas ML, Kuznar J, Viñuela E (1983) Effect of rifamycin derivatives and coumermycin A1 on in vitro RNA synthesis by African swine fever virus. Arch Virol 77:77–80

12. Salas ML, Rey J, Almendral JM, Viñuela E (1986) (to be published)

13. Salas J, Salas ML, Viñuela E (1986) (to be published)

14. Viñuela E (1985) African swine fever virus. Curr Top Microbiol Immunol 116: 151–170

15. Colgrove G, Haelterman EO, Coggins L (1969) Pathogenesis of African swine fever virus in young pigs. Am J Vet Res 30:1343–1359

16. Casal I, Enjuanes L, Viñuela E (1984) Porcine leukocyte cellular subsets sensitive to African swine fever virus in vitro. J Virol 52:37–46

17. Mebus CA, McVicar JW, Dardiri AH (1981) Comparison of the pathology of high and low virulence African swine fever viral infections. CEC/FAO Expert Consultation on African swine fever virus, pp 23–25

18. Maurer FD, Griesemer RA, Jones TC (1958) The pathology of African swine fever—a comparison with hog cholera. Am J Vet Res 19:517–539

19. Edwards JF, Dodds WJ, Slauson DO (1984) Coagulation changes in African swine fever virus infection. Am J Vet Res 45:2414–2420

20. Edwards JF, Dodds WJ, Slauson DO (1985) Megakaryocytic infection and thrombocytopenia in African swine fever. Vet Pathol 22:171–176

21. DeBoer CJ (1967) Studies to determine neutralizing antibody in sera from animals recovered from African swine fever and laboratory animals inoculated with African virus with adjuvants. Arch Ges Virusforsch 20:164–179

22. Stone SS, Hess WR (1965) Separation of virus and soluble non-infectious antigens in ASFV by isoelectric precipitation. Virology 26:622–629

23. Vigario JD, Terrinha AM, Moura Nunes JF (1974) Antigenic relationships among strains of African swine fever virus. Arch Ges Virusforsch 45:272–277

24. Norley SG, Wardley RC (1982) Complement-mediated lysis of African swine fever virus-infected cells. Immunology 46:75–82

25. Blasco R, Agüero M, Almendral JM, Viñuela E. (1986) (to be published)
26. García-Barreno B, Sanz A, Nogal ML, Viñuela E, Enjuanes E (1986) Monoclonal antibodies of African swine fever virus: Antigenic differences among field virus isolates and viruses passaged in cell culture. J. Virol 58:385–392
27. Almendral JM, Blasco R, Viñuela E. (1986) (to be published)

CHAPTER 29
Theiler's Murine Encephalomyelitis Virus (TMEV) Infection in Mice as a Model for Multiple Sclerosis

HOWARD LIPTON, STEPHEN MILLER, ROGER MELVOLD, AND ROBERT S. FUJINAMI

Theiler's murine encephalomyelitis viruses (TMEV) are enteric pathogens of mice and members of the Picornaviridae family. The nucleotide sequence of TMEV and the predicted amino acid sequences of the TMEV-encoded proteins have been compared with other picornaviruses, and a close relationship exists with encephalomyocarditis virus (EMCV), indicating that the TMEV belong to the cardiovirus subgroup of picornaviruses. However, the type of neurologic involvement the TMEV produce is quite distinct from that of other cardioviruses. Following intracerebral (IC) inoculation, certain TMEV strains produce a unique biphasic central nervous system (CNS) disease in their natural murine host [1,2], characterized by poliomyelitis during the first month post infection, and a chronic, inflammatory demyelinating disorder weeks to months later in surviving animals. Mice infected with tissue culture-adapted TMEV do not develop clinical polio but do develop demyelinating disease after a prolonged (30–60 days) incubation period [3]. The demyelination is related to persistent infection wherein for many months low levels of infectious virus can be recovered from the target organ, the CNS [1,4].

Two experimental animal models for multiple sclerosis (MS) have been described—autoimmune (experimental allergic encephalomyelitis, EAE) and viral; they form the basis of current hypotheses regarding the cause of this disease. Although these models are not exact replicas of the human disease, they do share many features in common with it. Of the several available animal models of chronic virus-induced demyelination, TMEV infection is perhaps one of the most promising because:

1. Chronic pathologic involvement is limited to the CNS white matter.
2. Primary demyelination is accompanied by mononuclear cell inflammation.
3. Demyelinating lesions of different ages are simultaneously present, and

recurrent episodes of myelin breakdown can be reconstructed from pathological material.
4. Demyelination clearly leads to clinical disease, e.g., upper motor neuron signs (limb spasticity, extensor spasms, and neurogenic bladder).
5. Demyelination appears to be immune-mediated and/or related to persistent infection of CNS cells.
6. Demyelinating disease is under multigenic control.
7. Strong linkage to certain major histocompatibility complex (H-2 in the mouse) genotypes exists.

The TMEV model also has the following advantages:

1. The use of inbred mouse strains allows for greater reproducibility of results and the potential for genetic manipulation.
2. Since the mouse is the natural host for TMEV infection, deductions from experimental studies may likely be relevant to human disease.
3. A small animal host enables research to be conducted at a lower cost.

Pathologic Features

Extensive but patchy areas of leptomeningeal and white matter mononuclear cell infiltration with concomitant primary demyelination are the pathologic hallmarks in this model [2,5]. As in EAE, stripping of myelin lamellae by invading mononuclear cell processes, and vesicular disruption of myelin are seen. At later times after infection, astrocytic gliosis and remyelination become conspicuous. In some instances, Schwann cells, rather than oligodendrocytes, are the major contributors to new myelin formation in the outer portions of the spinal cord, whereas proliferating oligodendrocytes remyelinate axons in the interior portions of the spinal cord [6].

Recent immunohistochemical studies have shown a different rate of disappearance of myelin basic protein (MBP), P_0, and myelin associated glycoprotein (MAG). In fresh lesions, MBP disappears sooner than MAG, whereas in recurrent lesions, where there is Schwann cell remyelination, P_0 is lost before MAG [7]. This suggests that myelin destruction does not represent a dying back phenomenon, but rather a direct attack on the myelin sheath because MAG is present in the inner oligodendroglial component of the myelin sheath whereas P_0 and MBP are in the entire thickness of the sheath.

Multigenic Control of Demyelinating Disease

The susceptibility to TMEV-induced demyelinating disease differs among inbred mouse strains. In studies with tissue-culture adapted virus, SJL, SWR, DBA/2, and PL represent susceptible strains, whereas C57BL/6, C57BL/10, BALB/c, and C57L are resistant ([8,9]; unpublished observations). The availability of such inbred strains and a variety of defined variants has permitted analysis of the genetic basis for susceptibility to a much finer degree than in humans, where many genetic variables cannot be controlled. Comparisons of the resistant C57BL/6 and susceptible SJL strains indicate that multiple genes are involved in determination of susceptibility, at least one in the *H-2* complex and at least one that segregates independently of *H-2* [8]. The *H-2* gene involved has been localized to the class I locus *H-2D* [9,10], but the non-*H-2* gene remains unidentified.

Differences at *H-2* genes, however, do not always appear to be crucial to determination of susceptibility. Recently, Melvold, Knobler, and Lipton [11] noted that in some strain combinations, such as the susceptible SJL and the resistant BALB/c, *H-2* genotypes of segregating backcross animals do not correlate well with susceptibility, which appears instead to be primarily determined by multiple non-*H-2* loci. In comparisons of the susceptible DBA/2 and the resistant BALB/c, the entire genetic basis for susceptibility must rely on non-*H-2* genes because both strains carry the $H-2^d$ haplotype. The predominant role of different loci in different strain comparisons probably reflects the involvement of many genes, and it may be well to think of the disease process as resembling a metabolic pathway with several stages that can be influenced by different gene products. In making particular strain comparisons, analysis is then affected by the facts that (a) only loci that are functionally different between the two strains under comparison can be identified (loci at which the strains are functionally identical will have no detectable effects), and (b) the activity of some genes may vary according to the "genetic environment" in which they exist, being influenced by the presence or absence of other genes. Thus, a satisfactory description of the genetic control of susceptibility probably requires a composite of numerous strain comparisons to identify a significant portion of the loci involved. This also provides an analogous situation to human studies, where particular HLA genes (for example, DR2 and Drw2) have positive associations with MS, but the relative risks are so low (ranging between 2 and 3) that genetic factors other than HLA phenotype must also be involved.

Sites of Virus Persistence

TMEV persistence clearly involves ongoing virus replication because infectious virus can be readily isolated from the CNS [1,4]. Brahic and co-workers [12,13] have found that TMEV multiplication is restricted at the level of RNA replication, but the exact kinetics of the persistent infection in the CNS remains unknown. By ultrastructural immunocytochemistry and in situ hybridization, TMEV has been found to persist in the CNS white matter for essentially the life of the mouse. Acutely (days 0–14), virus antigens and virus RNA are mainly localized to neurons and their processes, whereas chronically (days 15–>360), both virus antigens and virus RNA have been detected in a number of different types of cells, including macrophages, mononuclear inflammatory cells, astrocytes, and oligodendrocytes [14,16]. Different notions exist regarding the predominant cell type in the white matter that is infected. On the one hand, Dal Canto and Lipton [14] reported that the majority of TMEV antigens are in macrophages during persistence. Alternatively, Rodriguez and co-workers [15] found virus antigens only in oligodendrocytes 45 days post infection, and Brahic and colleagues [16,17] noted that demyelinating lesions were associated with viral infection of local nervous system cells and not with occurrence of inflammatory cells. The extent of oligodendrocyte infection is a crucial issue that still remains unresolved between these groups of investigators. Because of the possibility of sampling error in ultrastructural studies, newer light microscopic techniques, such as the combination of immunocytochemistry for the identification of cell types and in situ hybridization for the identification of virus nucleic acid, need to be applied to this model [12]. Further, there may be differences among the strains of TMEV used.

Immune-mediated Mechanism of Demyelination

The effector mechanism by which a nonbudding virus, such as TMEV, might lead to immune-mediated tissue injury is unknown. Because TMEV antigens have been found in macrophages [14,15], it has been proposed that myelin breakdown may result from an interaction between virus-specific, sensitized lymphocytes, trafficking into infected areas in the CNS, and the virus. Thus, myelinated axons may be nonspecifically damaged as a consequence of a virus-specific immune response, i.e., an "innocent-bystander" response. Clatch et al. [9,18] have recently shown that high levels of TMEV-specific delayed-type hypersensitivity (DTH), but not TMEV-specific antibody responses, correlate with the temporal onset of demyelinating disease, as well as with disease incidence among susceptible and resistant congenic recombinant mice. Thus, in this system, lymphokines produced by MHC class II (I-A) restricted, TMEV-specific T_{DTH} cells primed by interaction with TMEV-infected macrophages, would lead to the recruitment and activation of additional macrophages in CNS tissue, resulting in nonspecific macrophage-mediated demyelination. This hypothesis is consistent with the CNS pathologic changes observed in mice exhibiting TMEV-induced demyelinating disease [5]. This hypothesis is also supported by the classic observations of MacKaness [19] which demonstrated nonspecific resistance to *Mycobacterium* in naive recipients of *Listeria*-immune T cells infected with both viable *Mycobacterium* and *Listeria*. Regarding demyelination, both antigen-specific in vivo-derived T cells and in vitro-propagated T cell lines have been shown to cause bystander CNS damage via macrophage activation. Wisniewski and Bloom [20] showed that CNS and peripheral nervous system myelin can be damaged as a nonspecific consequence of a specific DTH reaction directed at non-nervous-tissue antigens, namely purified protein derivative (PPD) of tuberculin. More recently, Holoshitz et al. [21] showed that encephalitis could be produced in mice by the intravenous transfer of PPD-specific T cell lines following IC inoculation of PPD. It should be pointed out, however, that a recent study designed to evaluate bystander demyelination in peripheral nervous tissue was unsuccessful [22].

Alternatively to or concomitant with TMEV-specific DTH, if there is extensive infection of oligodendrocytes, demyelination may result from immune injury to these cells, as they may express TMEV antigens in conjunction with H-2 class I determinants. Because H-2 class I region determinants (e.g., H-2D) restrict the development and expression of Lyt-2$^+$ cytotoxic T cells (Tc) to allogeneic and virus-infected syngeneic cells, Tc could kill infected oligodendrocytes. This may seem unlikely because TMEV is a member of the picornavirus family, and no picornavirus is known to bud from, or insert virus-encoded proteins into, the host cell membrane. However, if one considers recent experiments showing that Tc can kill virus-infected cells in which glycosylated virus proteins are not readily detected on the cell membrane [23–26], and a recent report [27] describing Tc activity against coxsackievirus B3, this may be a feasible mechanism. However, the pathogenic role of TMEV-specific Tc responses remains in question, as such responses have not been reported in susceptible mice and widespread destruction of oligodendrocytes is not apparent histologically [5,14,15].

Finally, TMEV infection might trigger an autoimmune reaction comparable to that occurring in EAE, and this reaction may contribute, in part, to the demyelinating process. Although peripheral T cell proliferative responses against MBP and mouse whole-cord homogenate have not been demonstrated in TMEV infection [18,28], their contribution cannot be ruled out. Such an autoimmune response against CNS antigens could result from one of several mechanisms: (a) direct damage to CNS constituents resulting from cytopathologic effects of TMEV, (b) damage to CNS constituents as a result of TMEV-specific cell-mediated immunity (CMI) responses, and/or (c) cross-reactive immune responses between viral and CNS antigens. Although it has been recognized that viruses may share antigenic sites with normal host-cell components, homology between viral and nervous system myelin antigens has only been demonstrated recently [29,30]. Fujinami and Oldstone [30], who found amino acid sequence homology between the encephalitogenic site of rabbit myelin basic protein and the hepatitis B virus polymerase showed not only that immune responses were generated in rabbits by the viral peptide that cross-reacted with the self protein, but also that mononuclear cell infiltration was present in the CNS of animals immunized with the viral peptide. This concept of molecular mimicry, i.e., sharing of host sequences with viral sequences has been reviewed by Oldstone and Notkins (Chapter 23, this volume). In support of a molecular mimicry pathogenesis, Fujinami (unpublished results) has recently identified a monoclonal antibody from TMEV-infected animals that both neutralizes TMEV and reacts with galactocerebroside, a surface component on myelin. Further, the antibody on inoculation causes demyelination.

Acknowledgments

This work was supported by U.S. Public Health Service Grants NS-21913, A1-18755, and A1-16919 and by the Earl M Bane Fund. RSF is a Harry Weaver Scholar of the National Multiple Sclerosis Society—JF2009 and GRNMSF 1780 A-2.

References

1. Lipton HL (1975) Theiler's virus infection in mice: An unusual biphasic disease process leading to demyelination. Infect Immun 11:1147–1155
2. Lehrich JR, Arnason BGW, Hochberg F (1976) Demyelinative myelopathy in mice induced by the DA virus. J Neurol Sci 29:149–160
3. Lipton HL, Dal Canto (1979) The T0 strains of Theiler's viruses cause "slow viruslike" infections in mice. Ann Neurol 6:25–28
4. Theiler M (1937) Spontaneous encephalomyelitis of mice, a new virus disease. J Exp Med 65:705–719
5. Dal Canto MC, Lipton HL (1976) Primary demyelination in Theiler's virus infection. J Lab Invest 33:626–637
6. Dal Canto MC, Lipton HL (1980) Schwann cell remyelination and recurrent demyelination in the central nervous system of mice infected with attenuated Theiler's virus. Am J Pathol 98:101–110
7. Dal Canto MC, Barbano RL (1985) Immunocytochemical localization of MAG, MBP, P_0 and P_0 protein in acute and relapsing demyelinating lesions of Theiler's virus infection. J Neuroimmunol 10:129–140
8. Lipton HL, Melvold RM (1984) Genetic analysis of susceptibility to Theiler's virus-induced demyelinating disease in mice. J Immunol 132:1821–1825
9. Clatch RJ, Melvold RW, Miller SD, Lipton HL (1985) Theiler's murine encephalomyelitis virus (TMEV)-induced demyelinating disease in mice is influenced by the H-2D region: Correlation with TMEV-specific delayed-type hypersensitivity. J Immunol 135:1408–1414
10. Rodriguez M, David CS (1985) Demyelination induced by Theiler's virus: Influence of the H-2 haplotype. J Immunol 135:2145–2148
11. Melvold RM, Knobler RL, Lipton HL (1986) (submitted)
12. Brahic M, Haase AT, Cash E (1984) Simultaneous in situ detection of viral RNA and antigens. Proc Natl Acad Sci USA 81:5445–5448
13. Cash E, Chomorro M, Brahic M (1985) Theiler's virus RNA and protein synthesis in the central nervous system of demyelinating mice. Virology 144:290–294
14. Dal Canto MC, Lipton HL (1982) Ultrastructural immunohistochemical localization of virus in acute and chronic demyelinating Theiler's virus infection. Am J Pathol 106:20–29
15. Rodriguez M, Liebowitz JL, Lampert PW (1983) Persistent infection of oligodendrocytes in Theiler's virus-induced encephalomyelitis. Ann Neurol 13:426–433

16. Brahic M, Stroop WG, Baringer JR(1981) Theiler's virus persists in glial cells during demyelinating disease. Cell 26:123–128
17. Chamarro M, Aubert C, Brahic M (1986) The demyelinating lesions due to Theiler's virus are associated with ongoing CNS infection. J Virol 57:992–997
18. Clatch RJ, Lipton HL, Miller SD (1986) Characterization of Theiler's murine encephalomyelitis virus (TMEV)-specific delayed-type hypersensitivity responses in TMEV-induced demyelinating disease: Correlation with clinical signs. J Immunol 136:920–927
19. MacKaness GB (1964) The immunologic basis of acquired cellular resistance. J Exp Med 120:105–120
20. Wisniewski HM, Bloom BR (1975) Primary demyelination as a nonspecific consequence of a cell-mediated immune reaction. J Exp Med 141:346–359
21. Holoshitz J, Naparstek Y, Ben–Nun A, Marquardt P, Cohen IR (1984) T lymphocyte lines induce autoimmune encephalomyelitis, delayed hypersensitivity, and bystander encephalitis or arthritis. Eur J Immunol 14:729–734
22. Powell HC, Braheny SL, Hughes RAC, Lampert PW (1984) Antigen-specific demyelination and significance of the bystander effect in peripheral nerves. Am J Pathol 114:443–453
23. Reddehase MJ, Koszinowski UH (1984) Significance of herpesvirus immediate early gene expression in cellular immunity to cytomegalovirus infection. Nature 131:369–371
24. Gooding LR, O'Connell KA (1983) Recognition by cytotoxic T lymphocytes of cells expressing fragments of the SV40 tumor antigen. J Immunol 131:2580–2586
25. Tevethia SS, Tevethia MJ, Lewis AJ, Reddy UB, Weissman SM (1983) Biology of simian virus 40 (SV40) transplantation antigen (TrAg). IX. Analysis of TrAg in mouse cells synthesizing truncated SV40 large T antigen. Virology 128:319–330
26. Townsend ARM, McMichael AJ, Carter NP, Huddleston JA, Brownlee CG (1984) Cytotoxic T cell recognition of the influenza nucleoprotein and hemagglutinin expressed in tranfected mouse L cells. Cell 39:13–25
27. Huber SA, Job LP, Woodruff JF (1984) In vitro culture of coxsackievirus group B, type 3 immune spleen cells on infected endothelial cells and biologic activity of the cultured cells in vivo. Infect Immun 43:567–573
28. Barbano RL, Dal Canto MC (1984) Serum and cells from Theiler's virus-infected mice fail to injure myelinating cultures or to produce in vitro transfer to disease. J Neurol Sci 66: 283–293
29. Jahnke U, Fischer EH, Alvord EC Jr (1985) Sequence homology between certain viral proteins and proteins related to encephalomyelitis and neuritis. Science 229:282–284
30. Fujinami RS, Oldstone MBA (1985) Amino acid homology between the encephalogenic site of myelin basic protein and virus: A mechanism for autoimmunity. Science 230:1043–1045

Evolving Concepts in Viral Pathogenesis Illustrated by Selected Diseases in Humans

CHAPTER 30
Human T Cell Lymphotropic Viruses and Their Family of Diseases

L. RATNER AND ROBERT C. GALLO

Three classes of exogenous human retroviruses have been isolated, and designated human T-cell lymphotropic virus (HTLV) types I, II, and III. HTLV-I and -II are immortalizing and weakly cytopathic viruses, whereas HTLV-III is a nonimmortalizing, strongly cytopathic virus. HTLV-I is etiologically related to adult T-cell leukemia/lymphoma (ATLL). HTLV-III is the cause of the acquired immunodeficiency syndrome (AIDS). HTLV-II has been isolated on three occasions, from patients with a T cell variant of hairy cell leukemia, an intravenous drug abuser, and a hemophiliac. Because of the rarity of its detection, no clear disease association has been established for HTLV-II.

The isolation of HTLV-I depended on the establishment and maintenance of malignant T lymphocytes in culture. This process utilized the addition to the media of T cell growth factor (TCGF) also known as interleukin 2 (IL-2). By means of these techniques, HTLV-I can be readily cultured from malignant T cell lines established from tissues derived from patients with ATLL. This disease is generally considered to be a form of non-Hodgkin's lymphoma, characterized at the time of diagnosis by frequent infiltration of lymph nodes, liver, spleen, bone marrow, and peripheral blood. The most characteristic findings, however, are dermal and epidermal malignant infiltrates and hypercalcemia. The disease is clinically aggressive; it is poorly responsive to chemotherapy, and the median survival is about six months. The tumor cells in vivo and in vitro are pleomorphic, often with a multilobulated nucleus, and are generally OKT4(+), though they function as suppressor cells or lack immunologic function.

Patients with ATLL almost universally possess antibodies directed against *gag* and *env* protein products of HTLV-I. However, only 0.05–0.1% of individuals infected with HTLV-I develop ATLL. The prevalence of antibodies is three- to four-fold higher in family members of patients with ATLL than in the general population. HTLV-I infection is endemic in southern Japan, the Caribbean basin (including parts of the southeastern United States, Central America, and South America), and many parts of Africa. There is epidemiologic and sociologic data to implicate Africa as the origin of this retrovirus. This is supported by the identification of a virus structurally and antigenically related to HTLV-I in Old World primates; this virus is designated simian T-cell lymphotropic virus (STLV) type I.

HTLV-I can rapidly transform in vitro normal T lymphocytes derived from umbilical cord blood, bone marrow, or peripheral blood. The predominant infected cell in vivo is the T lymphocyte although HTLV-I does have the potential to infect some B lymphocytes and endothelial cells. One to three copies of the proviral form of HTLV-I are found in the ATLL tumor cells and are integrated in a mono- or oligoclonal fashion. However, no common integration site has been defined in malignant cells taken from different patients.

The integrated provirus of HTLV-I is 9032 base pairs (bp) in length. The LTR sequence is 754 bp in length and has several distinctive features compared to other retroviruses, including (a) a large R region of 221 bp, and (b) a polyadenylation signal located 276 bp upstream of the polyadenylation site. The genome includes gag, pol, and env genes, as well as an additional 3′ region of about 1500 bp, designated pX. A 42-kilodalton (kd) protein encoded by this region, designated X-lor, is translated from a double-spliced mRNA. The first amino acid is donated from the first codon of the env gene, and the remainder of the open reading frame corresponds to the 3′ 1067 bp of the pX region. The X-lor product encodes or induces a factor that activates 100 fold the transcriptional activity of the HTLV-I LTR. This activity is likely to be important in virus replication and cellular transformation.

Several other genes are activated by HTLV-I infection, including two designated HT3 and JD15, as well as c-sis and that for the TCGF receptor. The role of these gene products in transformation remains to be defined.

HTLV-I is one of the clearest examples of a viral cause of human cancer. This claim is based on (a) the clear epidemiological links of this virus to ATLL, (b) the in vitro transformation of primary human T cells, (c) the molecular biological results showing clonal integration of the provirus in the tumor cells, and (d) the association of a closely related virus with malignant lymphomas in monkeys.

Though HTLV-II, on the nucleotide level, is only 65% homologous to HTLV-I, it is functionally quite similar. The general structure of the viral genome parallels that described for HTLV-I. Of note is the similarity in structure and function of bovine leukemia virus, the cause of bovine lymphosarcoma, to that of HTLV-I and -II.

HTLV-III, or the lymphadenopathy associated virus (LAV), has been shown to be the cause of AIDS and the AIDS-related complex (ARC) of syndromes. This virus has been isolated by use of a permissive immortalized

OKT4(+) cell line that is relatively resistant to the cytopathic effects of the virus. Antibodies to HTLV-III may be detected by an enzyme-linked immunosorbent assay (ELISA) using detergent-lysed virus, and is confirmed by the demonstration, by Western blot, of reactivity to individual HTLV-III proteins. These specific antibodies have been demonstrated in 85–95% of patients with AIDS, ARC, or mothers of juvenile AIDS patients, 60–80% of hemophiliacs and intravenous drug abusers, 30–50% of asymptomatic homosexuals, and less than 1% of asymptomatic heterosexuals. The lag period for positive conversion of HTLV-III antibodies is two to 60 months. A small proportion of HTLV-III-infected individuals are antibody negative.

HTLV-III infects OKT4 antigen-bearing lymphoid cells, and is cytopathic for normal T lymphocytes. Viral sequences can be detected in lymphoid organs of infected individuals, though not reproducibly because of the lack of sensitivity of the current Southern blot technique. HTLV-III sequences have also been identified in brain samples from AIDS patients, most likely glial tissues.

The integrated proviral genome is 9729–9749 bp in length and varies by about 1–10% among different isolates. The tRNA primer binding site following the 5' long terminal repeat (LTR) is unusual in that it, unlike most other mammalian retroviruses, is complementary to the 3'-end of tRNA lysine. The genome includes genes for *gag, pol,* and *env* as well as two other open reading frames, *sor* and 3' *orf,* with 203 and 216 codons, respectively, whose function remains to be determined. An additional viral protein encodes or induces the expression of a factor that is responsible for trans-direct gene expression activation of the LTR. This protein is encoded by a 2.0-kb double-spliced mRNA. The open reading frame is derived partly from a region between *sor* and *env,* together with an alternative reading frame in the 3' portion of the *env* gene.

Variants of HTLV-III differ from each other more in the *env* gene than other regions. The most notable differences are in the extracellular portion of the viral envelope, which may result in decreased antigenic cross-reactivity of different HTLV-III isolates.

A retrovirus isolated from macaques with simian AIDS, designated STLV-III, shows antigenic cross-reactivity with HTLV-III and is functionally similar. Structural homology has also been demonstrated between HTLV-III and visna, a cytopathic retrovirus in sheep and goats. Thus, the retroviruses discussed in this chapter are structurally, functionally, and/or phylogenetically related to one another.

References

1. Robert-Guroff M, Sarngadharan NG, Gallo RC (1982) T-cell growth factor. *In* Growth and Maturation Factors. Wiley & Sons, New York, pp 267–309
2. Saxinger WC, Gallo RC (1985) Human T-cell growth factor (TCGF): Its discovery, properties, and some basic and applied uses in the long term propagation of human mature T-cells. *In* Human Cancer Immunology. North Holland Press (in press)

3. Gallo RC (1984) Human T-cell leukemia–lymphoma virus and T cell malignancies in adults. Cancer Surveys 3:113–159
4. Wong-Staal F, Gallo RC (1985) The family of human T-lymphotropic leukemia viruses: HTLV-I as the cause of adult T-cell leukemia and HTLV-III as the cause of acquired immunodeficiency syndrome. Blood 65:253–267
5. Wong-Staal F, Gallo RC (1985) Human T-lymphotropic retroviruses (HTLV). Nature 317:395–403

CHAPTER 31
Pathogenesis of Parvovirus-induced Disease in Humans

M.J. ANDERSON, J.R. PATTISON, AND NEAL S. YOUNG

Molecular Aspects

In 1975, Cossart and colleagues [1] described parvovirus-like particles in human sera. The subsequent indentification of the genome as a single piece of single-stranded DNA of about 5.5 kilobases [2] encapsidated within two or three proteins (MW 48,000, 80,000, and possibly 68,000) ([3], and unpublished) has confirmed that this virus belongs to the family Parvoviridae. More recent works suggests that although the virus has not been propagated in tissue culture, it is likely to be capable of autonomous replication; the terminal palindromic sequences of the 3' end are not complementary to those at the 5' end, contrary to findings in dependoviruses [4].

Clinical Epidemiology

Infection with this parvovirus is common (60–70% of adults are seropositive), and it has been found to occur wherever it has been sought in Europe, America, Africa, Australia, and the Middle and Far East. For many years the clinical consequences of infection remained obscure. At first the virus was found in occasional patients suffering minor nonspecific febrile illnesses with leukopenia, but infection often seemed to be wholly asymptomatic. However, one or two of the original viremic blood donors did develop a rubella-like illness a week or so after donation.

In 1983, the significance of these isolated reports of rash illness became clear when an outbreak of erythema infectiosum (fifth disease) in London, England, was shown to be due to parvovirus infection [5]. Erythema infectiosum is a common disease in children aged 7–14 years for which a viral etiology has long been suspected. The first sign of illness is often marked erythema of the cheeks (slapped cheek appearance), followed by an erythematous maculopapular rash. Following the initial description of the association of human parvovirus with erythema infectiosum in London, outbreaks in Canada, Japan, the United States, and Sweden have been found to share this etiology. However, sporadic cases also occur, affecting both children and adults throughout the year. In adults the disease is similar but joint involvement (arthritis/arthralgia) is more common (70–80% of cases as opposed to 10% of childhood cases). Infections without a rash occur in both children and adults, and, because joint involvement is more common in the latter, adults with parvovirus infection are sometimes first seen in rheumatology clinics with joints symptoms [6].

Paradoxically, a complication of human parvovirus infection was found two years before the common manifestation of infection. Transient aplastic crisis is a common complication of chronic hemolytic anemias, characterized by a sudden but temporary disappearance of erythroblasts from the bone marrow and of reticulocytes from the peripheral blood, and a sharp drop in hemoglobin leading to presentation with symptoms of severe anemia. Like erythema infectiosum, aplastic crisis had long been suspected of being infectious in origin but had come to be regarded as the common result of a variety of infections. The association between human parvovirus infection and aplastic crisis was initially demonstrated in sickle cell anemia in 1981 in both London, England and Jamaica [7,8] by reports confirming the presence of the virus in the acute-phase serum specimens or showing seroconversion at the time of the crisis. With the development of sensitive means of parvovirus diagnosis, it has become apparent that human parvovirus infection is responsible for more than 90% of aplastic crises in such diverse conditions as hereditary spherocytosis, pyruvate kinase deficiency and β-thalassemia intermedia, as well as in sickle cell anemia.

Thus evidence to date indicates a spectrum of disease associated with human parvovirus infection. It may be asymptomatic or associated with either a mild febrile illness with flulike symptoms, an erythematous rash complicated by arthralgia, arthralgia alone, or aplastic crisis in individuals with chronic hemolytic anemia. Experiments both in volunteers and in vitro are beginning to resolve the problems of pathogenesis posed by this array of possible clinical outcomes of infection.

In Vivo Studies in Volunteers

The sequence of events occurring in human parvovirus infection has been examined by studying infection in adult volunteers [9]. Following intranasal inoculation (Figure 31.1), an intense viremia developed, reaching peak titers corresponding to 10^{11} genome copies/ml serum, comparable to those found in asymptomatic blood donors or patients with aplastic crisis. At the time

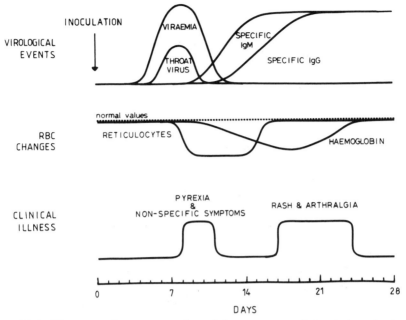

Figure 31.1. Diagrammatic representation of the sequence of events in volunteers following intranasal inoculation of human parvovirus.

of the viremia, virus was also detectable in upper respiratory tract secretions, but not in urine or feces. It would seem, therefore, that virus enters by and is shed from the upper respiratory tract.

Hematological changes occurred in the second week after inoculation (Figure 31.1). Reticulocytes disappeared from the peripheral blood for 7–10 days, and hemoglobin levels fell by an average of 0.18 g/dl/day. These changes are consistent with interruption of erythropoiesis in the bone marrow, as is seen in aplastic crisis. Lymphocyte, neutrophil, and platelet levels also showed a transient but significant drop; the pathogenesis of these changes remains to be determined, but in vitro studies (see below) suggest no direct effect of the virus on the progenitors of these cells.

Infected volunteers became ill at two quite distinct times following inoculation. The first phase of illness consisted of fever accompanied by a combination of malaise, myalgia, chills, or itching. These symptoms were temporally associated with viremia and especially, circulating IgM/virus immune complexes. Circulating interferon could not be detected at this time. The second phase of illness commenced 17–18 days after inoculation: A fine, pink, maculopapular rash appeared on the limbs and/or trunk, and arthralgia and some joint swelling commenced on the following day—all typical of erythema infectiosum. The rash faded after 2–3 days, and the joint symptoms resolved a day or two later.

This sequence of events accords well with observations of naturally occurring cases of erythema infectiosum and aplastic crisis. The hematological changes commence during the viremia, and virus is often found when cases of aplastic crisis first present [10]. The rash illness, however, occurs 5–7 days after the clearance of virus, at a time when virus-specific IgM is already detectable. This accords well with the failure to detect virus in cases of erythema infectiosum and suggests that the specific diagnosis can be made only by detecting parvovirus-specific IgM.

Two Diseases Occurring in a Natural Outbreak

The above findings in volunteers have recently been reflected in an extensive outbreak of parvovirus in Cleveland and surrounding Cuyahoga county, in the United States [11,12]. In a six-month period during 1984, there were 26 cases of transient aplastic crisis and more than 450 cases of erythema infectiosum. Sera collected from cases and controls at the same time and in the same geographic area clearly showed an association between parvovirus infection and the occurrence of both transient aplastic crisis and erythema infectiosum. About one-fourth of the patients with transient aplastic crisis also had a skin rash. The disease spread among household contacts, and virus was detected in urine and throat washings of affected individuals. Viremia occurred in black patients with sickle cell disease and in white patients with hereditary spherocytosis who later developed aplastic crises. High-concentration viremia in the patients with transient aplastic crisis was present during a limited period, from five days before to one day following the reticulocyte nadir; low-concentration viremia was detected as long as nine days after the reticulocyte nadir. IgM antibody levels were highest at the time of the reticulocyte nadir and for the following 10 days, whereas IgG titers rose later and remained elevated for at least three to four months. Titers of antigen or IgM antibody were markedly higher in patients with transient aplastic crisis than in those with fifth disease, consistent with amplification of virus proliferation in the active erythroid bone marrows of patients with hemolytic anemias, but this may have been related to the time of blood sampling.

In Vitro Studies with Human Bone Marrow

The cellular basis of the hematological changes associated with parvovirus infection has been investigated by studying the interaction of virus with bone marrow cells in vitro. The cells of the bone marrow actively proliferate to provide the circulating blood elements and to replenish the pool of hematopoietic stem cells. Aspects of this process can be studied in tissue culture systems where discrete colonies of cells committed to erythroid, granulocytic, monocytic, or megakaryocytic pathways of differentiation are formed from hematopoietic progenitors.

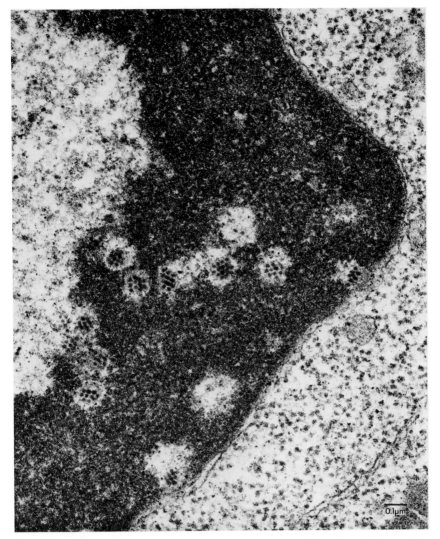

Figure 31.2. Parvovirus in human erythroid progenitor cell nucleus 48 hours after in vitro infection (Magnification, 100,000 ×).

In early experiments, sera from the original viremic blood donors [13] were found to selectively inhibit erythropoiesis (burst forming unit-erythroid [BFU-E] colony formation by the primitive progenitor as well as colony forming unit-erythroid [CFU-E] colony formation by the more mature progenitor). Myelopoiesis was unaffected. This inhibitory effect could be prevented by the addition of convalescent-phase sera containing anti-parvovirus IgM antibody, but not by antisera to either interferon or human immunoglobulin. Moreover, when virus was separated from serum proteins by ultracentrifugation, only the virus-containing fraction exhibited inhibitory activity.

The development of erythroid cells in culture is dependent on the presence of certain growth factors, some of which are provided by accessory cells within the heterogeneous bone marrow cell mix. To determine whether the virus acted directly upon erythroid progenitor cells or on an accessory cell, erythroid progenitors in the form of immature bursts formed from BFU-E were removed from primary cultures and replated together with parvovirus. In the presence of virus, secondary erythroid colony formation was inhibited [14]. Moreover, 48 hours after the addition of virus to these secondary cultures, viral antigen was demonstrable by immunofluorescence, and electron microscopy revealed parvovirus-like particles in crystalline assays within the nuclei (Figure 31.2) [15]. Infected cells showed ultrastructural signs of lysis, including vacuolization, pseudopod formation, abnormal mitochondrial structure, nucleolar degeneration, and chromatin margination. The ability of parvovirus-containing serum to inhibit erythropoiesis was first demonstrated by means of sera from asymptomatic blood donors. However, all serum specimens containing the virus, revealed either by radioimmunoassay for viral antigen or Southern analysis of DNA for parvovirus sequences, also inhibit erythroid colony formation.

Studies of the interaction of parvovirus with bone marrow erythroid progenitor cells have permitted the elucidation of certan biological properties of the virus. Inhibition of erythropoiesis is unaffected by lipid solvents but is abrogated by proteinase treatment, and the activity is resistant to heating (50°C for 30 minutes). These properties are those expected of a member of the Parvoviridae.

Comment

Evidence to date indicates that the same human parvovirus causes a rash illness and transient aplastic crisis. The precise pathogenesis of the rash has not yet been investigated, but it is likely to be immune mediated because it occurs after the intense viremia and at a time when a specific immune response has developed. Transient aplastic crisis, on the other hand, is a consequence of lytic infection of erythroid progenitors in the bone marrow. To date, there is no evidence of recent infection with this virus in other bonemarrow failure states including aplastic anemia, transient erythroblastopenia of childhood, paroxysmal nocturnal hemaglobinuria, and adult pure red-cell aplasia [14]. However, parvoviruses can persist by integration of viral DNA into their host cell's genome [16], and in animals persistent parvovirus infection can lead to pathogenic immune responses, for example, Aleutian disease of mink [17]. Thus the human parvovirus remains an excellent candidate as a causative agent of chronic erythropoietic failure states and chronic polyarthropathies. With respect to the latter a second, apparently unrelated, parvovirus, designated RA-1, has been described in association with rheumatoid arthritis [18]. Experiments to detect sequences homologous to cloned parvovirus in the peripheral blood and bone marrow cells of normal individuals and patients with hematologic and rheumatic disease are in progress.

Two characteristics of human parvovirus infection, the intense viremia and requirement of the virus for dividing host cells, suggest a poor fetal outlook if infection occurs during pregnancy. The virus has been shown to cross the placenta [19,20], and the frequency of parvovirus-specific IgM is significantly lower in pregnant women than in nonpregnant women of the same ages. Moreover, among those women found to have evidence of recent parvovirus infection, spontaneous abortion is not an infrequent occurrence. Numbers of cases studied are as yet small but suggest that fetal loss rather than physical abnormality may result from in utero infection with this virus.

References

1. Cossart JE, Field AM, Cant B, Widdows D (1975) Parvovirus-like particles in human sera. Lancet i:72–73
2. Summers J, Jones SE, Anderson MJ (1983) Characterisation of the agent of erythrocyte aplasia as a human parvovirus. J Gen Virol 64:2527–2532
3. Clewley JP (1984) Biochemical characterisation of a human parvovirus. J Gen Virol 65:241–244
4. Cotmore S, Tattersall P (1984) Characterisation and molecular cloning of a human parvovirus genome. Science 226:1161–1165
5. Anderson MJ, Lewis E, Kidd IM, Hall SM, Cohen BJ (1984) An outbreak of erythema infectiosum associated with human parvovirus infection. J Hyg (Camb) 93:85–93
6. White DG, Woolf AD, Mortimer PP, Cohen BJ, Blake DR, Bacon PA (1985) Human parvovirus arthropathy. Lancet ii:419–421
7. Pattison JR, Jones SE, Hodgson J, Davis LR, White JM, Stroud CE, Murtaza L (1981) Parvovirus infections and hypoplastic crises in sickle cell anaemia. Lancet i:664–665
8. Serjeant GR, Mason K, Topley JM, Serjeant BM, Pattison JR, Jones SE, Mohammed R (1981) Outbreak of aplastic crisis in sickle cell anaemia associated with parvovirus-like agent. Lancet ii:595–597
9. Anderson MJ, Higgins PG, Davis LR, Willman JS, Jones SE, Kidd IM, Pattison JR, Tyrrell DAJ (1985) Experimental parvovirus infection in humans. J Infect Dis 152:257–265
10. Anderson MJ, Jones SE, Minson AC (1985) Diagnosis of human parvovirus infection by dot-blot hybridization using cloned viral DNA. J Med Virol 15:163–172
11. Chourba TL, Coccia P, Holman RC, Tattersall P, Anderson LJ, Sudman J, Young NS, Kurcyznski E, Palmer E, Jason J, Evatt B (1986) Parvovirus B19—the cause of epidemic aplastic crisis and erythema infectiosum (fifth disease). J Infect Dis (in press)

12. Saarinen UM, Chourba TL, Tattersall P, Young NS, Anderson LJ, Coccia P. (1986) Human parvovirus-induced epidemic red cell aplasia in patients with hereditary hemolytic anemia. Blood 67:1411–1417
13. Mortimer PP, Humphries RK, Moore JG, Purcell RH, Young NS (1983) A human parvovirus-like virus inhibits hematopoietic colony formation in vitro. Nature 302:426–429
14. Young NS, Mortimer PP, Moore JG, Humphries RK (1984) Characterization of a virus that causes transient aplastic crisis. J Clin Invest 73:224–230
15. Young NS, Harrison M, Moore JG, Mortimer PP, Humphries RK (1984) Direct demonstration of the human parvovirus in erythroid progenitor cells infected in vitro. J Clin Invest 74:2024–2032
16. Berns KI, Cheung AK-M, Ostrove JM, Lewis M (1982) Adeno-associated virus latent infection. In Mahy BWJ, Minson AC, Darby GK (eds) Virus Persistence. Cambridge University Press, Cambridge, p 249
17. Porter DD, Ho HJ (1980) Aleutian disease of mink: A model for persistent infection. In Fraenkel-Conrat H, Wagner RR (eds) Comprehensive Virology, Vol 1:Virus-Host Interactions. Plenum, New York, p 233
18. Simpson RW, McGinty L, Simon L, Smith CA, Godzeski CW, Boyd RJ (1984) Association of parvoviruses with rheumatoid arthritis of humans. Science 223:1425–1428
19. Knott PD, Welply GAC, Anderson MJ (1984) Serologically proven intrauterine infection with parvovirus. Br Med J 289:1660
20. Brown T, Anand A. Ritchie LD, Clewley JP, Reid TMS (1984) Intrauterine human parvovirus infection and hydrops fetalis. Lancet ii:1033–1034

CHAPTER 32
Hepatitis B Virus and Cancer

MARK A. FEITELSON

There is much evidence suggesting a close association between hepatitis B virus (HBV) infection and the development of primary hepatocellular carcinoma (PHC), although the mechanism by which HBV participates in the development of PHC remains to be elucidated. The importance of this relationship to public health can be appreciated by the fact that PHC is the most common cancer in the world (at least 250,000 new cases annually), that there are approximately 200 million HBV carriers worldwide, and that persistent infection with HBV may be required for the development of PHC [1–3]. The latter point is supported by much epidemiological data, including a study showing that the relative risk of HBV carriers developing PHC as compared to noncarriers is more than 200, which is one of the highest relative risks known for a human cancer [4]. In some populations, the lifetime risk of chronic carriers developing PHC is as high as 50% in males, and almost half of all deaths in carriers 40 years or older have been attributed to PHC [4,5]. A better understanding of this relationship at cellular and molecular levels, then, will have a significant impact upon public health.

The lengthy period of HBV carriage (as much as 40 years or more) prior to the appearance of PHC is consistent with a multistep mechanism leading to carcinogenesis. Although known hepatic cocarcinogens such as aflatoxin B, alcohol, vinyl chloride, hormones, and cottonseed oil may alter the frequency in which PHC occurs in different HBV carrier populations [1,6], they alone do not fully explain the epidemiological data. However, the predominance of PHC in HBV carrier males as compared to females is consistent with a significant role for sex-related factors that may contribute to the outcome of infection [4]. The finding of a glucocorticoid responsive element in HBV DNA [30] may partially explain the sex differences in the outcome of HBV infection if such an element is differentially sensitive to the action of male and female sex hormones. The appearance of most PHC in carriers with long-term HBV infection suggests that one or more characteristics of persistent infection may be contributing to the risk of developing liver cancer. For example, because an important characteristic of chronic liver disease is the continual replacement of damaged or destroyed hepatocytes by regeneration, the actions of HBV, liver cocarcinogens, or both during this period may be crucial in establishing the genetic changes in hepatocytes prerequisite for PHC development. Thus, the ability of infected hepatocytes to effectively repair DNA damage prior to replication may be relevant to the initiation stage of carcinogenesis, as it is with chemically induced PHC [6]. Alternating waves of necrosis and regeneration, seen in chronic liver disease over many years, often result in hyperplastic foci of phenotypically dedifferentiated liver cells [6]. It is possible that these changes may alter cellular responsiveness to the actions of HBV, cocarcinogens, or both [6,7]. In addition, the role of the delta agent in the pathogenesis of HBV associated PHC needs to be evaluated because coinfection of HBV carriers with the delta agent often results in exacerbation of chronic liver disease [8]. However, the results thus far suggest little correlation between delta superinfection and the frequency of PHC [18].

The following observations suggest that host-virus interaction may depend upon the state of hepatocyte differentiation:

1. Only some cells in an HBV-infected liver produce enough of one or more viral antigens to be detected immunochemically [9].
2. PHC cells rarely demonstrate virus gene expression, whereas surrounding nontumorous tissue often does [9].
3. Duck embryos infected with an HBV-like duck hepatitis B virus (DHBV) do not replicate the virus until a particular stage of liver development in the embryo [10].

A recent proposal suggesting the presence of at least two types of liver cells, one type being a mature hepatocyte capable of supporting viral replication, and the other type being an immature cell unable to support HBV replication, provides a cellular model for the pathogenesis of HBV-associated PHC as well as putative explanations for a variety of other phenomena observed in HBV infection [2]. In this model, mature hepatocytes expressing viral gene products may be eliminated by the immune system during active hepatitis, whereas immature cells may escape elimination and establish a persistent infection with the virus. Such expansion of immature cells may ultimately give rise to clonally derived tumors characteristic of PHC.

The close association of HBV and PHC is also observed in woodchucks chronically infected with an HBV-like agent, i.e., woodchuck hepatitis virus (WHV) [11]. Most infected animals retained in captivity develop chronic active hepatitis and eventually die of PHC. In both cases, viral DNA or DNA fragments have been found integrated into the host genome in many chronically infected liver tissues and most PHC nodules examined [12,13].

A number of cell lines derived from HBV-infected patients with PHC also demonstrate integrated HBV DNA [12]. Although the number of viral integration sites as well as the cellular DNA sequences flanking each site seems to vary among the different tumors and cell lines studied, analysis of the DNA by Southern blot hybridization strongly suggests that PHC nodules are clonally derived [12]. In chronic hepatitis, the inserts appear to be colinear with the viral genome, whereas in PHC the viral sequences often appear to be highly rearranged [13]. Although there seems to be no specificity with regard to integration into host DNA sequences, there appears to be a restricted region of viral DNA where integration preferentially occurs [19]. Whether this restricted region of integration with respect to the viral DNA is important in the mechanism by which HBV and WHV are involved in liver carcinogenesis is not known. However, the presence of hepatitis B surface antigen (which includes the viral encoded envelope proteins and glycoproteins in a lipid bilayer and which is secreted as 22-nm diameter particles in the blood of many HBV-associated PHC patients and in the culture supernatants of HBV-associated human hepatoma cells lines) demonstrates that HBV DNA integrated into tumor cells often retains some capacity for gene expression. Perhaps integration is restricted to a certain region in the viral DNA in order to preserve the presence of recently identified glucocorticoid responses [30] and/or enhancer sequences [15] of HBV (which may stimulate or turn on the expression of nearby viral and/or host genes) in the integrated state.

These findings are important because many DNA and RNA tumor viruses integrate into host DNA and produce one or more viral gene products that participate(s) in the initiation and/or maintenance of the transformed state [20]. The finding that HBV and HBV-like viruses replicate through an RNA intermediate via reverse transcription suggests that their life cycle, in part, may be a temporally permuted version of the retroviral life cycle [12]. The question remains whether there are also analogies in the mechanism(s) by which HBV and retroviruses induce neoplasia. Despite extensive characterization of the integrated viral DNAs in infected human and woodchuck livers as well as in several human hepatoma cell lines, the role of these integrated sequences in the etiology of PHC remains obscure. It is unlikely that HBV encodes its own transforming gene product in a manner analogous to that documented for acutely transforming retroviruses or papovaviruses because (a) there is a very lengthy incubation time between infection and appearance of PHC, and (b) no consistent portion of the viral genome is present in all the cloned isolates so far analyzed. On the other hand, the data thus far do not exclude the participation of one or more HBV gene products in the initiation phase of liver carcinogenesis [6]. One possibility involves expression of all or a portion of the HBV nucleocapsid (HBcAg or core) polypeptide in small amounts in tumors. The major HBV nucleocapsid polypeptide is a DNA binding protein [21] that can be phosphorylated at one or more serine and threonine residues [22]. In a hepatoma cell line (PLC/PRF/5), the core gene is highly methylated, as compared to the same gene in the virus, suggesting that its expression in integrated DNA may be regulated this way [23]. Treatment of these cells with 5-azacytidine has apparently resulted in core gene expression [24]. Although conventional immunological staining rarely demonstrates any HBV expression in PHC nodules, and intact core gene sequences are often lacking in cloned inserts, it is still possible that small amounts of core alone, or as a fusion polypeptide with viral or host sequences, may be important in transformation. A role for protein serine–theonine kinases in transformation (by the demonstrable action of the c-*ras* gene products upon the activity of adenyl cyclase and consequently serine phosphorylation) has recently been suggested [20]. At this writing, however, such a model is purely speculative.

Published data concerning the structural characterization of integrated HBV and WHV DNAs suggest some similarities with a number of other models of viral carcinogenesis. For example, some retroviruses that do not possess an oncogene give rise to cancers after integration and a lengthy incubation period by means of a promotion–insertion mechanism by which a cellular oncogene is expressed abnormally under viral promotion [16]. Although both virus promotor and enhancer sequences have been identified [15] and are present in a number of cloned inserts, and although fused virus-host transcripts have been found in HBV-associated human hepatoma cell lines [16,25], the apparent lack of consensus among the flanking host DNA sequences suggests the absence of a single or small number of cellular genes whose expression would be correlated with the transformed state. There are also many chromosomal rearrangements associated with PHC [12,17] that might result in the rearrangement of one or more oncogenes accompanying insertions of HBV DNA. However, no oncogene rearrangements have been found [16]. The further finding of increased c-*ras*[H] and c-*myc* expression in liver tumors is provocative [14], but it is not known whether HBV infection or integration results in the altered expression of these cellular oncogenes nor whether these genes are important in the establishment or maintenance of PHC. The same arguments are true concerning changes in expression of a variety of other host-cell products that seem to be specifically associated with PHC [26,27]. It has not been ascertained whether the expression of growth factors, growth factor receptors, or other host components (such as p53 [20]) known to be linked to carcinogenesis in some instances, plays a role in PHC. In summary, although the present data appear to be inconsistent with a promotion–insertion mechanism, the information is too preliminary and incomplete to rule it out definitively.

When the putative oncogenic potential of HBV DNA is assayed by the transfection of NIH-3T3 cells, a small number of foci (only a few percent of that observed with SV40 DNA) are found [27,28]. The absence of HBV DNA from the transformants and from some PHC samples recently analyzed [28] raises the question of whether HBV DNA really makes a genetic contribution to PHC [16]. When one considers the genetic alterations that have been reported for host and viral DNA in PHC [5,12,13,17,19,25,27,28], these data are then consistent with the hypothesis that HBV may act as an insertional mutagen. In this way, integration per se would not be directly involved in transformation, save via its role in creating genetic instability (deletions, translocations, etc.) over a long period of time. Clones of expanding cells that survive in chronic active hepatitis or cirrhotic livers could inherit mutations in one or more host genes important in growth control, and thereby become "initiated." This may take the form of one or more mutated oncogenes, as recently described in mice with spontaneous PHC [31]. Whether the tumorigenic phenotype ultimately arises from loss of a chromosome or chromosome fragment [12,17] that suppresses cell transformation, and that cellular genes associated with such a phenotype are expressed because the suppressing genetic loci are lost [29], remains an enigma. However, the loss of chromosome 11p in PHC [17, C. Rogler, personal communication] compared to normal tissue from several patients, suggests that such a deletion is one of the events leading to tumor formation, as with other recessive tumors [32]. It is possible that the mechanism of transformation is elusive because it is "hit and run." Such a mechanism may require HBV DNA integration and expression of one or more viral genes early on. Combined with genetic rearrangements in the host DNA, and the possible trans-acting effects of one or more HBV gene products, the new pattern of host gene expression may be able to sustain a phenotype whose expression is independent of whether HBV is present. In such a model, HBV need only rewire some of the elaborate circuitry responsible for the state of cellular differentiation and patterns of growth [20] and then drop out of the picture. Although such a model is purely speculative at this time, the development of tissue culture or animal models in which HBV-associated PHC could be experimentally induced will go far to further our understanding of the relationship between HBV and cancer.

Acknowledgments

This work was supported by United States Public Health Service grants CA-06551, RR-05539, and CA-06927 from the National Institutes of Health and by an appropriation from the Commonwealth of Pennsylvania.

References

1. Vyas GN, Blum HE (1984) Hepatitis B virus infection. Western J Med 140:754–762
2. Blumberg BS, London WT (1982) Hepatitis B virus: Pathogenesis and prevention of primary cancer of the liver. Cancer 50:2657–2665
3. Szmuness W (1978) Hepatocellular carcinoma and hepatitis B virus: Evidence for a causal association. Prog Med Virol 24:40–69

4. Beasley RP, Hwang L-Y (1984) Epidemiology of hepatocellular carcinoma. *In* Vyas GN, Dienstag JL, Hoofnagle JH (eds) Viral Hepatitis and Liver Disease. Grune and Stratton, New York, pp 209–224

5. Robinson WS, Miller RH, Marion PL (1983) Hepatitis B virus as an environmental carcinogen. *In* Milman HA, Sell S (eds) Application of Biological Markers to Carcinogen Testing. Plenum Publishing Corp, New York, pp 465–473

6. Smuckler EA, Ferrell L, Clawson GA (1984) Proliferative hepatocellular lesions, benign and malignant. *In* Vyas GN, Dienstag JL, Hoofnagle JH (eds) Viral Hepatitis and Liver Disease. Grune and Stratton, New York, pp 201–207

7. Berenblum I (1985) Challenging problems in cocarcinogenesis. Cancer Res 45:1917–1921

8. Rizzetto M (1983) The delta agent. Hepatology 3:729–737

9. Ray MB (1979) Hepatitis B Virus Antigens in Tissues. University Park Press, Baltimore, MD, pp 1–59

10. Urban MK, O'Connell AP, London WT (1985) Sequence of events in natural infection of Pekin duck embryos with duck hepatitis B virus. J Virol 55:16–22

11. Snyder RL, Summers J (1980) Woodchuck hepatitis virus and hepatocellular carcinoma. *In* Essex M, Todaro E, zur Hausen H (eds) Virus in Naturally Occurring Cancers. Cold Spring Harbor Conferences on Cell Proliferation. Cold Spring Harbor Laboratory, Cold Spring Harbor, New York, pp 447–457

12. Feitelson MA (1985) *In* Becker Y, Hadar J (eds) Molecular Components of Hepatitis B Virus. Martinus Nijhoff Publishing, Boston, pp 1–273

13. Shafritz DA, Rogler CE (1984) Molecular characterization of viral forms observed in persistent hepatitis infections, chronic liver disease and hepatocellular carcinoma in woodchucks and humans. *In* Vyas GN, Dienstag JL, Hoofnagle JH (eds) Viral Hepatitis and Liver Disease. Grune and Stratton, New York, pp 225–243

14. Fausto N, Shank PR (1983) Oncogene expression in liver regeneration and hepatocarcinogenesis. Hepatology 3:1016–1023

15. Shaul Y, Rutter WJ, Laub O (1985) A human hepatitis B viral enhancer element. EMBO J 4:427–430

16. Varmus HE (1984) Do hepatitis B viruses make a genetic contribution to primary hepatocellular carcinoma? *In* Vyas GN, Dienstag JL, Hoofnagle JH (eds) Viral Hepatitis and Liver Disease. Grune and Stratton, New York, pp 411–414

17. Rogler CE, Sherman M, Su CY, Shafritz DA, Summers J, Shows TB, Henderson A, Kew M (1985) Deletion in chromosome llp associated with a hepatitis B integration site in hepatocellular carcinoma. Science 230:319–322

18. Kew MC, Dusheiko GM, Hadziyannis SJ, Patterson A (1984) Does delta infection play a part in the pathogenesis of hepatitis B virus related hepatocellular carcinoma? Br Med J 288:1727

19. Dejean A, Sonigo P, Wain-Hobson S, Tiollais P (1984) Specific hepatitis B virus integration in hepatocellular carcinoma DNA through a viral 11-base-pair direct repeat. Proc Natl Acad Sci USA 81:5350–5354

20. Bishop JM (1985) Viral oncogenes. Cell 42:23–38

21. Petit M-A, Pillot J (1985) HBc and HBe antigenicity and DNA binding activity of major core protein p22 in hepatitis B virus core particles isolated from the cytoplasm of human liver cells. J Virol 53:543–551

22. Feitelson MA, Marion PL, Robinson WS (1982) The core particles of HBV and GSHV. II. Characterization of the protein kinase reaction associated with ground squirrel hepatitis virus and hepatitis B virus. J Virol 43:741–748

23. Miller RH, Robinson WS (1983) Integrated hepatitis B virus DNA sequences specifying the major viral core polypeptide are methylated in PLC/PRF/5 cells. Proc Natl Acad Sci USA 80:2534–2538

24. Yoakum GH, Korba BE, Lechner JF, Tokiwa T, Gazdar AF, Seeley T, Siegel M, Leeman L, Autrup H, Harris CC (1983) High-frequency transfection and cytopathology of the hepatitis B virus core antigen gene in human cells. Science 222:385–389

25. Ou JH, Rutter WJ (1985) Hybrid hepatitis B virus–host transcripts in a human hepatoma cell. Proc Natl Acad Sci USA 82:83–87

26. Wen Y-M, Mitamura K, Merchant B, Tang ZY, Purcell RH (1983) Nuclear antigen detected in hepatoma cell lines containing integrated hepatitis B virus DNA. Infect Immun 39:1361–1367

27. Koshy R, Freytag von Loringhoven ABL, Koch S, Marquardt O, Hofschneider PH (1984) Structure and function of integrated HBV genes in the human hepatoma cell line PLC/PRF/5. *In* Vyas GN, Dienstag JL, Hoofnagle JH (eds) Viral Hepatitis and Liver Disease. Grune and Stratton, New York, pp 265–273

28. Robinson WS, Miller RH, Klote L, Marion PL, Lee S-C (1984) Hepatitis B virus and hepatocellular carcinoma. *In* Vyas GN, Dienstag JL, Hoofnagle JH (eds) Viral Hepatitis and Liver Disease. Grune and Stratton, New York, pp 245–263

29. Klein G, Klein E (1985) Evolution of tumors and the impact of molecular oncology. Nature 315:190–195

30. Tur-Kasta R, Burk RD, Shaul Y, Shafritz DA (1986) Hepatitis B virus DNA contains a glucocorticoid-responsive element. Proc Natl Acad Sci USA 83:1627–1631

31. Reynolds SH, Stowers SJ, Marondot R, Anderson MW, Aaronson SA (1986) Detection and identification of activated oncogenes in spontaneously occurring benign and malignant hepatocellular tumors of the B6C3F1 mouse. Proc Natl Acad Sci USA 83:33–37

32. Koufos A, Hansen MF, Copeland NG, Jenkins NA, Lampkin BC, Cavenee WK (1985) Loss of heterozygosity in three embryonal tumours suggests a common pathogenetic mechanism. Nature 316:330–334

CHAPTER 33
Progressive Epstein–Barr Virus Infection

WARREN A. ANDIMAN, BEN Z. KATZ, AND GEORGE MILLER

Within the past dozen years, the spectrum of diseases that has been associated with the Epstein–Barr virus (EBV) has increased. Among the clinical conditions that have provoked the most interest are those in which EBV infection progresses far beyond that normally observed in the course of infectious mononucleosis. At the heart of these relapsing, progressive, and, oftentimes, fatal syndromes are specific acquired or inherited disorders of immunity that permit an abnormal degree of viral replication, unchecked proliferation of EBV-transformed B cells, or failure to recognize important EBV associated antigens. We will review some salient features of the immunopathogenesis of infectious mononucleosis and then describe several important nosologic entities in which immunologic control of the infection is disordered to such an extent that the disease progresses, sometimes resulting in neoplasia and death.

Mononucleosis in Normal Individuals

Primary infection with Epstein–Barr virus usually occurs in childhood and is generally asymptomatic. Thirty percent to 50% of primary infections that are delayed until adolescence or early adulthood are recognized as heterophile-positive infectious mononucleosis (IM). IM is a paradigm of a self-limited lymphoproliferative disease, and therefore its immunopathogenesis is the object of intensive investigation. Nevertheless, the immunologic processes involved are still incompletely understood [1–3].

EBV infection is usually acquired via salivary exchange between a sero-negative susceptible and a virus-shedding contact. Approximately 15% of seropositive individuals shed EBV in their oropharyngeal secretions in low titers at any given time. Productive infection in the susceptible subject probably begins in epithelial cells of the buccal mucosa or salivary gland, whence the virus gains access to B lymphocytes in the lymphoid tissues of the pharynx and then disseminates to the entire lymphoid system. In early IM, nearly 20% of circulating B cells are infected with the virus. Many of the B lymphocytes infected by EBV are capable of proliferating indefinitely in vitro; this process is termed *immortalization,* and EBV is felt to be latent within these transformed cells. Latently infected B cells produce little or no infectious virus; nevertheless, EBV DNA is present within the nucleus of the B cell, usually as circularized, nonintegrated molecules.

Most of the atypical lymphocytes that are characteristic of IM are T8 (suppressor/cytotoxic) cells. In early IM, these lymphocytes may outnumber B cells 50 to 1. Cytotoxic/suppressor T cells, both specifically and nonspecifically, kill EBV-infected B lymphocytes, prevent their further outgrowth, and inhibit their immunoglobulin secretion. In addition, natural killer (NK) cells are capable of lysing nonspecifically EBV-infected B cells. Interferons, which are secreted by T cells and induce NK cell activity, may play a role in nonspecific lysis as well.

During acute IM, a general depression of cellular immunity has been observed. This has been demonstrated in vivo by anergy to skin test antigens, and in vitro by decreased lymphocyte responses to plant mitogens and soluble antigens, and in mixed leukocyte reactions.

The humoral arm of the immune system also plays a role in limiting IM. There are at least three specific classes of EBV antigens that stimulate antibody responses: (a) viral capsid antigen (VCA), (b) early antigens (EA; felt to be a complex of viral enzymes associated with the early phases of the viral replicative cycle), and (c) the EB nuclear antigens (EBNA, of which there are several subcomponents). There are also viral envelope antigens that elicit neutralizing antibodies.

The first antibodies detected during acute EBV infection are directed against the structural components of the virus, first VCA and then EA. IgM antibodies to VCA and IgG antibodies to EA appear transiently during the acute and early convalescent phases of IM. Anti-EBNA antibodies usually begin to appear one to two months after infection, possibly as a result of cell death and release of EBNA from within infected cells. The origin of the transient heterophile antibodies that develop during the course of IM are not at all understood; they may arise as a consequence of nonspecific immunoglobulin secretion by EBV-stimulated B cells.

The continued increase in virus-specific antibodies and in T- and NK-mediated responses are thought to bring the disease under control in the normal host. Although most of the EBV-infected B cells are destroyed, some survive. Antibodies to some of the virus-associated antigens, especially neutralizing antibody, and T cells primed to recognize antigens expressed on the B cell surface, keep this low-grade infection in check. However, when a seropositive subject is iatrogenically immunosuppressed, reactivation of latent infection may occur. Such patients begin to excrete high titers of EBV, and lymphoproliferative disorders may arise as a result of renewed proliferation of transformed B cells. In contrast, in some congenitally immunodeficient individuals, immunologic surveillance of primary infection may be defective, and the infection may persist, recur, or progress inexorably.

EBV-Induced Lymphoproliferative Disease in Genetically Impaired Individuals

Rare individuals, because of a discrete (usually X-linked) but uncharacterized genetic impairment, are incapable of mounting a normal immune response to primary infection with EB virus. A registry of kindreds where multiple males of different ages succumbed to infection was begun in the mid-1970s and now contains over 25 families and more than 100 patients [4]. Sporadic cases in females are more unusual but well described [5]. A specific immunologic defect has not been identified, and it is likely that more than one is involved because the syndrome has variable phenotypic expression.

Clinical Syndromes

It has become customary to classify patients with the so-called X-linked lymphoproliferative syndrome (XLPS) according to the two major disease processes to which they succumb. The first accounts for nearly three quarters of the patients who die as a result of a "proliferative" disorder in which the pathology reveals either extreme mononucleosis-like changes (i.e., organs infiltrated with vast numbers of lymphocytes and plasma cells), or widespread malignancies such as Burkitt's lymphoma or immunoblastic sarcoma. The second group of diseases occurs in the minority of patients who survive their initial infection with EBV; they acquire an "aproliferative" disease. The most common is hypogammaglobulinemia, but aplastic anemia and agranulocytosis have also been described. An increased susceptibility to bacterial infection, and the late occurrence of lymphoproliferative disorders also are seen in these survivors. In patients with both early- and late-onset lymphoproliferative diseases, EBNA and EBV DNA can be readily identified in tissues that are affected pathologically.

Pathogenesis of Fatal Infection and Lymphoproliferation in Genetically Predisposed Individuals

Most patients with XLPS have a normal clinical response to routine childhood immunizations and other viral and bacterial infections. Furthermore, studies done in EBV-seronegative persons in affected kindreds reveal normal cell-mediated immune (CMI) responses to skin test antigens and plant mitogens, normal T and B cell numbers, and normal quantitative immunoglobulins. These facts strongly suggest that the defective response to EBV is unique and is not preceded by some generalized immunologic abnormality.

Immunologic studies done in these patients during the course of their acute infection reveal the same depression in CMI observed in IM and, in some, exaggerated NK cell activity against a wide variety of target cells. Unfortunately, most of the immunologic evaluations in this group of individuals have been performed in long-term survivors, thereby muddling the issue as to which phenomena are primary and which are a consequence of the infection. Moreover, the immune defect present in those who survive may be different from that present in those who succumb to their primary infection. In the survivors, T and B cell numbers are normal, but there are global deficiencies in T cell function, including decreased response to plant lectins and to specific antigens, and poor performance in mixed leukocyte reactions [6]. Spontaneous activity of NK cells against target cells is also markedly diminished and cannot be augmented by treatment with interferon [7]. Finally, the great majority of survivors either lack all antibody to EBV, or fail to produce or to lose antibodies to EBNA.

No single theory to date has successfully explained both the proliferative and aproliferative disorders associated with XLPS. On the basis of what is known of the immunopathogenesis of IM, it is logical to assume that the critical defect in patients who fail to respond to EBV infection is one of disordered immunoregulation, especially involving NK and cytotoxic T cells. However, there are patients with the proliferative form of XLPS who demonstrate normal humoral responses and exaggerated NK responses, indicating that the defect may well lie elsewhere.

In contrast, the aproliferative diseases that often develop in survivors may result from an uncontrolled virus-mediated lytic infection of antibody-producing B cells or may be explained by an overabundant suppressor T cell response that progressively limits the further outgrowth of B cells and their progeny. This suppressor cell response would have to be far in excess of that normally occurring during recovery from IM in normal individuals. Whether the necrosis seen in the lymph nodes and reticuloendothelial organs of some of these patients is a morphologic manifestation of viral-induced B cell lysis, or whether it reflects an uncontrolled response on the part of cytotoxic T cells and NK cells to the massive ingress of EBV-transformed B cells is unknown. Although agranulocytosis and aplastic anemia are very rare but well-described complications of IM in normal individuals and XLPS, the mechanism by which these sequelae occur is unknown.

Future clues to the pathogenesis of XLPS may be forthcoming from study of two other heritable disorders in which lymphoproliferative disease is a frequent complication—Chediak–Higashi disease and ataxia-telangiectasia. In the former, the defect appears to lie in inadequate killing by NK cells; in the latter, a specific chromosomal abnormality has been observed in the neoplastic tissue.

EBV-associated Lymphoproliferations and Malignancies Occurring After Organ Transplantation

Recently, there has been increased recognition and reporting of progressive lymphoproliferative disease and lymphoid neoplasms in recipients of organ transplants. In patients receiving renal allografts, cancer rates are approximately 100 times greater than in the general population, and 20% of the neoplasms that occur are malignant lymphomas. An unusually large number involve the central nervous system, usually considered to be an immunologically privileged site. In patients who receive heart transplants, the incidence of lymphoreticular neoplasia has ranged from 9% to 13%. The immunosuppressive regimens employed in such patients and, most recently, the widespread use of cyclosporin A have been implicated as factors critical to the genesis of these lymphoproliferative lesions. Because most of the tumors are of B cell origin, it had been hypothesized from the outset that they might well be associated with EBV.

Clinical Syndromes

Because the kidney is the most frequently transplanted organ, more is known of the sequelae that occur in this group of transplant patients than in any other. In renal allograft recipients who develop lymphoma, two clinical syndromes have been recognized [8]: Younger patients present with signs and symptoms suggesting mononucleosis. The lymphoproliferative disorder usually arises during periods of increased immunosuppression, often after treatment for rejection. The disease progresses rapidly and is widely disseminated, and the mean interval from diagnosis to death is three months. Older patients more often present with localized tumor masses, sometimes confined to the CNS, and the disease usually occurs more than three years following transplantation. Although the tumors progress more slowly and disseminate less widely in the older individuals, the mortality rate approaches 70%. Interestingly, the lymphomas that develop almost always occur in extranodal sites which, besides the brain, usually include the gut, the lung, and soft tissue sites where anti-lymphocyte globulin injections were given as part of an immunosuppressive regimen.

In one study that examined the incidence of EBV infections and lymphoproliferative disease among recipients of different organs, including kidney, liver, heart, and lung, an IM-like illness occurred in 8 of 14 patients at an interval of two to eight months before the discovery of the tumor [9]. In the majority of instances, this viral illness was accompanied by serologic evidence of primary EB virus infection, although some of the patients also had concurrent evidence of primary or reactivated cytomegalovirus infection. The risk of EBV-associated lymphomas is greater with primary than reactivated infection.

Pathologic and Immunologic Studies of Transplant-associated Lymphoproliferations

Various histopathologic, immunologic, and molecular biologic techniques have been employed to characterize the lymphoproliferative disorders that occur in transplant patients. Morphologically these lymphoproliferations have been classified as polymorphic diffuse hyperplasias of B cells or as frank B cell lymphomas, immunoblastic sarcomas, or diffuse histiocytic lymphomas. In a number of instances, the polymorphic B cell proliferations have evolved into monoclonal tumors that are clinically more aggressive. In striking contrast, Hodgkin's disease is by far the most common lymphoma seen in the general population.

A seminal controversy surrounding lymphoproliferative lesions that occur in immunosuppressed transplant recipients is whether they are true malignancies or viral-associated hyperplasias. According to purists, the former should be monoclonal and the latter polyclonal. Two general techniques have been used to address the question of clonality: (a) immunologic cell-typing, using a variety of monospecific fluoresceinated antibodies; and (b) analysis of tumor cell DNA by the Southern blot hybridization technique for rearranged immunoglobulin genes. Only monoclonal B cell proliferations contain sufficiently abundant rearranged genes of an identical genotype to be clearly detected on a gel [10].

Analysis of frozen sections of tumor tissue from renal transplant patients by fluorescence microscopy for the presence of surface and intracytoplasmic immunoglobulin light and heavy chains, complement, and Fc receptors, Ia (HLA-DR), and T cell surface antigens has revealed that both polyclonal and monoclonal B cell proliferations may occur in renal transplant patients, sometimes at different sites in the same patient. Furthermore, polyclonality has been observed both in polymorphic B cell hyperplasias (where it is to be expected) and in tumors classified as true lymphomas on histologic grounds. Sometimes surface marker studies indicate that the lymphomas arise from immature B cells or an even more undifferentiated cell population. None of the tumors appears to arise from T cells. Because the proliferating cells that occur in immunosuppressed heart transplant recipients do not regularly express surface or cytoplasmic immunoglobulin, the clonality of these tumors can be ascertained reliably only by use of Southern blot analysis to identify rearranged immunoglobulin genes. The results of one study of 10 cardiac transplant recipients who developed lymphoproliferative disorders indicate that all contained substantial monoclonal cell populations typical of convential B cell lymphomas at the earliest stages of detectable disease [11]. No data are available on the possible existence of cytogenetic abnormalities in this group of tumors. However, the results of analysis of 19 lymphoproliferations occurring in renal transplant patients revealed clonal translocations in 3 of 19. Two of these had polyclonal lesions, according to immunologic cell typing data suggesting the existence of a clone of genetically aberrant cells within the proliferation [12].

In summary, these data suggest that lymphoproliferative disorders in different groups of transplant patients may present clinically at various stages of development: In renal transplant recipients, the lymphoproliferations may appear at an early, polyclonal stage in their ontogeny, whereas in most cardiac transplant patients, true neoplasia and concomitant monoclonality are already present when the tumor is first discovered. These variations may be associated with the grafted organ itself, differences in immunosuppressive therapy, or differences in the degree of preexisting disease. Interestingly, the polyclonal tumors appear to respond clinically to acyclovir, whereas the monoclonal ones do not [12]. This would imply that viral replication regularly accompanies polyclonal B cell tumors, because acyclovir affects only the viral replicative cycle. In contrast, reduction of dose or termination of therapy with cyclosporin A has an apparent salutary effect on the progress of most tumors, irrespective of clonality [13].

Virologic and Serologic Studies in Transplant Recipients

The evidence supporting an association between EBV and the B cell lymphoproliferative lesions that occur in transplant recipients is abundant. Several investigators have observed seroconversions or significant rises in antibody titer to the capsid and early antigens in the vast majority of patients who develop lymphoproliferative lesions, and twice as often in cardiac and liver as in renal transplant patients. Furthermore, patients who develop primary infection while immunosuppressed seem to be at greatest risk for developing neoplastic disease; nearly half the patients who develop tumors have had recent primary EBV infections [9]. Finally, when tests for the presence of transforming virus in the oropharynx of transplant patients have been performed, nearly all patients have been shown to shed virus.

There is also evidence for the presence of viral genomes in the tumors themselves. With rare exceptions, the polyclonal lymphoproliferations and lymphomas that occur in the wake of organ transplant contain either EBNA or EBV DNA or, usually, both. Utilizing cRNA/DNA filter hybridization, DNA/DNA reassociation kinetics, or Southern blot analysis, researchers have found multiple copies of the genome in nearly all specimens examined. These findings are particularly striking when viewed against the background that positive hybridizations with EBV DNA are *not* found in any lymphomas or leukemias that occur spontaneously in nonimmunosuppressed individuals, with the exception of those whose primary locus is the central nervous system [14].

EBV-associated Lymphoproliferative Disorders in AIDS

Lymphoproliferative Disease in Infants and Children with AIDS.
The clinical manifestations of AIDS in children are somewhat different from those that occur in adults. Although the pediatric patients acquire opportunistic infections with *Pneumocystis carinii, Candida,* cytomegalovirus, and

atypical mycobacteria, as do adults, approximately half develop an unusual form of chronic, diffuse interstitial pneumonia with a highly characteristic follicular lymphocytic infiltrate [15]. Central nervous system lymphoma has also been recognized. Both diseases are associated with the presence of EBV DNA in affected tissues, and there is evidence that simultaneous infection with more than one strain may occur [16]. In addition, EBV may participate in the hypergammaglobulinemia and abnormal B cell immunoregulation characteristic of AIDS.

Virtually all pediatric AIDS patients with lymphocytic interstitial pneumonia (LIP) have extremely high antibody titers to the replicative antigens of EBV. In contrast, some of them fail to produce antibody to EBNA, and most do not recognize that specific component of the nuclear antigen complex encoded by the Bam HI K region of the genome. This antigen is universally recognized by normal individuals [17]. The very high antibody titers to EBV found in patients with LIP suggest that the lung may be the site of extensive viral replication. Thus, some cell in the lung, either epithelial or lymphoid, permits viral growth; in the case of AIDS, HTLV-III may be a co-pathogen.

LIP is not a lymphoma. Instead there is striking mature lymphoid follicle formation in the lung, including germinal centers, and infiltration with plasmacytoid cells. The results of one study indicate that over three-quarters of the biopsies showing LIP contain EBV DNA, whereas none of those containing CMV, *P. carinii,* or mycobacterial infection contains the EB viral genome [15]. At present, LIP has rarely been associated with the presence in the lung of any coinfecting pathogens. In a few instances, the use of immunoglobulin gene probes has shown the pulmonary proliferations to be polyclonal.

Lymphoid Neoplasms in Adult AIDS Patients

Of the lymphoid tumors that develop in adult AIDS patients (and in some children with AIDS), almost all are of the Burkitt's lymphoma (BL) type. They often present in extranodal sites, particularly the CNS, bone marrow, and gastrointestinal tract, and, as is usual in BL, both EBNA and EBV DNA are frequently found in the tumor tissue. The chromosomal translocations that occur in classical, endemic BL have also been observed in the lymphomatous tissue of AIDS patients. Thus the relative roles of the virus and oncogenes in the genesis of these neoplasms are unclear, as is true for BL in general.

Chronic Mononucleosis

Chronic mononucleosis (chronic IM) may well be a newly emerging, discrete clinical entity [18–22]. As a group, chronic IM patients have complaints of fatigue, fever, myalgia, and pharyngitis of at least six months' duration; there is often an overlay of depression and/or neurosis. Generalized lymphadenopathy and hepatosplenomegaly are commonly seen. The general signs and symptoms resemble those of IM, but their duration is exceedingly long, and the specific findings of lymphocytosis with atypia and the heterophile antibody response so characteristic of classical IM are often absent.

Some patients have had documented IM in the past, occasionally of a severe nature, whereas others have not. In some individuals, symptoms are always present, albeit with varying intensity; in others, they recur, sometimes at regular intervals. These patients rarely have evidence of any underlying illness or classical immunodeficiency. Moreover, they almost never develop the lymphoproliferation seen in patients with XLPS, nor do they develop any other serious viral infections. The amount of EBV present in the secretions of these patients is the same as in healthy seropositive controls. Therefore, if this syndrome is truly due to EBV, the immunologic abnormality is very subtle, such that it allows EBV (but not other viruses) to persist, and causes chronic debilitating illness of varying degree, but no lymphoproliferation.

Many types of immunologic abnormalities have been sought in this group of patients. The most consistent finding is an unusual EBV serologic profile, which includes prolonged and/or elevated antibody responses to EA (usually to its so-called restricted component) [19,20,22], persistently elevated IgM antibodies against viral capsid antigen [19], and the failure to produce antibody to EBNA and/or its Bam HI K component [17,21,22]. Other immunologic abnormalities sometimes found in these patients include mild hypoglobulinemia and, occasionally, increased numbers of helper and suppressor T cells, autoimmune phenomena, abnormalities of the interferon system, and decreased skin test sensitivity. Some of these patients also appear to be simultaneously infected with cytomegalovirus [20,21].

As far as can be surmised at present, chronic IM patients constitute a heterogeneous group clinically, serologically, and immunologically. Although some patients are felt to be neurotic and to complain of generalized symptoms out of proportion to any objective findings, many have abnormal serologic responses to EBV which suggest an immune impairment. But even in one family in which a common underlying immunodeficiency, i.e., common variable hypogammaglobulinemia, was recognized, two siblings presented with different immunologic abnormalities (e.g., B cell vs. T cell) and with different EBV-associated clinical syndromes (e.g., recurrent vs. persistent IM) [18].

In the future, a more precise clinical and serologic classification of the syndrome into several subtypes may emerge. At that time, it is expected that the specific immune deficiencies in each subset of patients will be delineated.

Acknowledgments

This work was partly supported by grants from the National Institutes of Health (AI 14741 and AI 21186) and by a grant from the Charles H. Hood Foundation. Dr. Katz is a Pfizer postdoctoral fellow.

References

1. Epstein MA, Achong BG (1977) Pathogenesis of infectious mononucleosis. Lancet ii:1270–1272
2. Sugden B (1982) Epstein–Barr virus: A human pathogen inducing lymphoproliferation in vivo and in vitro. Rev Infect Dis 4:1048–1061
3. Sullivan JL (1983) Epstein–Barr virus and the X-linked lymphoproliferative syndrome. Adv Pediatr 30:365–399
4. Purtilo DT, DeFlorio D, Hutt LM, Bhawan J, Yang JP, Otto R, Edwards W (1977) Variable phenotypic expression of an X-linked recessive lymphoproliferative syndrome. N Engl J Med 297:1077–1081
5. Robinson JE, Brown N, Andiman WA, Halliday K, Francke U, Robert MF, Andersson-Anvret M, Horstmann D, Miller G (1980) Diffuse polyclonal B-cell lymphoma during primary infection with Epstein–Barr virus. N Engl J Med 302:1293–1297
6. Sullivan JL, Byron KS, Brewster FE, et al (1983) X-linked lymphoproliferative syndrome: Natural history of the immunodeficiency. J Clin Invest 71:765
7. Sullivan JL, Byron KS, Brewster FE, Purtilo, DT (1980) Deficient natural killer cell activity in X-linked lymphoproliferative syndrome. Science 210:543–545
8. Hanto DW, Sakamoto K, Purtilo DT, Simmons RL, Najarian JS (1981) The Epstein–Barr virus in the Pathogenesis of posttransplant lymphoproliferative disorders. Surgery 90:204–213
9. Ho M, Miller G, Atchison RW, Breinig M, Dummer JS, Andiman WA, Starzl TE, Eastman R, Griffith BP, Hardesty RL, Bahnson HT, Hakala TR, Rosenthal JT (1985) Epstein–Barr virus infections and DNA hybridization studies in post-transplantation lymphoma and lymphoproliferative lesions: The role of primary infection. J Infect Dis 152:876–886
10. Cleary ML, Chao J, Warnke R, Sklar J (1984) Immunoglobulin gene rearrangement as a diagnostic criterion of B cell lymphoma. Proc Natl Acad Sci USA 81:593–597
11. Cleary ML, Warnke R, Sklar J (1984) Monoclonality of lymphoproliferative lesions in cardiac-transplant recipients. N Engl J Med 310:477–482
12. Hanto DW, Gajl-Peczalska KJ, Frizzera G, Arthur DC, Balfour H, McClain K, Simmons RL, Najarian JS(1983) Epstein–Barr virus (EBV) induced polyclonal and monoclonal B-cell lymphoproliferative diseases occurring after renal transplanation. Ann Surg 198:356–369
13. Starzl TE, Nalesnik MA, Porter KA, Ho M, Iwatsuki S, Griffith BP, Rosenthal JT, Hakala TR, Shaw BW, Hardesty RL, Atchison RW, Jaffe R, Bahnson HT (1984) Reversibility of lymphomas and lymphoproliferative lesions developing under cyclosporin–steriod therapy. Lancet i: 583–587
14. Andiman WA, Gradoville, L, Heston L, Neydorff R, Savage ME, Kitchingman G, Shedd D, Miller G (1983) Use of cloned probes to detect Epstein–Barr viral DNA in tissues of patients with neoplastic and lymphoproliferative diseases. J Infect Dis 148:967–977
15. Andiman WA, Eastman R, Martin K, Katz BZ, Rubinstein A, Pitt J, Pahwa S, Miller G (1986) Opportunistic lymphoproliferations associated with Epstein–Barr viral DNA in infants and children with AIDS. Lancet i:1390–1393
16. Katz BZ, Andiman WA, Eastman R, Martin K, Miller G (1986) Infection with two genotypes of Epstein–Barr virus in an infant with AIDS and lymphoma of the central nervous system. J Infect Dis 153:601–604

17. Miller G, Grogan E, Fischer D, Niederman JC, Schooley RT, Henle W, Lenoir G, Liu C-R (1985) Antibody responses to two Epstein–Barr virus nuclear antigens defined by gene transfer. N Engl J Med 312:750–755
18. Ballow M, Seeley J, Purtilo DT, St Onge S, Sakamato R, Rickles FR (1982) Familial chronic mononucleosis. Ann Int Med 97:821–825
19. Tobi M, Ravid Z, Feldman-Weiss V, Ben-Chetrit E, Morag A, Chowers I, Michaeli Y, Shalit M, Knobler H (1982) Prolonged atypical illness associated with serological evidence of persistent Epstein–Barr virus infection. Lancet i:61–63
20. DuBois RE, Seeley JK, Brus I, Sakamoto K, Ballow M, Harada S, Bechtold TA, Pearson G, Purtilo DT (1984) Chronic mononucleosis syndrome. South Med J 77:1376–1382
21. Straus SE, Tosato G, Armstrong G, Lawley T, Preble OT, Henle W, Davey R, Pearson G, Epstein J, Brus I, Blaese RM (1985) Persisting illness and fatigue in adults with evidence of Epstein–Barr virus infection. Ann Int Med 102:7–16
22. Jones JF, Ray CG, Minnich LL, Hicks MJ, Kibler R, Lucas DO (1985) Evidence for active Epstein–Barr virus infection in patients with persistent, unexplained illness: Elevated anti-early antigen antibodies. Ann Int Med 102:1–7

CHAPTER 34

Human Papillomaviruses: Why Are Some Types Carcinogenic?

HARALD ZUR HAUSEN

Almost 80 years ago, papillomaviruses were described as filtrable infectious agents in humans [1]; thus these represent the first oncogenic viruses analyzed. Although numerous clinical studies on human papillomas and the occasional conversion of some of these tumors into malignant growth were reported subsequently, no serious experimental studies were carried out until the early 1930s. In this period, Shope descibed an infectious papillomatosis in cottontail rabbits that converted in some of these animals, into squamous cell carcinomas [2]. He also noted that these infections could be transmitted to domestic rabbits, resulting in a higher rate of malignant conversions. Within the following decade, Rous and his coworkers (see ref. 3 for a review) revealed an interesting interaction between cottontail rabbit papillomavirus (CRPV) infections and the application of chemical carcinogens. His group demonstrated a synergistic or cooperative effect between these agents, resulting in tumors that uniformly contained papillomavirus sequences [4].

These early studies implicated two important factors in the etiology of carcinomas as a consequence of papillomavirus infections:

1. Genetic control of malignant conversion, as indicated by the differences in the conversion rates between cottontail rabbits and domestic rabbits
2. The role of chemical cofactors in significantly increasing the risk of carcinoma development

Progress made recently has resulted from two basic observations: (a) the development of an in vitro system for cell transformation by bovine papillomaviruses [5–8], and (b) the demonstration of the plurality of human papillomaviruses (HPV) [9–12].

Transformation studies established the causative role of bovine papillomavirus in malignant conversion of murine and hamster cells and permitted identification of those viral gene functions involved in the transformation process [13–15]. The expression of these genes is a prerequisite for fibroma or fibrosarcoma development in hamsters and mice. Their loss results in reversion of the malignant phenotype [16].

The demonstration of the plurality of human papillomavirus types provided a satisfying explanation for differences in growth patterns, histopathology, and prognosis of various types of human papillomas. To the time of writing, more than 40 individual genotypes have been identified (see ref. 17 for review), and in all likelihood this number will increase further.

The obvious involvement of some of these types in human carcinogenesis raises new interesting questions.

Papillomaviruses in Human Carcinomas

At least two types of human carcinomas are consistently associated with papillomavirus infections: (a) squamous cell carcinomas of the skin arising in a rare condition called epidermodysplasia verruciformis, and (b) anogenital cancer (including cervical, vulvar, penile, and perianal carcinomas). Epidermodysplasia verruciformis (EV) is a hereditary disorder, characterized by an extensive verrucosis which usually develops during childhood or early adolescence. A remarkable susceptibility to specific types of HPV infections is demonstrated along with a virtual absence of others. Close to 20 distinct types of HPVs have been identified from these patients (see refs. 3 and 18 for review).

About one-third of EV patients develop squamous cell carcinomas within a period of 5–10 years, almost exclusively at light-exposed sites. The analysis of these carcinomas revealed the presence of specific types of HPV, predominantly of HPV-5, occasionally HPV-8, rarely HPV-14 and HPV-17, and possibly a few additional types. At present, with very few exceptions, no evidence has been obtained for integration of viral DNA within the host cell genome. The majority of cells contain large amounts of episomal HPV-5 or -8 molecules.

In two respects, epidermodysplasia verruciformis infections resemble Shope papillomavirus infections:

1. EV represents a hereditary disorder with increased susceptibility to HPV infections and a clearly enhanced risk for subsequent malignant conversion. This points again to a genetic control of papillomavirus infections.
2. Carcinogenic cofactors, in this case most likely the ultraviolet rays of sunlight, contribute significantly to cancer development.

Moreover, squamous cell carcinomas in EV patients are clearly associated with specific types of HPV infections which seem to represent high-risk types for malignant conversion. Thus, EV seemingly is a most interesting model for cooperative effects between genetically controlled host factors, specific types of HPV infections, and environmental carcinogenic effects.

In anogenital cancer, again specific types of HPV infections appear to result in a higher risk of malignant conversion. Approximately 70% of all cervical, vulvar, and penile cancers contain the DNA of HPV types 16 and 18 (see ref. 19 for review). A few additional types have recently been isolated from cervical cancers, bringing the percentage of positive biopsies to about 80%. Because the majority of negative tumors cross-hybridize with other genital HPV prototypes under conditions of lowered stringency, it is likely that the majority of them, if not all, contain HPV sequences of related but as yet unidentified types.

During the past 30 years, a number of cell lines have been isolated for cervical cancer biopsies. As in the in vivo situation 60–70% of these lines contain HPV-16 or -18 sequences, among them the well-characterized HeLa line. The availability of these lines, in addition to a number of primary biopsies, permitted careful analysis of the state of viral DNA within the tumor cells and its genetic activity. The data demonstrate that the majority of tumors contains the viral DNA in an integrated state. This observation contrasts with findings in Shope papillomavirus-induced carcinomas and carcinomas in EV patients [20]. Moreover, the integrational pattern reveals a remarkable degree of specificity. The circular viral molecules are regularly opened within the E1–E2 open reading frames (ORFS). This clearly interrupts the early region and excludes ORF E5, which has been identified in bovine papillomavirus as one of the transforming regions [21] and as a genetically active ORF in carcinoma cells. Similarly, trans activation by E2 of ORFs E6 and E7, as demonstrated in bovine papillomavirus-infected cells [22], cannot be operative in cervical cancer cells.

Active transcription of the remaining early region (ORFs E6 and E7) is readily demonstrable in cervical cancer lines and in the majority of cervical cancer biopsies [20]. An exceptional negative report [23] revealing transcription in one out of four biopsies only, may have resulted from admixture of nontumorous tissue: All these biopsies contained only one genome copy per cell, which is unusual for primary tumors. The demonstration of actin message as a marker for the sensitivity of the test system includes also actin mRNA derived from normal cells and therefore does not permit comparative statements on HPV expression in a low percentage of carcinoma cells.

Besides transcription of E6 and E7, fusion transcripts between RNA from these ORFs and adjacent host cell DNA are readily demonstrable in HPV-containing cell lines [20]. Their functional significance remains to be determined.

Today there is little doubt that transcription of E6 and E7 leads to synthesis of proteins. A 15.5K protein has been identified as a product of E6 transcription in bovine papillomavirus-transformed cells [24]. Monoclonal antibodies raised against a bacterially expressed E7 fusion protein of HPV 16 specifically recognize cytoplasmic and perinuclear regions in immunoperoxidase-stained HPV-16-containing cell lines (Oltersdorf et al., unpublished observations).

It is important to point out that HPV-16 and -18 are found in a high percentage of precursor lesions of genital cancer. These are cervical dysplasias with significant nuclear atypia and aneuploidy but also (at external genital sites) Bowenoid proliferations, most frequently diagnosed as Bowenoid papulosis [25,26]. It appears that these proliferative changes are the primary manifestations of HPV-16 and HPV-18 infections. They differ substantially from typical genital warts and koilocytotic "flat" condylomata of the cervix, which are most frequently caused by HPV types 6 and 11. The induction of koilocytotic dysplasias in heterografted cervical tissue by HPV-11 infections [27] appears to be the first step in proving directly that these lesions are caused by HPV infections.

It is intersting to note that the state of viral DNA in dysplasias and Bowenoid lesions containing HPV-16 or -18 is episomal [28]. Thus, integration seems to be a specific event in carcinomas.

As in EV infection, infections by specific HPV types appear to pose a high risk for malignant conversion in anogenital cancer. In contrast to EV infection, however, a genetic predisposition for anogenital cancer is not apparent in HPV infection. What about carcinogenic cofactors? A role for cofactors has been suggested by two sets of observations:

1. Cervical cancer is approximately 20 fold more frequent worldwide than vulvar or penile cancer, although males should be as frequently infected as females and in all likelihood transmit the viruses to their sexual partners.
2. Penile cancer occurs at high frequency only in noncircumcised populations with low standards of sexual hygiene.

The cervix is certainly much more exposed to potentially carcinogenic metabolites resulting from acute and chronic inflammations by bacterial and protozoal infections. In addition, the demonstration of cotinine and nicotine in the vaginal fluid of heavy smokers can probably be attributed to elevated exposure of cells in continuous contact with this fluid, to chemical carcinogens absorbed during smoking. Indeed, a slightly higher risk for cervical cancer has been demonstrated in heavy smokers (reviewed in ref. 29). Viral infections with initiating properties, like herpes simplex virus and probably also cytomegalovirus [30], commonly affect the mucosa of internal genital sites much more extensively than external genital sites and may also represent important cofactors.

A similar situation seems to account for the high rate of penile cancer in the noncircumcised male who practices no penile hygiene. Chronic inflammations and smegma production should provide an environment favorable for cancer development. On the basis of these observations, the speculation that cofactors play a significant role in the etiology of anogenital cancer seems to be justified.

Specific types of HPVs are also found increasingly in oral, tongue, laryngeal, and lung cancers [31–34]. Because these types of cancers are known to be associated with exposure to chemical carcinogens (from smoking, tobacco chewing, heavy alcohol consumption), a similar interaction between specific virus infections and chemical factors in the etiology of these cancers may deserve special interest. Among many other aspects, this would permit the design of new strategies in preventing some of the most common human cancers.

Models for the Carcinogenicity of Human Papillomaviruses

On the basis of the discussion of papillomavirus–host interactions, a few common characteristics become evident:

1. Specific types are involved in carcinogenesis.
2. Viral DNA persists in the carcinoma cells, and specific early functions are consistently expressed.
3. Chemical and physical carcinogens appear to play a significant role as cofactors.

Integration of viral DNA reveals a remarkable specificity in genital carcinomas, but no similar observations have yet been made in other systems.

What experimental approaches now available can definitively prove the role of HPV in human carcinogenesis and aid our understanding of the mechanism of their interaction with chemical and physical carcinogens? One possible approach has recently been developed by Kreider [27]. Heterografting of human tissue beneath the renal capsule of nude mice permits the observation of human cells in an animal host for at least several months. Infection of these cells by HPV-11 has already allowed its role to be established in condyloma induction and in the induction of koilocytotic dysplasias in human cervical tissue. It is anticipated that this system will permit the demonstration of specific histological patterns induced by other types of HPVs. It may also represent a useful in vivo system to establish the interaction between HPV infections and other carcinogenic factors in malignant conversion. Such experiments have been initiated very recently.

At the molecular level, prevention of the expression of the E6–E7 ORFs may corroborate their function in the maintenance of the malignant phenotype. If suppression of E6–E7 expression in cervical cancer cell lines by antisense mRNA [35] leads to a reversion of the malignant phenotype, this would strongly support a causative role of these viral functions for the malignant growth. These approaches have become feasible rather recently, and are presently being studied in several laboratories.

A new concept originated from studies originally devised by Stanbridge et al. [36]. Cell fusion of HeLa cells with normal human fibroblasts or keratinocytes resulted in hybrids that were not carcinogenic upon inoculation into nude mice. Malignant revertants had uniformly lost chromosomes, presumably those of the nonmalignant donor. These data suggested that malignant functions are controlled by specific chromosomes of the normal fusion partner.

The demonstration of the presence of HPV-18 DNA within HeLa cells [37] and of its genetic activity [20] permitted an analysis of HPV-18 transcription in malignant and nonmalignant HeLa hybrids.

Both cell types show no differences in their in vitro growth patterns and in their cloning efficiency in soft agar. Under these conditions, no differences were noted in HPV-18 transcription, which was readily demonstrated in tumorigenic as well as in nontumorigenic hybrid lines (Schwarz and zur Hausen, unpublished observations). Preliminary data obtained after implantation of the respective hybrids in diffusable chambers into nude mice indicated a dramatic reduction or shutoff of HPV-specific mRNA synthesis in one nontumorigenic hybrid, whereas in the corresponding tumorigenic subline, production of HPV-18-specific mRNA continued (Schwarz, Fusenig, and zur Hausen, unpublished observations). This finding is supported by observations that additions of mouse embryonic extract to the tissue culture medium significantly reduced growth of the nontumorigenic subline by not affecting the tumorigenic line (Boukamp and zur Hausen, unpublished observations).

These studies seem to strongly support a concept presented a few years ago [38], suggesting that intracellularly a cellular interfering factor (CIF) regulates early functions of tumor viruses. Obviously, cellular genes suppress early papillomavirus expression. This is supported by other experimental data showing that cycloheximide treatment of bovine papillomavirus-infected cells increases transcription of early genes five to ten fold [39].

In the HeLa hybrid system the suppressing function contributed by the chromosomes from the normal cells is inactive in vitro. However, it is activated in vivo, strongly implicating a humoral factor provided by another cell compartment. A scheme of a concept outlining these interactions is given in Figure 34.1. If confirmed, this concept will have a number of important implications:

1. The malignant phenotype of HeLa cells depends on expression of HPV-18 early functions.
2. Genes regulating HPV-18 expression in HeLa cells are functionally inactive and cannot be activated in vivo; most likely they are structurally modified.
3. Chromosomes of normal cells fused with HeLa cells contain genes that, upon activation by a humoral factor, suppress HPV-18 early functions.
4. Chemical and physical carcinogens may interact with HPV infections by modifying the cellular suppressor genes (CIF genes) or the genes responsible for binding and transporting the humoral CIF-activating factor.
5. The CIF-activating factor should not be species specific.

The experimental basis for this concept is still preliminary. The data clearly require additional controls; yet results are encouraging.

In this respect it will be interesting to analyze the transcription of early functions in normal basal cells infected by HPV-16, -18, or other types of HPV before they start to differentiate. According to the model, a reduced transcription rate in comparison to that in carcinomas is expected. Expression of RNA in Shope papillomavirus-induced carcinomas is elevated in com-

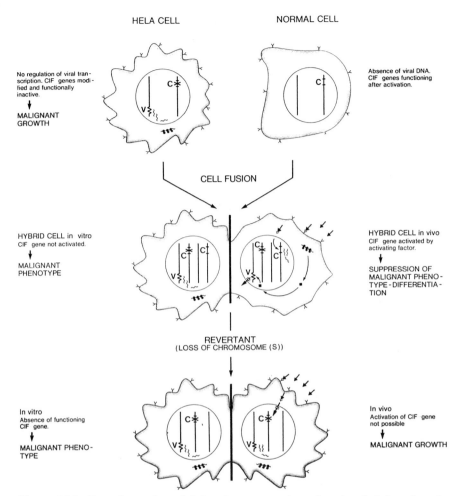

Figure 34.1. Experimental model for the suppression of early viral functions by trans-acting cellular interference factor(s) (CIF). The fusion of HeLa cells containing integrated HPV-18 sequences and structurally modified CIF genes, to normal cells results in non-tumorigenic hybrids. The "malignant phenotype" in vitro leading to growth patterns and cloning efficiencies similar to those observed in the parental HeLa cell is explained by noninduction of CIF genes in chromosomes contributed by the normal cell. In vivo, however, a humoral factor (arrows) activates these genes, resulting in synthesis of a protein that interferes with HPV-18 transcription. Loss of nonrearranged CIF genes during subsequent replications results in restoration of the characteristics of malignant growth.

parison to non-virus-producing papillomas [40], suggesting leaky control of early viral functions within the latter as the reason for the high susceptibility of these rabbits to malignant conversion.

According to this concept, EV results from a genetic defect of CIF genes for suppressing specific types of papillomaviruses, and less likely from defects in genes responsible for binding the CIF-activating factor or from a failure

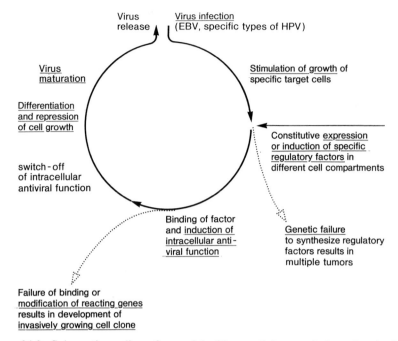

Figure 34.2. Schematic outline of a model of intracellular regulation of early functions of EBV and HPV. The infecting virus stimulates growth of specific target cells.
This in turn provokes the constitutive expression or induction of regulatory factors
originating from different cell compartments. Binding and further processing of the
factor induce intracellularly the cellular interfering factor (CIF) which suppresses
the viral information. Differentiation of the respective cells leads to a regulated
switchoff of CIF genes and permits expression of viral functions and virus maturation,
eventually resulting in release of infectious virus. Genetic failure to synthesize CIF-
activating factors, modification of receptor genes, or structural changes within the
CIF genes would result in malignant growth of the infected cells.

to produce the latter factor, because this would probably affect additional
viral infections.

In genital cancer, modification of CIF genes controlling widespread potentially carcinogenic HPVs should result in cancer development, Similarly,
inactivation of genes responsible for binding and transporting the CIF-activating factor in cells infected by the respective HPVs should have the same
consequence. These events should be substantially facilitated by other mu-
tagenic factors. The relative risks for malignant conversion should increase
with the number of infected cells.

This concept may also be applicable to other human tumor virus infections,
most notably to Epstein–Barr virus (EBV) infection. Although cell-mediated
immune functions are clearly involved in the control of unrestricted lymphoproliferation of EBV-transformed lymphoblasts [41], the failure to induce

malignant growth after heteroinoculation into nude mice (in contrast to Burkitt's lymphoma cells) may suggest in vivo activation of intracellular control. The X-chromosome-linked lymphoproliferative syndrome (XLPS) (reviewed in ref. 41) may represent less an immunological failure, and rather a genetic defect in the intracellular control of EBV infections. The basic features of the postulated interactions are depicted in Figure 34.2.

Human cells, in contrast to rodent cells, can barely be transformed in vitro by chemical and physical carcinogens, and thus seem to have developed specific controls during evolution which protect them against carcinogenic functions of some of their common and persisting viral infections. The first examples are now emerging of a fairly indirect mode of action of chemical and physical carcinogens, by modification of cellular control functions for such viral oncogenes. If this mechanism holds true for other common human cancers, it will certainly have a deep impact on our strategies in cancer prevention and cancer therapy.

References

1. Ciuffo G (1907) Innesto positivo con filtrato di verrucae volgare. Giorn Ital Mal Venereol 48:12–17
2. Shope RE (1933) Infectious papillomatosis of rabbits, with a note on the histopathology. J Exp Med 58:607–624
3. Orth G, Favre M, Breitburd F, Croissant O, Jablonska S, Obalek M, Jarzabek-Chorzelska M, Rzesa G (1980) Epidermodysplasia verruciformis: A model for the role of papillomaviruses in human cancer. In Viruses in Naturally Occurring Cancers. Cold Spring Harbor Laboratory Press, Cold Spring Harbor, New York, pp 259–282
4. Ito Y (1975) Papilloma-myxoma viruses. In Cancer, vol 2. Plenum Press, New York
5. Black PH, Hartley JW, Rowe WP, Huebner RJ (1963) Transformation of bovine tissue culture cells by bovine papillomavirus. Nature 199:1016–1018
6. Boiron M, Levy JP, Thomas M, Friedman JC, Bernard J (1964) Some properties of bovine papilloma virus. Nature 201:423–424
7. Thomas M, Boiron M, Tanzer J, Levy JP, Bernard J (1964) In vitro transformation of mice cells by bovine papillomavirus. Nature 202:709–710
8. Olson C, Gordon DE, Robl MD, Lee KP (1969) Oncogenicity of bovine papilloma virus. Arch Environ Health 19:827–837
9. Gissmann A, zur Hausen H (1976) Human papillomaviruses: Physical mapping and genetic heterogeneity. Proc Natl Acad Sci USA 73:1310–1313
10. Gissmann L, Pfister H, zur Hausen H (1977) Human papillomaviruses (HPV): Characterization of four different isolates. Virology 76:569–580
11. Orth G, Favre M, Croissant O (1977) Characterization of a new type of human papillomavirus that causes skin warts. J Virol 24:108–120
12. Orth G, Breitburd F, Favre M, Croissant O (1977) Papillomaviruses: Possible role in human cancer. In Origins of Human Cancer. Cold Spring Harbor Laboratory Press, Cold Spring Harbor, New York, 1043–1068

13. Geraldes A (1969) Malignant transformation of hamster cells by cell-free extracts of bovine papillomas (in vitro). Nature 222:1283–1284

14. Dvoretzky I, Shober R, Chattopadhy SK, Lowy DR (1980) A quantitative in vitro focus forming assay for bovine papillomavirus. Virology 103:369–375

15. Breitburd F, Favre M, Zoorob R, Fortin D, Orth G (1981) Detection and characterization of viral genomes and search for tumoral antigens in two hamster cell lines derived from tumors induced by bovine papillomavirus type 1. Int J Cancer 27:693–702

16. Turek LP, Byrne JC, Lowy DR, Dvoretzky I, Friedman RM, Howley PM (1982) Interferon induces morphologic reversion with elimination of extrachromosomal viral genomes in bovine papillomavirus-transformed mouse cells. Proc Natl Acad Sci USA 79:7914–7918

17. zur Hausen H, Schneider A (1986) The role of papillomaviruses in human anogenital cancer. In The Papillomaviruses (in press)

18. Pfister H (1984) Biology and biochemistry of papillomaviruses. Rev Physiol Biochem Pharmacol 99:111–181

19. zur Hausen H (1986) Evidence for an association between human papillomaviruses and neoplasia. Int Med (in press)

20. Schwarz E, Freese UK, Gissman L, Mayer W, Roggenbuck B, Stremlau A, zur Hausen H (1985) Structure and transcription of human papillomavirus sequences in cervical carcinoma cells. Nature 314:111–114

21. Schiller JT, Vass WC, Lowry DR (1984) Identification of a second transforming region in bovine papillomavirus DNA. Proc Natl Acad Sci USA 82:7880–7884

22. Spalholz BA, Yang Y-C, Howley H (1985) Transactivation of a bovine papillomavirus transcriptional regulatory element by the E2 gene product. Cell 42:183–191

23. Lehn H, Krieg P, Sauer G (1985) Papillomavirus genomes in human cervical tumors: Analysis of their transcriptional activity. Proc Natl Acad Sci USA 82:5540–5544

24. Androphy EJ, Schiller JT, Lowy DR (1985) Identification of the protein encoded by the E6 transforming gene of bovine papillomavirus. Science 230:442–445

25. Ikenberg H, Gissmann L, Gross G, Grussendorf E-I, zur Hausen H (1983) Human papillomavirus type 16-related DNA in genital Bowen's disease and in Bowenoid papulosis. Int J Cancer 32:563–565

26. Crum CP, Mitao M, Levine RU, Silverstein S (1985) Cervical papillomaviruses segregate within morphologically distinct precancerous lesions. J Virol 54:675–681

27. Kreider JW, Howett MK, Wolfe SA, Bartlett GL, Zaino RJ, Sedlacek TV, Mortel R (1985) Morphological transformation in vivo of human uterine cervix with papillomavirus from condylomata acuminata. Nature 317:639–641

28. Dürst M, Kleinheinz A, Hotz M, Gissmann L (1985) The physical state of human papillomavirus type 16 DNA in benign and malignant genital tumors. J Gen Virol 66:1515–1522

29. zur Hausen H, Peto R (1986) Viral origin of cervical cancer. Cold Spring Harbor Laboratory Press, Cold Spring Harbor, New York (in press)

30. zur Hausen H (1982) Human genital cancer: Synergism between two virus infections or synergism between a virus infection and initiating events. Lancet ii:1370–1372

31. Löning T, Ikenberg H, Becker J, Gissmann L, Hoepfer I, zur Hausen H (1985) Analysis of oral papillomas, leukoplakias and invasive carcinomas for human papillomavirus-related DNA. J Invest Dermatol 84:417–420

32. de Villiers E-M, Weidauer H, Otto H, zur Hausen H (1985) Papillomavirus DNA in human tongue carcinomas. Int J Cancer 36:575–579

33. Kahn T, Schwarz E, zur Hausen H (1986) Molecular cloning and characterization of the DNA of a new human papillomavirus (HPV 30) from a laryngeal carcinoma. Int J Cancer (in press)

34. Stremlau A, Gissmann L, Ikenberg H, Stark M, Bannasch P, zur Hausen H (1985) Human papillomavirus type 16-related DNA in an anaplastic carcinoma of the lung. Cancer 55:737–740

35. Izant JG, Weintraub H (1985) Constitutive and conditional suppression of exogenous and endogenous genes by anti-sense RNA. Science 228:345–352

36. Stanbridge E, Der CJ, Dorsen C-J, Nishimii RY, Peehl DM, Weissman BE, Wilkinson JE (1982) Human cell hybrids: Analysis of transformation and tumorigenicity. Science 215:252–259

37. Boshart M, Gissmann L, Ikenberg H, Kleinheinz A, Scheurlen W, zur Hausen H (1984) A new type of papillomavirus DNA, its presence in genital cancer biopsies and in cell lines derived from cervical cancer. EMBO J 3:1151–1157

38. zur Hausen H (1980) The role of viruses in human tumors. Adv Cancer Res 33:77–107

39. Kleiner E, Dietrich W, Pfister H (1986) Differential regulation of papillomavirus early gene expression in transformed fibroblasts in carcinoma cell lines. (Submitted for publication)

40. Nasseri M, Wettstein FO (1984) Differences exist between viral transcripts in cottontail rabbit papillomavirus-induced benign and malignant tumors as well as non-virus-producing and virus-producing tumors. J Virol 51:706–712

41. Purtilo D (1984) Immune Deficiency and Cancer. Plenum Press, New York

New Trends in Diagnosis and Epidemiology

CHAPTER 35
Nucleic Acid Probes to Detect Viral Diseases

DOUGLAS D. RICHMAN AND GEOFFREY M. WAHL

The presence of a virus infection can be documented either by isolation of the virus by culture or by detection of a virus-specific component. These viral components include structural proteins (which are usually identified on the basis of their antigenicity), virus-induced enzymes, and viral nucleic acids.

There are several features of nucleic acid hybridization that have prompted investigation of this technology for the detection of viral diseases. First, nucleotide sequences can be identified as unique for the genome of specific agents and therefore can constitute probes of very high specificity. Second, the high avidity of complementary nucleic strands for each other should permit highly sensitive hybridization assays. Finally molecular cloning and oligonucleotide synthesis make possible the availability of virtually unlimited amounts of standardized reagent for the detection of any organism.

Procedures

Nucleic acid from infected tissue can be detected either in situ to permit histologic and cytologic localization, or after extraction of the nucleic acid from tissues. The theory and practice of nucleic hybridization are detailed elsewhere [1–3], and in situ hybridization is discussed by Haase (Chapter 36, this volume). Although kinetic liquid phase hybridization can best be applied to the quantitative determination of concentration, complexity, and homology of nucleic acid preparations [2], fixation of nucleic acid to a solid filter support permits much less cumbersome assays which are semiquantitative and amenable to the examination of a large number of samples [3].

DNA or RNA is placed on a nitrocellulose or nylon filter either directly by spotting [4,5] or by filtration through a vacuum manifold [3]. The filter is then baked [3] or irradiated [6] to ensure bonding of the nucleic acid to the solid support, or in some procedures the filter may be used directly [7]. First, DNA must be partially purified before placement on the solid support. The protein and lipid contaminants in samples must be removed prior to immobilization. If filtration is used, contaminants will clog the filter. More importantly, contaminants can compete with either target or probe binding to reduce specific reactions; also, protein coimmobilization with target sequences can lead to spuriously high backgrounds. Techniques for the extraction of RNA for assay on filters have also been developed [7,8]. In general, less than 10μg of total nucleic acid or the contents of 10^5 to 10^6 cells are applied on a surface area of approximately 5 mm^2. Larger quantities of material will not bind with 100% efficiency on such a small area and can lead to high background hybridization signals.

In addition to the examination of partially purified DNA with the dot-blot format, DNA may be digested with restriction endonucleases, fractionated by agarose gel electrophoresis, and then transferred to filters (Southern blot) [1,3,9]. Similarly, electrophoretic fractionation of RNA and transfer to filters may be performed ("Northern" blot) [1,3]. These blotting techniques combine the high resolution of gel electrophoresis with the high sensitivity and specificity of nucleic acid hybridization. Such blots may increase the specificity of hybridization by eliminating or diluting nonspecific reactions. Analysis of size-fractionated nucleic acid by these methods also permits the characterization of restriction endonuclease fragments of DNA, grouping of the size classes and genome mapping of RNA transcripts, determination of the relative abundance of genes and subgenomic fragments or of transcripts, and the possibility of integration of viral genes in host DNA (see below).

The specificity of a DNA probe may vary greatly with the portion of the genome comprising the probe. Several subgenomic fragments of both herpes simplex virus (HSV) and cytomegalovirus (CMV) have been shown to hybridize with mammalian DNA [10–12]. Whether this cross-reactivity proves to be due to true sequence homology or to guanosine–cytosine rich (GC) sequences will not alter the fact that such fragments are undesirable as probes for viral nucleic acid in the presence of human cells. The virus specificity of a probe can also depend on the fragment selected. The 3' end of enterovirus genomes can be used to detect most enteroviruses [13,14], whereas a poliovirus probe consisting only of the first 220 5'-nucleotides reacts only with poliovirus RNA [14].

The BamH1-A fragment of HSV type 1 is only two- to fourfold less sensitive for detecting HSV type 1 DNA as compared to HSV type 2 DNA [15]. Fragments from the junction regions of HSV-1 and HSV-2 yield absolutely type-specific probes under stringent hybridization and washing conditions [16]. These probes can then be used for the detection, identification, and typing of HSV DNA in specimens from a number of epidemiologically unrelated patients.

Synthetic oligonucleotide probes which are usually 14–30 nucleotides in length can be designed to be much more selective. For example, a synthetic oligonucleotide utilizing sequences from the thymidine kinase gene of HSV can be designed to be absolutely type-specific for type 1, absolutely type-specific for type 2, or completely cross-reactive. In addition, synthetic oligonucleotide probes can be prepared from sequence data without the need for a molecular clone. Unfortunately, sensitivity is roughly proportional to the length of the sequence from which the probe is prepared. A much shorter span of target DNA is available for detection with these oligonucleotide probes. For example, synthetic ^{32}P-labeled oligonucleotide probes are approximately $1/100$th as sensitive as subgenomic cloned HSV DNA, which is usually 1,000–10,000 bp in length (Redfield et al., unpublished observations).

The most frequently utilized method for preparing labeled probe is *nick translation* of double-stranded DNA with deoxynucleotide $5'$-[^{32}P]triphosphates [17]. Several alternative methods of probe preparation have been investigated [18]. In one alternative method, the probe is prepared by cloning the viral sequences in the single-stranded DNA phage M13, and then annealing a commercially available oligonucleotide primer to M13 sequences adjacent to the insert. The M13–primer hybrid is then reacted with T4 DNA polymerase and radioactive and nonradioactive deoxynucleotide triphosphates, and the primer is extended away from the insert. Stopping the reaction at an appropriate time should yield a significant proportion of molecules in which the extended primer spans almost the entire length of the M13 cloning vector, but leaves the inserted sequence single-stranded.

Another method of probe preparation generates single-stranded RNA (ssRNA) probes. In this method, the probe sequence is inserted adjacent to an RNA polymerase promoter that is recognized specifically by the polymerase encoded by either *Salmonella* phage SP6 or *Escherichia coli* phage T7. Although these and several other alternative approaches to probe preparation have theoretical advantages, in practice they are no more sensitive and are often less specific than the standard nick-translated probes [18]. For example, ssRNA probes from Epstein–Barr virus (EBV) will bind to target RNA from cells infected with influenza A virus or human T-cell leukemia/ lymphoma virus type I (HTLV-I) [18]. Northern blotting has demonstrated that this nonspecific reactivity of ssRNA probes (in contrast to nick-translated DNA probes prepared from the same subgenomic clone of EBV) results from the binding of the GC-rich (67%) probe to the large amount of ribosomal RNA in unfractionated RNA extracts of cells (Richman and Wahl, unpublished observations). Even very stringent hybridization and washing conditions could not prevent these tenacious RNA–RNA hybrids from forming, thus leading to a significant background in uninfected samples.

Once the probe is prepared, the target DNA that is fixed to the solid-phase support is then incubated in a solution containing probe; incubation is followed by washes under conditions of buffer and temperature determined to permit maximal specific hybridization and minimal nonspecific "background" binding of the probe. The quantity of specific target sequences is rarely more than 1 ng in the most positive samples, and is typically in the picogram or subpicogram range. In order to minimize the hybridization time required to detect picogram quantities of target sequences, hybridization reactions employ the highest probe concentrations that will not give unacceptable nonspecific background hybridization. Agents that accelerate hybridization rates are also routinely employed [19,20]. The parameters governing reaction rate and hybrid stability, enabling one to derive conditions for obtaining the most rapid, sensitive, and specific assay, have been reviewed in some detail [2,3,18]. Probe bound to the solid support is then detected by autoradiography or enzymatic reaction.

Applications

The uses of nucleic acid probes to detect viral diseases can be divided into two categories: (a) development of techniques for viral diagnosis and (b) studies of viral pathogenesis. The diagnostic applications have not attained practical utility but hold promise for the future. Nucleic acid hybridization has provided a powerful tool for the investigations of viral pathogenesis.

Currently available nucleic acid detection techniques can detect approximately 10^5 copies of target DNA. With studies to detect HSV DNA, this level of sensitivity translates to 1 pg of purified HSV DNA, 10^4 plaque-forming units (PFU), and one to four HSV-infected cells in a population of 10^5 uninfected cells [15]. When an HSV-1 probe is used to analyze male or female genital lesions, 100% of those are detected with an infectivity titer of $> 10^{2.5}$ tissue culture median infective dose ($TCID_{50}$); 90% with $> 10^{2.0}$ $TCID_{50}$; and 40% with $< 10^{2.0}$ $TCID_{50}$ [15]. When the HSV-1 probe is used to detect HSV2, the overall sensitivity is 78% that of viral culture. A 90% sensitivity relative to viral culture is seen when the HSV-1 probe is used to analyze eye swabs from HSV-1-infected rabbits [16]. Use of the homologous probe thus makes a substantial difference in the sensitivity of the assay. However, even with the homologous probe, the DNA detection method is no more sensitive than that of a sensitive enzyme immunoassay [16].

The use of techniques to detect viral components such as DNA is especially useful for agents that are slow or difficult to grow. For example, Spector et al. [12], using the technique just described with cloned restriction endonuclease fragments of cytomegalovirus (CMV), succeeded in detecting cytomegaloviral DNA in urine and buffy-coat specimens from a number of patients, including neonates and bone marrow transplant recipients. Hybridization yielded positive results quickly and often with specimens from a number of immunosuppressed patients, including some whose viral culture results were negative but who were subsequently shown to have significant cytomegaloviral disease. The availability of this technology for an agent like CMV, for example, promises a number of potentially clinically useful applications. With the advent of chemotherapeutically promising nucleoside analogues for CMV and EBV, the identification of patients at high risk for lethal infections could provide an important application of this new diagnostic approach. In addition, DNA hybridization should permit a method for quantitating antiviral effects both in vitro and in vivo; such quantitation is slow and difficult when cell culture techniques are employed [21].

These findings, and similarly encouraging results with EBV [22], adenovirus [23], rotavirus [24], varicella zoster virus [25], human parvovirus [26,27], enteroviruses [14], and hepatitis B virus [28,29], demonstrate the feasibility of using nucleic acid hybridization for viral diagnosis. There are, however, a number of major limitations which have to be overcome before nucleic acid hybridization becomes generally useful. First, the extraction of nucleic acid from clinical materials can be quite cumbersome. The preparation of DNA from eye swabs or from herpetic vesicles and ulcers is a simple process [15]; however, DNA present in buffy coats or swabs of materials that contain large amounts of mucus, such as respiratory secretions or cervical swabs, can be more difficult to prepare. The preparation of DNA from these materials often requires phenol extraction to eliminate the protein that competes with nucleic acid for binding sites on filters. In addition, the hybridization and washing steps remain slow and laborious compared to antigen detection techniques.

A second limitation to the practical application of nucleic acid hybridization for diagnosis is the necessity for radiolabeled probes, especially ^{32}P. Progress in the development of nonradioactive detection systems are reviewed by Haase (Chapter 36, this volume). For the purposes of diagnosis, nonradioactive detection systems will need to be *at least* as sensitive as ^{32}P-labeled probes because these probes are consistently less than 100% sensitive. Whether improved nucleic and detection techniques will make this approach generally more useful than antigen detection remains to be determined; however, certain viruses that have numerous non-cross-reactive serotypes (rhinoviruses, enteroviruses), and viroids, which have no antigens at all, may prove amenable to rapid detection only by nucleic acid hybridization.

The contributions of nucleic acid hybridization to the study of viral pathogenesis, in contrast to diagnosis, is now quite evident. The value of in situ hybridization for the histologic localization of viral infection (for example, in demonstrating EBV replication in the oropharyngeal epithelial cells during infectious mononucleosis [30]) is discussed in detail by Haase in Chapter 36. Filter hybridization permits several additional questions to be addressed. The presence of viruses in malignancies of suspected viral etiology can be documented. For example, DNA hybridization has demonstrated the involvement of hepatitis B virus in hepatocellular carcinoma [21,32], human papillomavirus (HPV) in invasive cervical and cutaneous carcinoma [33–38], and HTLV-I in certain leukemias [39,40]. This is particularly useful for many of those agents that are difficult or impossible to propagate in culture. The physical state of the viral genome in the malignant cell can be examined with restriction endonuclease digestion and Southern blotting. Hepatitis B virus

DNA is randomly integrated in the host cell DNA in hepatocellular carcinoma [30,31], and HPV DNA is episomal in most carcinomas examined to date [33–37], although integrated HPV-18 has been detected in several cervical carcinomas [38]. Mouse mammary tumor virus apparently integrates randomly, but integration of viral DNA in numerous clones of mouse tumor tissue is restricted to one or two sites [39]. In contrast, with another transforming retrovirus, human T-cell leukemia virus type I, the malignant cells in each individual represent a clonal expansion of a cell having a single integration site, with that site being one of multiple, possibly random, sites [40].

Nonintegrated DNA from JC papovavirus has been demonstrated in the organs of patients with progressive multifocal leukoencephalopathy, and is especially concentrated in the diseased areas of brain [42]. The presence of HSV DNA in murine neuronal tissue has been demonstrated with similar techniques, and, yet of unexplained significance, the genome termini become much less prevalent in latently infected tissues [43]. These observations have been extended in mice and confirmed in human trigeminal ganglia [44]. In addition, the presence and size distribution of RNA transcripts in infected tissues can be assessed with Northern blots of RNA extracted from infected tissue.

Conclusion

The utility of nucleic acid hybridization for the study of viral pathogenesis is well documented. The theoretical advantages of detecting viral DNA in clinical material for diagnosis have been considered, and studies have demonstrated the feasibility of this approach. The actual adaptation of nucleic acid hybridization for clinical diagnosis awaits faster and simpler methods for specimen preparation and hybridization, and more sensitive detection systems that do not utilize radioisotopes.

Acknowledgment

This work was supported by the Veterans Administration and the National Institutes of Health (AI-2268 and HL-32471).

References

1. Maniatis T, Fritsch EF, Sambrook J (1982) Molecular Cloning: A Laboratory Manual. Cold Spring Harbor Laboratory, Cold Spring Harbor, New York
2. Minson AC, Darby G (1982) Hybridization techniques. In New Developments in Practical Virology. Alan R. Liss Inc, New York pp 67–82
3. Meinkoth J, Wahl GM (1984) Hybridization of nucleic acids immobilized on solid supports. Anal Biochem 138:267–284

4. Kafatos FC, Weldon-Jones C, Efstratiadis A (1979) Determination of nucleic acid sequence homologies and relative concentrations by a dot hybridization procedure. Nucleic Acids Res 7:1541–1552

5. Thomas PS (1980) Hybridization of denatured RNA and small DNA fragments transferred to nitrocellulose. Proc Natl Acad Sci USA 77:5201–5205

6. Church GM, Gilbert W (1984) Genomic sequencing. Proc Natl Acad Sci USA 81:1991–1995

7. Bresser J, Doering J, Gillespie D (1983) Quick-blot: Selective mRNA or DNA immobilization from whole cells. DNA 2:243–249

8. White BA, Bancroft FC (1982) Cytoplasmic dot hybridization. J Biol Chem 257:8569–8572

9. Southern EM (1975) Detection of specific sequences among DNA fragments separated by gel electrophoresis. J Mol Biol 98:503–517

10. Peden K, Mounts P, Hayward GS (1982) Homology between mammalian cell DNA sequences and human herpesvirus genomes detected by a hybridization procedure with a high-complexity probe. Cell 31:71–80

11. Puga A, Cantin EM, Notkins AL (1982) Homology between murine and human cellular DNA sequences and the terminal repetition of the S component of herpes simplex virus type 1 DNA. Cell 31:81–87

12. Spector S, Rau JA, Spector D, McMillan R (1984) Detection of human cytomegalovirus in clinical specimens by DNA–DNA hybridization. J Infect Dis 150:121–126

13. Hyypia T, Stalhandeske P, Vainionpaa R, Pettersson U (1984) Detection of enteroviruses by spot hybridization. J Clin Microbiol 19:436–438

14. Rotbart HA, Levin M, Villarreal LP (1984) Use of subgenomic poliovirus DNA hybridization probes to detect the major subgroups of enteroviruses. J Clin Microbiol 20:1105–1108

15. Redfield DC, Richman DD, Albanil S, Oxman NM, Wahl GM (1983) Detection of herpes simplex virus in clinical specimens by DNA hybridization. Diag Microbiol Infect Dis 1:117–128

16. Richman DD, Cleveland PH, Redfield DC, Oxman NM, Wahl GM (1984) Rapid viral diagnosis. J Infect Dis 149:298–310

17. Rigby PW, Dieckmann M, Rhodes C, Berg P (1977) Labeling deoxyribonucleic acid to high specific activity in vitro by nick translation with DNA polymerase. J Mol Biol 113:237–251

18. Wahl GM, Albanil S, Ignacio K, Richman DD (1985) Nucleic acid hybridization: A powerful technology useful for medical diagnosis. In Medical Virology IV, de la Maza, L.M., Peterson, E.M. (eds). L. Erlbaum Associates, Hillsdale, N.J., 1985

19. Wahl GM, Stein M, Stark GR (1979) Efficient transfer of large DNA fragments from agarose gels to diazobenzyloxymethyl–paper and rapid hybridization by using dextransulfate. Proc Natl Acad Sci USA 76:3683–3687

20. Renz M, Kurz C (1984) A colorimetric method for DNA hybridization. Nucleic Acids Res 12:3435–3444

21. Chou S, Merigan T (1982) Rapid detection and quantitation of human cytomegalovirus in urine through DNA hybridization. N Engl J Med 308:921–925

22. Andiman W, Gradoville L, Heston L, Neydorff R, Savage ME, Kitchingman G, Shedd D, Miller G (1983) Use of cloned probes to detect Epstein–Barr viral DNA in tissues of patients with neoplastic and lymphoproliferative diseases. J Infect Dis 148:967–977

23. Virtanen M, Laaksonen M, Soderlund H, Palva A, Halonen P, Ranki M (1983) Novel test for rapid viral diagnosis: Detection of adenovirus in nasopharyngeal mucus aspirates by means of nucleic-acid sandwich hybridization. Lancet i:381–383

24. Flores G, Purcell RH, Perez I, Wyatt RG, Boeggeman E, Sereno M, White L, Chanock RM, Kapikian A (1983) A dot hybridization assay for detection of rotavirus. Lancet i:555–558

25. Seidlin M, Takiff HE, Smith HA, Hay J, Straus SE (1984) Detection of varicella–zoster virus by dot-blot hybridization using a molecularly cloned viral DNA probe. J Med Virol 13:53–61

26. Anderson MJ, Jones SE, Minson AC (1985) Diagnosis of human parvovirus infection by dot-blot hybridization using cloned viral DNA. J Med Virol 15:163–172

27. Clewley JP (1985) Detection of human parvovirus using a molecularly cloned probe. J Med Virol 15:173–182

28. Berninger M, Hammer M, Hoyer B, Gerin JL (1982) An assay for the detection of the DNA genome of hepatitis B virus in serum. J Med Virol 9:57–68

29. Weller IVD, Fowler MJF, Monjardino J, Thomas HC (1982) The detection of HBV-DNA in serum by molecular hybridization: A more sensitive method for the detection of complete HBV particles. J Med Virol 9:273–280

30. Sixbey JW, Nedrud JG, Raab-Traub N, Hanes RA, Pagano JS (1984) Epstein–Barr virus replication in oropharyngeal epithelial cells. N Engl J Med 310:1225–1230

31. Shafritz DA, Shouval D, Sherman HI, Hadziyannis SJ, Kew MC (1981) Integration of hepatitis B virus DNA into the genome of liver cells in chronic liver disease and hepatocellular carcinoma: Studies in percutaneous liver biopsies and post-mortem tissue specimens. N Engl J Med 305:1067–1073

32. Brechot C, Hadchouel M, Scotto M et al. (1981) State of hepatitis B virus DNA in hepatocytes of patients with hepatitis B surface antigen-positive and -negative liver diseases. Proc Natl Acad Sci USA 78:3906–3910

33. Zachow KR, Ostrow RS, Bender M, Watts S, Okagaki T, Pass F, Faras AJ (1982) Detection of human papillomavirus DNA in anogenital neoplasias. Nature 300:771–773

34. Ostrow RS, Bender M, Niimura M, Seki T, Kawashima M, Pass F, Faras AJ (1982) Human papillomavirus DNA in cutaneous primary and metastasized squamous cell carcinomas from patients with epidermodysplasia verriciformis. Proc Natl Acad Sci USA 79:1634–1638

35. Green M, Brackmann KH, Sanders PR, Lowenstein PM, Freel JH, Eisinger M, Switlyk SA (1982) Isolation of a human papillomavirus from a patient with epidermodysplasia verruciformis: Presence of related viral DNA genomes in human urogenital tumors. Proc Natl Acad Sci USA 79:4437–4441

36. Gissmann L, Wolnik L, Ikenberg H, Koldovsky U, Schnurch HG, zur Hausen H (1983) Human papillomavirus types 6 and 11 DNA sequences in genital and laryngeal papillomas and in some cervical cancers. Proc Natl Acad Sci USA 80:560–563

37. Durst M, Gissmann L, Ikenberg H, zur Hausen H (1983) A papillomavirus DNA from a cervical carcinoma and its prevalence in cancer biopsy samples from different geographic regions. Proc Natl Acad Sci USA 80:3812–3815

38. Boshart M, Gissman L, Ikenberg H, Kleinheinz A, Scheurlen W, zur Hausen H (1984) A new type of papillomavirus DNA, its presence in genital cancer biop-

sies and in cell lines derived from cervical cancer. Eur Mol Biol Org J 3:1151–1157

39. Wong-Staal F, Hahn B, Manzari V, Colombini S, Franchini G, Gelmann E, Gallo RC (1983) A survey of human leukemias for sequences of a human retrovirus. Nature 302:626–628
40. Seiki M, Eddy R, Shows RB, Yoshida M (1984) Nonspecific integration of the HTLV provirus genome into adult T-cell leukemia cells. Nature 309:640–642
41. Peters G, Kozak C, Dickson C (1984) Mouse mammary tumor virus integration regions int-1 and int-2 map on different mouse chromosomes. Mol Cell Biol 4:375–378
42. Grinnell BW, Padgett BL, Walker DL (1983) Distribution of nonintegrated DNA from JC papovavirus in organs of patients with progressive multifocal leukoencephalopathy. J Infect Dis 147:669–675
43. Rock DL, Fraser NW (1983) Detection of HSV-1 genome in central nervous system of latently infected mice. Nature 302:523–525
44. Efstathiou S, Minson AC, Field HJ, Anderson JR, Wildy P (1986) Detection of herpes simplex virus-specific DNA sequences in latently infected mice and in humans. J Virol 57:446–455

CHAPTER 36

In Situ Hybridization and Covert Virus Infections

ASHLEY T. HAASE

Introduction: The Rationale for Using In Situ Hybridization to Detect Covert Infections

In the course of the productive life cycle of a virus, the macromolecular machinery of the cell is redirected to reproduce viral genomes and virion components. As a consequence, there is abundant evidence of infection detectable by conventional assays for infectious virus, virus particles, or subcomponents. In covert infections, however, viral particles and components are produced in insufficient quantities for detection by conventional diagnostic methods. In this situation, a search must be made for the viral genome itself.

Viral genomes can be detected by hybridization of virus-specific probes to nucleic acids either in extracts from the tissue or in individual cells in tissue sections (i.e., in situ hybridization). If viral genomes are widely dispersed in the tissue, analyses of extra nucleic acid isolated from the total population of cells will be successful. If, on the other hand, infection is focal, this kind of analysis will likely fail because virus-specific sequences will be diluted beyond the limits of detection by nucleotide sequences from uninfected cells. In situ hybridization will be needed to detect virus genes in covertly infected cells when infection involves less than about 5% of the cells.

Anticedents and Precedents for Using In Situ Hybridization to Reveal Viral Genes in Covert Infections

A growing body of evidence attests to the necessity and utility of in situ hybridization for discovering viral genes in tissues, because in fact in these situations infection is highly focal and viral gene expression is restricted. These general conclusions rest on analysis of slow and persistent infections of humans and animals, particularly visna and subacute sclerosing panencephalitis (SSPE).

Visna is a slowly evolving paralytic disease of sheep caused by a retrovirus in the subfamily of lentiviruses [1]. In productive infections in tissue culture, the virus multiplies to high titers in three or four days, and kills the infected cell by a process that involves cell fusion. Infected cells contain viral antigens, which are readily demonstrable by immunofluorescence; they also contain 100 copies of DNA and 10,000 copies of viral RNA, which are detectable by solution or solid-phase hybridization of specific probes to nucleic acids extracted from the population of cells [2]. By contrast, in infected animals there is little free virus after the animals mount a cellular and humoral immune response [3–5]: Infectious virus cannot usually be recovered directly from tissue homogenates; viral particles and antigens are only rarely detectable; and viral genomes cannot be detected in nucleic acid extracted from tissues. However, viral DNA and RNA are detectable by in situ hybridization in about 0.1% to 1% of the cells at copy numbers two orders of magnitude lower than in productive infections in tissue culture [6,7].

In visna then, most infected cells harbor the viral genome in a repressed state. Because of the decreased synthesis of viral RNA and proteins, few cells produce infectious virus particles, and there is therefore little overt evidence of infection in tissues examined directly for antigens, particles, or infectivity. Moreover, the 100- to 1000-fold dilution of viral DNA by DNA from uninfected cells during isolation for hybridization accounts for the failure of solution or blot hybridization to detect viral genomes in the tissues. Viral DNA or RNA is readily detected by in situ hybridization in 0.1% to 1% of the cells in a typical tissue section with 50,000 to 100,000 cells.

Subacute sclerosing panencephalitis (SSPE) provides another good example of the utility of in situ hybridization in uncovering unsuspected infection. SSPE is a rare neurological complication of measles virus infection in which several years separate a typical acute case of measles with apparent recovery, from a slowly progressive and fatal disease of the nervous system. In the terminal phase of the disease, the diagnosis can be made by finding inclusion bodies and measles antigens in tissues, and by virus isolation in explant cultures [8]. At an earlier stage of disease, however, the majority of infected cells lack sufficient antigen for detection by the most sensitive contemporary methods, and have only a few copies of viral RNA in each cell [9]. In this earlier stage of SSPE, infection again is focal, but cells with five to ten copies of measles RNA and comprising less than 1% of the population of cells can be detected by in situ hybridization in biopsies taken for diagnosis.

Restricted Gene Expression as a General Theme in Slow and Persistent Infections

In both visna and SSPE, in its earlier stages, the greatly diminished synthesis of viral genes and gene products provides a mechanism and an explanation for the persistence of virus and the slow evolution of infection. Because most infected cells produce little if any particles of antigens, the virus genome is able to persist in cells because infected cells are not detected and destroyed by immune surveillance. In addition, because the tissue-destructive effects of infection, whether directly or indirectly mediated, also depend on the level of synthesis of viral components, the same restriction in viral gene expression slows the tempo of pathological changes, and thus prolongs the incubation period preceding overt disease. There are now numerous examples, spanning the taxonomic diversity of viruses, of restricted synthesis and expression of viral genomes in slow and persistent infections [10–12].

The Search for Virus Genes in Chronic Diseases of Humans

One of the major implications of the foregoing analyses of slow and persistent infections is the possibility of unsuspected viral infections in chronic diseases of humans. Certainly, if restricted gene expression and focal infection are the rule in slow infections, there is no reason to be discouraged by negative results of assays for infectious virus, particles, or antigens; or by negative hybridization analyses of extracted nucleic acids.

Multiple sclerosis (MS) is of particular interest in this respect, because of epidemiological evidence consistent with remote viral infection as an important etiologic component of the disease, and because of the exaggerated immune response of these patients to measles and other viruses [13,14,15]. Measles virus antigens have not been detected in MS, but recently measles RNA has been detected by in situ hybridization in about half of the cases of MS, but also in an equivalent proportion of non-MS controls. These observations, and other recent reports [16,17] of polio- and herpesvirus genes in the CNS of humans, leave little doubt that viral genes may be deposited in the brain quite frequently in the course of common viral infections. The question as to the etiological role, if any, that viruses play in chronic diseases, however, remains unanswered. Viral genes in tissues could simply represent adventitious infection without pathological significance; or, alternatively, the virus gene expression could be the initiating stimulus for a pathological process in susceptible individuals. Resolution of this central issue is beyond the scope of any single experimental strategy.

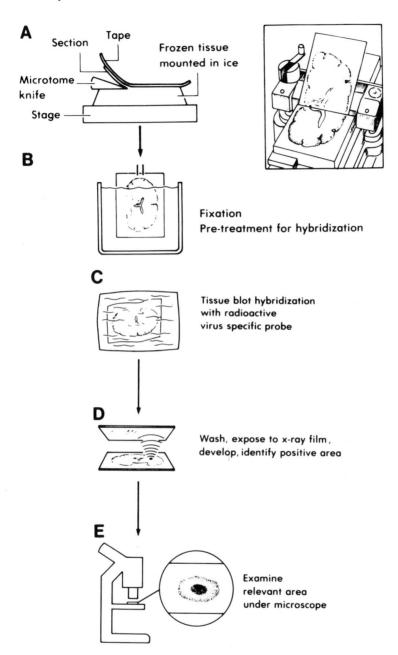

Figure 36.1. Principle of the combined macroscopic–microscopic hybridization assay. (Reprinted by permission from Haase et al., *Virology* 140: 201–206, 1985.)

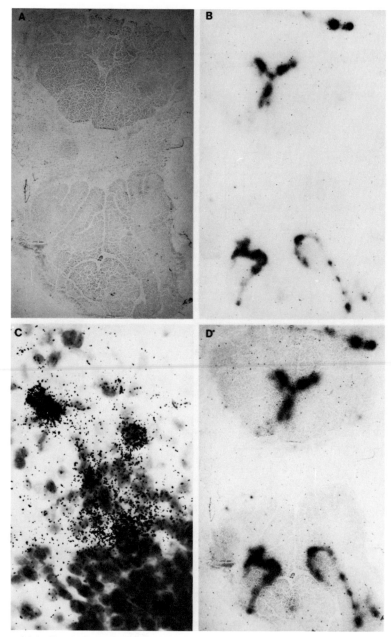

Figure 36.2. Detection of visna virus RNA by the combined macroscopic–microscopic hybridization assay. A visna virus-specific probe labeled with ^{125}I and ^{35}S was hybridized to two large coronal sections of brain from an infected animal. (**A**) Coronal sections after hybridization. (**B**) Macroscopic radioautograph. (**C**) Microscopic radioautograph. (**D**) Superimposed section and macroscopic radioautograph. Viral RNA is present in cells at the periphery of an inflammatory joint bordering the lateral ventricles (**B, C**). (Reprinted by permission from Haase et al, *Virology* 140: 201–206, 1985.)

Technological Underpinnings, Advances, and Prospects

Contemporary in situ hybridization methods [18] are capable of revealing even single copies of as small a genome as that of hepatitis B virus (3.2 bp) [19]. These impressive sensitivities have been achieved by (a) refinements in methodology that optimize diffusion and binding of probe to specific target sequences in cells; and (b) the high specific activities and efficiency of latent image formation of probes labeled with ^{35}S and ^{125}I. Probes have been labeled with ^{125}I and ^{35}S, to specific activities approaching 10^{10} dpm/μg [20]; with efficiencies of latent image formation that are two to five fold higher than ^{3}H, the theoretical sensitivities of detection are in the range of 10^{-18} g of virus nucleotide sequence per cell.

The ability to detect viral genes in single cells by in situ hybridization has recently been greatly augmented in sampling power by adapting methods of whole body radioautography for use in screening whole organs or animals (see Southern and Oldstone, Chapter 18, this volume). In the combined macroscopic–microscopic method, whole animals or organs are frozen, and sections are cut and transferred to clear tape [21] (Figure 36.1). The tape is hybridized to probes labeled with both ^{125}I and ^{35}S to generate signals detectable both with x-ray film and nuclear track emulsion. The anatomic distribution of viral genes is first determined at low resolution by aligning the section with x-ray film (Figure 36.2A, B, D) to locate regions that have bound virus-specific probe (Figures 36.1E, 36.2C). After the tape has been coated with emulsion, exposed, and developed, these regions are examined microscopically to assess the cellular distribution of viral genes. This and cognate methods should provide important new information about the distribution of viral or other genes in the tissues of humans and animals, and should greatly enhance our ability to detect viral genes in covert infections.

References

1. Brahic M, Haase AT (1981) Lentivirinae: Maedi/visna virus group infections. Comp Diag Viral Dis 4:619–643
2. Haase AT, Stowring L, Harris J, Traynor B, Ventura P, Peluso R, Brahic M (1982) Visna DNA synthesis and the tempo of infection in vitro. Virology 119:399–410
3. Narayan O, Griffin DE, Silverstein AM (1977) Slow virus infection: Replication and mechanisms of persistence of visna virus in sheep. J Infect Dis 135:800–806
4. Griffin DE, Narayan O, Adams RJ (1978) Early immune responses in visna, a slow viral disease of sheep. J Infect Dis 138:340–350
5. Petursson G, Nathanson N, Georgsson G, Panitch H, Palsson P (1976) Pathogenesis of visna. Lab Invest 35:402–412
6. Haase AT, Stowring L, Narayan O, Griffin D, Price D (1977) Slow persistent infection caused by visna virus: Role of host restriction. Science 195:175–177
7. Brahic M, Stowring L, Ventura P, Haase AT (1981) Gene expression in visna virus infection in sheep. Nature 292:240–242.

8. Agnarsdottir G (1977) Subacute sclerosing panencephalitis. Recent Adv Clin Virol 1:21–49
9. Haase AT, Gantz D, Eble B, Walker D, Stowring L, Ventura P, Blum H, Wietgrefe S, Zupancic M, Tourtellotte W, Gibbs Jr CJ, Norrby E, Rozenblatt S (1985) Natural history of restricted synthesis and expression of measles virus genes in subacute sclerosing panencephalitis. Proc Natl Acad Sci USA 82:3020–3024
10. Koch EM, Neubert WJ, Hofschneider J (1984) Lifelong persistence of a paramyxovirus sendai-6/94 in C129 mice: Detection of a latent viral RNA by hybridization with a cloned genomic cDNA probe. Virology 136:78–88
11. Brahic M, Stroop W, Baringer J (1981) Theiler's virus persists in glial cells during demyelinating disease. Cell 26:123–128
12. Lavi E, Gilden D, Kighkin M, Weiss S (1984) Persistence of mouse hepatitis virus A59 RNA in a slow virus demyelinating infection in mice as detected by in situ hybridization. J Virol 51:563–566
13. Weiner LP, Johnson RT, Herndon RM (1973) Viral infections and demyelinating diseases. N Engl J Med 288:1103–1110
14. Haase AT, Ventura P, Gibbs Jr CJ, Tourtellotte WW (1981) Measles virus nucleotide sequences: Detection by hybridization in situ. Science 212:672–675
15. Haase AT, Stowring L, Ventura P, Burks J, Ebers G, Tourtellotte W, Warren K (1985) Detection by hybridization of viral infection of the human central nervous system. Ann N Y Acad Sci 436:103–108
16. Brahic M, Haase AT, Cash E (1984) Simultaneous in situ detection of viral RNA and antigens. Proc Natl Acad Sci USA 81:544–5448
17. Brahic M, Simth RA, Gibbs Jr CJ, Garruto RM, Tourtellotte WW, Cash E (1985) Detection of picornavirus sequences in nervous tissue of amyotrophic lateral sclerosis and control patients. Ann Neurol (in press)
18. Haase AT, Brahic M, Stowring L, Blum H (1984) Detection of viral nucleic acids by in situ hybridization. Methods Virol 7:189–226
19. Blum H, Stowring L, Figus A, Montgomery C, Haase AT, Vyas G (1983) Detection of hepatitis B virus DNA in hepatocytes, bile duct epithelium, and vascular elements by in situ hybridization. Proc Natl Acad Sci USA 80
20. Blum H, Haase AT, Harris JD, Walker D, Vyas G (1984) Asymmetric replication of hepatitis B virus DNA in human liver: Demonstration of cytoplasmic minus-strand DNA by blot analyses and in situ hybridization. Virology 139:87–96
21. Haase AT, Gantz D, Blum H, Stowring H, Ventura P, Geballe A, Moyer B, Brahic M (1985) Combined macroscopic and microscopic detection of viral genes in tissues. Virology 140:201–206

CHAPTER 37

Oligonucleotide Fingerprinting in the Investigations of Outbreaks of Viral Disease

OLEN KEW AND BALDEV NOTTAY

Introduction

Oligonucleotide fingerprinting is a rapid and powerful method for determining sequence relatedness among RNA molecules [1]. Fingerprinting has found wide application in virology, particularly in monitoring the distribution of specific genotypes of RNA viruses in the natural environment [2,3]. The capacity of fingerprinting to identify different genotypes among strains of the same serotype has provided epidemiologists with a precise tool for independently tracking their epidemic transmission.

The precision of the epidemiologic data available from fingerprint analysis is directly related to the rate at which the viral genome evolves during replication in the natural host. With rapidly evolving viruses, such as influenza or polio viruses, significant evolution occurs during the course of an outbreak, such that epidemic transmission pathways can be reconstructed by analysis of the pattern of cumulative changes in the fingerprints of temporally ordered isolates [4,5]. In such systems, substantial similarities (e.g., >50% of the large oligonucleotide spots shared) in the fingerprints of independent isolates provide unambiguous evidence for epidemiologic linkage between infections.

In contrast, some RNA viruses, including many with arthropod vectors, evolve so slowly in nature that viruses with very similar fingerprints can be isolated over many years [6–9]. However, many of these have restricted geographic ranges, and are described as *topotypes* [8,9]. In these systems, fingerprinting is most useful in determining the geographic range of a topotype over time, in order to provide insights into the complex network of biological parameters controlling the distribution of specific viral genotypes in nature.

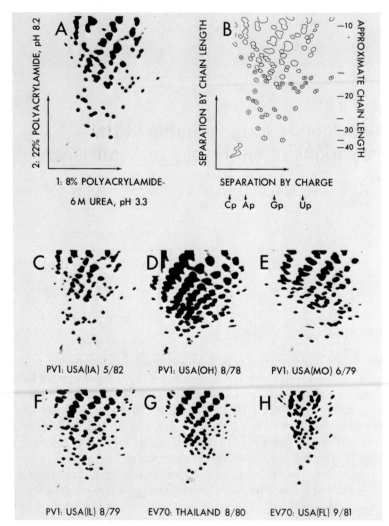

Figure 37.1. Oligonucleotide fingerprints of representative human enterovirus isolates.
(A, B): Fingerprint of the Sabin type 1 polio vaccine strain (A) and interpretive tracing
(B) illustrating the mechanisms of electrophoretic separation of the oligonucleotide
fragments in polyacrylamide gels. The large, characteristic oligonucleotides occupy
the lower half of each fingerprint. Relative first-dimension mobilities of each nucleoside
monophosphate are shown. Enumeration of oligonucleotide spots follows the nomenclature of Nomoto et al. [27]. **(C–H)** Panels C–F illustrate differentiation among
four poliovirus isolates of the same serotype. The fingerprint of a vaccine-related
isolate (C) is nearly identical to that of Sabin 1 (A). The reduced-intensity oligonucleotide spot 7 along with the presence of an additional spot immediately to its left
suggests that the virus population is heterogeneous, containing roughly equal ratios
of genomes having wild-type and mutant oligonucleotide 7. Three wild-type 1 polioviruses, isolated from poliomyelitis patients in the United States over a 12-month
period, had very different fingerprints, indicating separate epidemiologic origins of
disease. Earlier-case isolates from New York, the Netherlands, and Mexico had fin-

Principles of the Method

Viral RNA is digested to completion with ribonuclease T_1, which specifically cleaves after guanosine residues. To a first approximation, the guanosine residues are distributed randomly along the RNA molecule, such that digests of entire RNA virus genomes contain 2000–4000 oligonucleotide fragments varying in length from 1 to >30 base residues. Digests are resolved by two-dimensional electrophoresis: first, according to charge, which follows from base composition (8% polyacrylamide, 6 M urea–citrate, pH 3.3); and second, according to chain length (22% polyacrylamide, 50 mM Tris–borate, pH 8.2). The large oligonucleotides (≥ 12 base residues, representing 10% to 15% of the genome), having sequences represented only once in the genome, separate into easily recognizable patterns, or "fingerprints," of 30 or more spots that are highly characteristic for each RNA sequence (Figure 37.1). Because the characteristic oligonucleotides derive from all regions of the RNA molecule, sequence variations among strains may be surveyed over the entire genome, with the results obtained in a readily interpretable, pictorial form.

Detailed protocols for fingerprint analysis have been described recently [1–3,10]. Efficient methods exist for fingerprinting both single- and double-stranded RNAs [2,3,11], and for the separate analysis of each component of a segmented genome [3]. Viral RNAs can be labeled in vivo by the biosynthetic incorporation of [^{32}P]orthophosphate, or in vitro by radiophosphorylation of the 5′ ends of the oligonucleotide fragments [10]. Although most of the original fingerprinting procedures used large polyacrylamide gels (approximately 40 cm in the direction of migration), we routinely obtain high-resolution fingerprints by using smaller gels (16 × 16 cm), which are much easier to prepare and faster to run.

Fingerprinting is extremely sensitive to small mutational differences in viral genomes [12]. Any base change will alter the mobility of an oligonucleotide. If the change occurs in the approximately 10% of the genome encoding the largest oligonucleotides, then one of the characteristic spots will appear to have shifted in the fingerprint pattern. Therefore, approximately one of every 10 base changes will register as an altered characteristic spot. Because the characteristic spots are ≥ 12 base residues long, an overall base sequence divergence of 10% causes mobility shifts in approximately 75% of the characteristic spots [12]. As a result, quantitative estimates of the extent of overall sequence divergence are possible only when the fingerprint analyses compare very closely related strains.

---

gerprints very similar to those shown on panels D, E, and F, respectively, thereby confirming the independence of the later domestic cases. A single basic genotype of enterovirus 70 was the agent of a major pandemic of acute hemorrhagic conjunctivitis [21], as demonstrated in part by the similarities in the fingerprints of isolates from Thailand (G) and the United States (H).

Oligonucleotide Fingerprinting in Molecular Epidemiology

Representatives of nearly every RNA virus group have been fingerprinted. Comprehensive reviews of the many applications of oligonucleotide fingerprinting to the characterization of RNA viruses and to the monitoring of their epidemic and zoonotic transmission have appeared recently [2,3]. Space limitations allow only a few examples to be presented here.

Fingerprinting is now a standard method for following the transmission of influenza A viruses. The evolution of H1N1 and H3N2 viruses during pandemics was shown to occur both by selection of new substitution mutants [4,13] and by genetic reassortment [14,15]. In contrast, influenza C viruses from distant geographic regions isolated over a 32-year period were found to be very closely related [16].

Zoonotic isolates of foot-and-mouth disease virus have been widely studied by fingerprinting [17,18]. For example, a 1981 outbreak of foot-and-mouth disease in France and the Channel Islands was traced to the probable escape from containment of a 1965 strain used as seed virus in a vaccine production facility [18]. The epidemiology of polioviruses [19] and other enteroviruses [20,21] has also been investigated by fingerprinting. Isolates derived from the live poliovaccines can be readily distinguished from wild viruses by fingerprinting [19], providing information critical to outbreak control programs. Fingerprinting has been used to demonstrate links between cases and outbreaks of poliomyelitis that were not suspected from other epidemiologic data [5,19].

Because genomic variation is an essential prerequisite for distinguishing among related strains, fingerprinting cannot reliably recognize epidemiologic links between infections when the agents evolve very slowly. For example, independent isolates of vesicular stomatitis virus (a rhabdovirus [6]) and alphaviruses of the Western equine encephalitis complex [8], obtained over several decades from widely separated areas, had, within each group, very similar fingerprints. Such observations effectively preclude the use of fingerprinting in these cases to track the origins of isolates.

Fingerprinting Compared with Other Methods for Identifying Viruses

As an epidemiologic tool, fingerprinting is generally competitive with the best modern serologic methods, including analyses using panels of monoclonal antibodies [22]. Interpretations of fingerprint data are not complicated by problems commonly encountered in serologic studies: antigenic drift among closely related viruses, convergence of antigenic structure among more distantly related viruses, heterogeneity of immune responses of different experimental animals, or differences in the relative immunodominance of antigenic sites arising from variations in experimental immunization conditions. Further, fingerprinting does not require production of new reagents in order to positively characterize new isolates. Finally, because most natural mutations generate synonymous codons, fingerprinting has the additional advantage of being able to detect a predictable fraction of all genomic changes, regardless of whether the mutations alter polypeptide structure or biological activity.

The chief theoretical limitation to fingerprinting is its often short epidemiologic range. This is a consequence of the high mutability typical of RNA genomes [23] and the great sensitivity of fingerprinting to mutation. In contrast, restriction endonuclease analysis, frequently used to identify DNA viruses [24], is at least an order of magnitude less sensitive to substitution mutations [25], which become fixed into DNA genomes at frequencies many orders of magnitude lower than those generally observed with RNA genomes [23].

The major practical limitation to fingerprinting is that it is generally too slow for most routine diagnostic applications. Analyses generally take a minimum of two days for completion after preparation of labeled RNA. However, fingerprint studies provide valuable information on patterns of epidemic transmission, and establish a basis for application of other molecular techniques that overcome the primary limitations of fingerprinting. For example, sequence analysis of the oligonucleotides characteristic of an outbreak strain [2,3,26,27] opens the way for design of synthetic deoxyoligonucleotides complementary to specific sites along the genome [28]. The synthetic DNA molecules can serve as sequence-specific hybridization probes that allow identification of viral genotypes by rapid and simple blot hybridization methods [28,29], procedures that can easily be adapted to large-scale analyses. Genomic sequence analyses are able to detect genetic relationships over much longer intervals of evolutionary (or epidemiologic) time. The synthetic deoxyoligonucleotides may also be used as primers for reverse transcriptase-catalyzed primer extension sequence analyses [30] to establish a broader view of the patterns of infectious transmission [31]. Clearly, oligonucleotide fingerprinting, in conjunction with these supplementary techniques, will remain an indispensable method for monitoring the transmission of RNA viruses in human and animal populations.

References

1. DeWachter R, Fiers W (1982) Two-dimensional electrophoresis of nucleic acids. *In* Rickwood D, Hames BD (eds) Gel Electrophoresis of Nucleic Acids: A Practical Approach. IRL Press, Oxford, p 77
2. Clewley JP, Bishop DHL (1982) Oligonucleotide fingerprinting of viral genomes. *In* Howard CR (ed) New Developments in Practical Virology. Alan R. Liss, New York, p 231
3. Kew OM, Nottay BK, Obijeski JF (1984) Applications of oligonucleotide fingerprinting to the identification of viruses. *In* Maramorosch K, Koprowski H (eds) Methods in Virology, vol 8. Academic Press, New York, p 41

4. Young JF, Desselberger U, Palese P (1979) Evolution of human influenza viruses in nature: Sequential mutations in the genomes of new H1N1 isolates. Cell 18:73–83

5. Nottay BK, Kew OM, Hatch MH, Heyward JT, Obijeski JF (1981) Molecular variation of type 1 vaccine-related and wild polioviruses during replication in humans. Virology 108:405–423

6. Clewley, JP, Bishop DH, Kang CY, Coffin J, Schnitzlein WM, Reichmann ME, Shope RE (1977) Oligonucleotide fingerprints of RNA species obtained from rhabdoviruses belonging to the vesicular stomatitis virus subgroup. J Virol 23:152–166

7. Trent DW, Grant JA (1980) A comparison of New World alphaviruses in the Western equine encephalitis complex by immunochemical and oligonucleotide fingerprint techniques. J Gen Virol 47:261–282

8. Trent DW, Grant JA, Vorndam AV, Monath TP (1981) Genetic heterogeneity among Saint Louis encephalitis virus isolates of different geographic origin. Virology 114:319–332

9. El Said LH, Vorndam V, Gentsch JR, Clewley JP, Calisher CH, Klimas RD, Thompson WH, Grayson M, Trent DW, Bishop DHL (1979) A comparison of La Crosse virus isolates obtained from different ecological niches and an analysis of the structural components of California encephalitis serogroup viruses and other bunyaviruses. Am J Trop Med Hyg 28:364–386

10. Pedersen FS, Haseltine WA (1980) A micromethod for detailed characterization of high molecular weight RNA. Methods Enzymol 65:680–687

11. Walker PJ, Mansbridge JN, Gorman BM (1980). Genetic analysis of orbiviruses using RNase T_1 oligonucleotide fingerprints. J Virol 34:583–591

12. Aaronson RP, Young JF, Palese P (1982) Oligonucleotide mapping: Evaluation of its sensitivity by computer simulation. Nucleic Acids Res 10:237–246

13. Ortin H, Najera R, Lopez C, Davila M, Domingo E (1980) Genetic variability of Hong Kong (H3N2) influenza viruses: Spontaneous mutations and their locations in the viral genome. Gene 11:319–332

14. Young JF, Palese P (1979) Evolution of influenza A viruses in nature: Recombination contributes to genetic variation of H1N1 strains. Proc Nat Acad Sci USA 76:6547–6551

15. Nakajima K, Nakajima S, Suguira A (1982) The possible origin of H3N2 influenza virus. Virology 120:504–509

16. Meier-Ewert H, Petri T, Bishop DHL (1981) Oligonucleotide fingerprint analyses of influenza C virion RNA recovered from five different isolates. Arch Virol 67:141–147

17. Domingo E, Davila M, Ortin J (1980) Nucleotide sequence heterogeneity of the RNA from a natural population of foot-and-mouth disease virus. Gene 11:333–346

18. King AMQ, McCahon D, Slade WR, Newman JWI (1982) Biochemical identification of viruses causing the 1981 outbreaks of foot and mouth disease in the UK. Nature 293:479–480

19. Kew OM, Nottay BK (1984) Molecular epidemiology of polioviruses. Rev Infect Dis 6 (Suppl. 2):S499–S504

20. Harris TJR, Robson K, Brown F (1977) Molecular aspects of the antigenic variation of swine vesicular disease and coxsackie B5 viruses. J Gen Virol 35:299–315

21. Kew OM, Nottay BK, Hatch MH, Hierholzer JC, Obijeski JF (1983) Oligonucleotide fingerprint analysis of enterovirus 70 isolates from the 1980 to 1981 pandemic of acute hemorrhagic conjunctivitis: Evidence for a close genetic relationship among Asian and American strains. Infect Immun 41:631–635
22. Gerhard W, Koprowski H (1984) Monoclonal antibodies. *In* Notkins AL, Oldstone MBA (eds) Concepts in Viral Pathogenesis, vol 1. Springer-Verlag, New York, p 376
23. Holland JJ (1984) Continuum of change in RNA virus genomes. *In* Notkins AL, Oldstone MBA (eds) Concepts in Viral Pathogenesis, vol 1. Springer-Verlag, New York, p 137
24. Wadell G (1984) Molecular epidemiology of human adenoviruses. Curr Top Microbiol Immunol 110:191–220
25. Brown WM, George M, Wilson AC (1979). Rapid evolution of animal mitochondrial DNA. Proc Nat Acad Sci USA 76:1967–1971
26. Donis-Keller H, Maxam AM, Gilbert W (1977) Mapping adenines, guanines and pyrimidines in RNA. Nucleic Acids Res 4:2527–2538
27. Nomoto A, Kitamura N, Lee JJ, Rothberg PG, Imura N, Wimmer E (1981) Identification of point mutations in the genomes of the poliovirus Sabin vaccine LSc 2ab, and catalogue of RNase T_1- and RNase A-resistant oligonucleotides of poliovirus type 1 (Mahoney) RNA. Virology 112:217–227
28. Studencki AB, Wallace RB (1984) Allele specific hybridization using oligonucleotide probes of very high specific activity: Discrimination of the human β^A and β^s globin genes. DNA 3:7–15
29. Meinkoth J, Wahl G (1984) Hybridization of nucleic acids immobilized on solid supports. Anal Biochem 138:267–284
30. Zimmern D, Kaesberg P (1978) 3′-Terminal nucleotide sequence of encephalomyocarditis virus RNA determined by reverse transcriptase and chain-terminating inhibitors. Proc Natl Acad Sci USA 75:4257–4261
31. Skehel JJ, Daniels RS, Douglas AR, Wiley DC (1983) Antigenic and amino acid sequence variations in the hemagglutinin of type A influenza viruses recently isolated from human subects. Bull WHO 61:671–676

CHAPTER 38
Monoclonal Antibodies as Reagents

DAVID A. FUCCILLO AND JOHN L. SEVER

Immunization of an animal by the standard procedure produces a polyclonal antibody response to many antigenic structures on an antigen as well as to any other contaminating materials in the antigen preparation. Therefore, one of the preconditions for obtaining specific antisera has been to highly purify the antigen. Very often, however, this requirement is impossible to satisfy especially with biological antigens. This is the case with viral preparations where knowledge and separation of complex biological molecules without destruction of native structure usually cannot be accomplished. About ten years ago, Kohler and Milstein described a technique that could overcome these problems and produce specific antibodies against previously unknown antigens never before available in a purified or enriched state. The technique involved the fusion of mouse myeloma cells with normal mouse spleen cells to produce hybrid cells called *hybridomas* [1]. These cells have the growth characteristics of the mouse tumor cells along with the capability of antibody production found in normal spleen cells. A single clone of these hybrid cells will produce and secrete a homogeneous monoclonal protein, i.e., monoclonal antibody. If the donor animal of the spleen cells has been previously inoculated with an antigen, the hybridoma cells produced after fusion will synthesize and secrete large volumes of various monoclonal antibodies with high specificity to the antigen. The capacity to produce monospecific antibodies is proving to be invaluable for serologic diagnosis and typing of infectious agents, correlating neutralizing capacity with specific antibody, and defining the potential of antigens in protection against infection.

Production of Monoclonal Antibodies

The procedure used to produce monoclonals varies from laboratory to laboratory, but usually the BALB/c mouse strain is utilized for inoculation with the antigen of interest [2]. Different adjuvants, schedules, and routes of immunization have been tried, but most often three injections of antigen (with the first containing complete Freund's adjuvant) are used in the immunization procedure. The animals are rested for four to six weeks after the initial series of inoculations before they are given a final inoculation to stimulate high antibody production. During the rest period, animals should be bled in order for the circulating antibody and the eventual decrease in antibody levels to be detected before the final booster injection is given. Three to four days after the final injection, the mice are killed and spleens are taken for fusion with myeloma cells.

A suspension of spleen cells is prepared and mixed with myeloma cells in the presence of polyethylene glycol. Several myeloma cell lines have been tested. In early experiments the P3X63AG8 cells were used, but these cells were found to produce some of their own immunoglobulins. Later studies have used variant mouse myeloma cell lines such as the Sp2/OAG14 and P3/X63AG8.653, which do not produce immunoglobulins [3,4].

The myeloma cell lines used to produce hybridomas are deficient in the enzyme hypoxanthine phosphoribosyl transferase (HPRT), which permits cells to utilize exogenous purines for nucleic acid synthesis. The myeloma cells cannot survive in a medium containing hypoxanthine, aminopterin, and thymidine (HAT). The unfused spleen cells are not killed by the HAT medium, but they do not proliferate and eventually die under the culture conditions used. The hybrids formed between spleen and myeloma cells, however, contain the spleen cell HRPT and thus can utilize the HAT medium for cell growth.

The most critical step in the procedure of producing monoclonal antibodies is the screening of hybridoma cultures for the hybrids secreting useful antibodies. This will become quite evident in our later discussions. Because fusion generates hundreds of hybrids, it is crucial to have a relatively fast and sensitive method for identifying those hybrids producing antibodies with the desired characteristics. Because different screening methods can reveal different antibodies, it is advisable to use the method appropriate for the expected application of the monoclonal antibody.

The selected hybridomas are subcloned by plating in multi-well culture dishes at concentrations that will give less than one hybrid clone per well. Clones are then grown for two to three weeks until cells become overgrown and wells appear cloudy. The cell culture supernates are assayed again for antibody, and those clones producing desired antibody are then transferred to larger culture dishes. After cells have reached a high density, they are frozen and maintained in liquid nitrogen for future use. Supernatants from these cultures will contain the antibodies of interest.

An alternate method of producing antibody is to inoculate cultured cells into the peritoneal cavity of BALB/c mice and after a few days collect the ascites fluid. This fluid will usually contain concentrations of antibody a thousand times greater than that produced by cell culture.

Use of Monoclonal Antibodies

Compared to earlier methods of obtaining antisera after immunization of animals, this technique is superior because it permits production of a virtually unlimited supply of a well-defined specific antibody. This potential has attracted the interest of immunologists and researchers in many different scientific fields [5,6]. Monoclonal antibodies are being successfully applied to a number of diagnostic problems and, in particular, to the rapid identification of infectious agents. This was quite evident when we conducted a limited survey of major diagnostic laboratories (Table 38.1). It should be pointed out, however, that although monoclonal antibodies are very useful for research purposes, their utilization in routine diagnostic reagents or kits may not be as attractive. We will discuss some of these advantages and disadvantages arising from the substitution of traditional antisera with monoclonal antibodies.

Advantages and Disadvantages

Monoclonal antibodies offer a number of advantages over conventional antisera. One of the most important properties is their precise specificity for a single antigenic determinant or epitope in an antigen. The antibodies can be used to characterize different parts of a molecule with regard to its antigenicity, functional activity, or genetic variability. Their potential value in research appears to be unlimited. Their use in commercial viral antigen detection applications may be somewhat limited, however, by their great specificity. Monoclonal antibodies for commercial diagnostic kits must be carefully selected to allow for the antigenic heterogeneity inherent in biological systems. Antigenic determinants may vary slightly, because of methods of collection and of storage, or may be modified by fixation in tissue or by the action of enzymes [7,8]. Before any assay utilizing monoclonals can be put into widespread general diagnostic use, a large number of tests may be required to eliminate the potential of missing specimens containing slight antigenic changes. This disadvantage may be overcome by using mixtures of several monoclonals, each reacting with a different epitope.

As with antibodies produced in animals, monoclonal antibodies may cross-react. The epitope reactive with a monoclonal could be present in a similar

Table 38.1. Current use of monoclonal antibodies for medical virology.[a]

Laboratory[b]	Herpes I and II	CMV	Varicella	RSV	Chlamydia	Influenza A and B	Rabies	Parainfluenza 1, 2, 3	Papova BK
1	FA(c)	FA(c)	FA(c)		FA(c)		FA(c)	FA(n)	
2	FA(c)	FA(c)	FA(c)	FA(c)		FA(n)	FA(u)		
3	FA(u)			FA(u)	FA(n)				
4	FA(c)		FA(c)	FA(n)					
5	FA(u)	FA(u)	FA(u)			FA(u)			
6	FA(n) IA(n)	FA(c) AP(c)			FA(n) AP(n)				FA(n)
7	IA(n)			IA(n)					

[a] AP, antiperoxidase assay; FA, fluorescent antibody; IA, immunoassay (i.e., RIA, ELISA, etc.); (c) obtained commercially; (n) not obtained commercially but produced in-house or obtained from another investigator; (u) source unknown.

[b] Laboratory: 1, Dr. Stanley R. Plotkin, The Joseph Stokes, Jr. Research Inst., Philadelphia, PA; 2, Dr. Nathalie J. Schmidt, State Department of Public Health, Berkeley, CA; 3, Dr. Joseph M. Joseph, MD State Dept. of Health & Mental Hygiene, Baltimore, MD; 4, Dr. Kenneth MacIntosh, Children's Hospital, Boston, MA; 5, Dr. G.D. Hsiung, Veterans Administration Medical Center, West Haven, CT; 6, Dr. Max Chernesky, St. Joseph's Hospital, Hamilton, Ontario, Canada; 7, Dr. Pekka Halonen, University of Turku, Turku 52, Finland.

form in another completely unrelated virus. Selection of the original clones again is a very important step in the process of obtaining useful monoclonals.

The uniform type of antibody produced by monoclonals could present additional problems because each monoclonal synthesizes only a single class or subclass of immunoglobulin. These antibodies may have only limited biological capacity as compared to polyclonal antisera; they may not be usable in complement fixation, agglutination, cytotoxicity, or precipitation assays. Also, because they are a single type or class of antibody, they may not tolerate manipulations such as conjugation with fluorescein or enzyme labeling materials required to produce antibody reagents.

Monoclonal antibody of a single class will have a fixed affinity for a particular antigen. If this affinity is not high enough, the use of the monoclonal in many diagnostic assays may not be possible. An assay format that appears to have the greatest potential for detecting most viral antigens is the capture assay. In this format, an antibody against a particular antigen is attached to a solid phase. If the antigen is present in a sample, it will be bound by the capture antibody. A detector antibody can then be added and its presence amplified in a number of different ways, i.e., radioisotopes, fluorescence, luminescence, or enzymes. It is obvious that if the capture antibody does not have a high affinity for the antigen in the initial stage, then the assay will not function. The selection process becomes even more complicated because the affinity of various monoclonals must be checked.

The cost and time required to inoculate mice and produce cultures of hybridomas is minor when compared to the selection process necessary to obtain usable monoclonals for a particular application. In fact, cost can far exceed that necessary for antibody production by the classical method, including purification by affinity chromatography. Also hybridomas can be unstable and may be lost as antibody producers at any time. Therefore, where continued production of antibodies is essential—for instance, in commercial application—backup cells must be kept viable for long periods of time. In addition, if it is necessary to have a mixture of monoclonals for a final product, then all of these requirements are multiplied.

Current Use of Monoclonals for Medical Virology

As can be seen from our survey (Table 38.1), monoclonal antibodies are a very potent source of reagent for diagnostic and epidemiologic studies. Most of these applications involve the use of immunofluorescence. This technique has been shown to be a very sensitive and accurate diagnostic procedure. Its only drawback in the past has been the nonavailability of high-quality antiviral antisera. Monoclonal antibodies overcome this problem. In research, monoclonal antibodies have been used primarily to analyze or select various viral antigenic variants or to determine mutation rates of specific antigenic determinants. In the area of applying monoclonals to rapid diagnosis of viruses, reagents are available, some commercially, to detect herpes simplex virus types 1 and 2, cytomegalovirus, varicella virus, rabies, and respiratory syncytial virus (RSV). A recent study used monoclonals for detecting and characterizing various strains of RSV [9]. Such studies help to establish the antigenic diversity of viruses and produce information for epidemiologic studies. This is the case with RSV, where such information could prove important in studying the incomplete immunity seen following RSV infection.

Future Application

Future applications of monoclonal antibodies for diagnostic and epidemiologic studies will depend upon the solution of some of the problems previously stated. This will require the development of sensitive and specific selection processes. Monoclonal antibodies offer definite advantages for the simplicity of design to various immunoassays. One type of assay that has been described involves the use of anti-idiotype monoclonal antibodies [10]. With this particular immunoassay, the first monoclonal antibody is directed against a particular antigen. The second monoclonal antibody is directed against the first monoclonal, specifically in the Fab portion of the molecule. In an assay, if the antigen were present, the primary monoclonal antibody would react with the antigen and therefore block the binding of a second antibody to its reaction site. The difference found in the measurement of the bound versus the free second antibody would be related to the presence or absence of the antigen. This type of antigen detection assay would be very rapid and simple to perform.

Another approach using the capture format would involve two monoclonal antibodies directed against two different epitopes of an antigen. One can be bound to the solid phase as the capture antibody and the other labeled with an indicator (i.e., isotope, enzyme, etc.). An antigen-containing specimen can be placed with the monoclonal indicator antibody. This mixture could then be added to the solid phase, producing a rapid, single, one-step assay for antigen.

A potentially valuable development has been described, which uses the cell fusion technique for the production of bispecific antibodies [11]. If two antibody-producing hybridomas are fused together, the derived hybridoma will express the immunoglobulin chains of its parent. The product from these hybrid hybridoma cells could become valuable reagents for immunocyto-chemistry studies, antibody delivery of drugs and especially for development of rapid, simple immunoassays. For example, a bispecific antibody could be used in a capture format assay as the capture antibody. Activity of this antibody could be directed against two different epitopes on an antigen. The availability of such a reagent would eliminate some of the problems previously described. Another application of a bispecific antibody could be its use in crosslinking two antigens in a mixture such as viral antigen and an enzyme.

As previously indicated, very simple assays, using these types of reagents, could be designed.

Conclusion

Despite the previously described limitations of monoclonal antibodies, these reagents have become very potent tools for a wide spectrum of scientific studies. Eventually monoclonal antibodies will replace classical antisera, when they can be shown to have equivalent functional capability.

References

1. Kohler G, Milstein C (1975) Continuous cultures of fused cells secreting antibody of predefined specificity. Nature 256:495–497
2. Kennett TJ, McKearn TJ, Bechtol KB (1980) Monoclonal antibodies. Hybridomas: A new dimension in biological analysis. Plenum Press, New York, pp 1–420
3. Schulman M, Wilde CD, Kohler G (1978) A better cell line for making hybridomas secreting specific antibodies. Nature 276:269–270
4. Kearney JF, Radbruch A, Liesegang B, Rajewsky K (1979) A new mouse myeloma cell line that has lost immunoglobulin expression but permits the construction of antibody-secreting hybrid cell lines. J Immunol 123:1548–1550
5. McMichael AJ, Bastin JM (1980) Clinical applications of monoclonal antibodies. Immunol Today 1:56–60
6. Falkenberg FW, Pierard D, Mai U, Kantwerk G (1984) Polyclonal and monoclonal antibodies as reagents in biochemical and in clinical-chemical analysis. J Clin Chem Clin Biochem 22:867–882
7. Haaijman JJ, Deen C, Krose CJM, Zijlstra JJ, Coolen J, Radl J (1984) Monoclonal antibodies in immunocytology. A jungle full of pitfalls. Immunol Today 5:56–58
8. Stein H, Gater K, Asbahr H, Mason DY (1985) Methods in laboratory investigation. Use of freeze-dried paraffin-embedded sections for immunohistologic staining with monoclonal antibodies. Lab Invest 52:676–683
9. Anderson LJ, Hierholzer JC, Tsou C, Hendry RM, Fernie BF, Stone Y, McIntosh K (1985) Antigenic characterization of respiratory syncytial virus strains with monoclonal antibodies. J Infect Dis 151:626–633
10. Yolken RH (1983) Monoclonal antibodies in microbiologic immunoassay. Lab Management 21:49–55
11. Milstein C, Cuello AC (1984) Hybrid hybridomas and the production of bispecific monoclonal antibodies. Immunol Today 5:299–304

CHAPTER 39

Anti-peptide Antibodies: Some Practical Considerations

THOMAS M. SHINNICK AND RICHARD A. LERNER

Introduction

The observation that antibodies to almost any region of a protein can be induced by a short peptide corresponding to that region has led to the development of a general technology for producing protein-reactive antibodies as well as generating novel insights into the structure of proteins in solution.

Uses of Peptide Immunogens

The potential uses of peptides and peptide-elicited antibodies have been recently reviewed [1,2] and will not be discussed in detail here. However, most of the uses of the peptide-elicited antibodies in basic research are reflected in two facets of the synthetic peptide immunogen approach. First, essentially all one needs in order to produce antibodies that specifically react with a given protein is the amino acid sequence of that protein. This is particularly important nowadays because most amino acid sequences are being generated by the translation of nucleotide sequences, and indeed, occasionally all one knows about a putative gene product is its amino acid sequence. The protein-reactive peptide-elicited antibodies are excellent reagents for establishing the identity between a protein sequence and the protein itself. Indeed, positive, coincident, specific reactivity with antibodies elicited by two or more adjacent nonoverlapping peptides from an amino acid sequence may be stronger evidence that a particular protein is encoded by a given gene than a series of successful genetic experiments. Besides identifying the protein product, the peptide-elicited antibodies are useful reagents for determining the cellular location of the protein products, for analyzing the possible en-

zymatic activities of the products, and for purifying the products by immunoaffinity chromatography. The anti-peptide antibodies are particularly
good for immunoaffinity chromatography because specifically bound proteins
can be eluted very gently and often in enzymatically active form by competition with peptide [3].

The second facet of the synthetic peptide immunogen approach is that the
antibodies react with a small region of the protein that is chosen in advance
by the investigator by the process of selecting which peptide immunogens
to use. As such, the antibodies can be said to have *predetermined specificities*
or *targets*. Such site-directed antisera are excellent reagents for following
the fate of particular portions of a protein through protein processing pathways; for following, during DNA or RNA rearrangements, alternative exon
usage such as that occurring during immunoglobulin production [4,5] or adenovirus transcription [6]; or for producing antibodies that can distinguish
closely related proteins. For example, Alexander et al. [7] used peptide immunogens to elicit antibodies that could easily distinguish Thy-1 and Thy-
2, two glycoproteins that differ by a single amino acid [8]. Also, the ability
to target antibodies to a particular region of a protein allows one to investigate
structure–function relationships. For example, Schneider et al. [9] have used
site-directed antibodies to determine the orientation of the low-density lip-
roprotein (LDL) receptor in the membranes of fibroblasts. Additional examples and discussion of the possible uses of the peptide-elicited antibodies
may be found in the cited reviews.

A large part of the power and utility of this approach comes from the
ability to employ sets of antibodies and to do controls that ensure that one
is studying the desired antibody–antigen reaction and not some fortuitous
cross-reactivity. For example, as mentioned above, if antibodies elicted by
two or more nonoverlapping peptides from a given amino acid sequence
immunoprecipitate the same protein or generate the same staining pattern
in immunocytochemical reactions, one can be quite confident that one is
detecting the authentic protein that carries that amino acid sequence. Furthermore, the desired, protein-specific reactivity was elicited by a peptide.
Therefore, one can determine, by a competition experiment, whether the
observed reactivity is due to the peptide-elicited antibodies, as opposed to
antibodies elicited by the carrier protein or present in preimmune sera. That
is, if the observed reactivity is due to the peptide-elicited antibodies, then
that reactivity should be specifically abolished or greatly reduced by pretreatment of the antibodies with an excess of the peptide used to elicit the
antibodies. In short, these controls plus the ability to preselect the target of
the antibodies allow one to generate a powerful set of reagents to investigate
the molecular biology and architecture of a protein.

Finally, we should note that this technology holds promise for several
medical applications. Here the exquisite specificity of the anti-peptide antibodies might be exploited to produce reagents for immunodiagnostic assays
to detect and identify pathogenic infections. Indeed, peptide-elicited anti-

bodies have been used to distinguish between closely related serotypes of hepatitis B virus [10] and foot-and-mouth disease virus [11]. The peptides themselves might form the basis of safe, chemically defined vaccines. The advantages of such a vaccine would be

1. One does not have to grow or handle large quantities of the pathogenic agent, as is required for the production of the conventional vaccine.
2. The peptides show excellent stability at room temperatures.
3. The immunogen can be precisely defined chemically, allowing a vaccine free of any biological contamination to be produced.

Practical Considerations

Because one rarely knows the structure of a protein to sufficient degree to make meaningful structural predictions, how does one select peptides for use as immunogens to elicit protein-reactive antibodies? A number of studies from our laboratories and those of others have generated a set of elementary guidelines for choosing peptides. These guidelines should be considered only as helpful hints because what we learn from one peptide may not always be relevant to another peptide. Also, as experiments move away from require- ment for *any* protein-reactive antibodies to a requirement for *site-directed* antibodies, the selection of the peptide immunogen will largely be determined by the question being asked and the antibody specificity required to answer it.

The first general guideline is that the peptide should be longer than a pen- tamer because shorter peptides tend not to be immunogenic. Peptides of 10– 20 residues in length appear to have a high probability of eliciting protein- reactive antibodies. Longer peptides have been used successfully but are not inherently better than peptides 10–20 residues long. Second, for ease of handling, the peptide should be soluble in aqueous solutions. For this reason, one should avoid highly hydrophobic stretches of amino acid sequences and should confirm that the peptide contains a number of hydrophilic or charged residues. Third, we expect that in order for an antibody to react with an intact protein, the antibody must react with a site on the surface of the protein. To identify such sites, a number of computer programs such as those of Chou and Fasman [12], Hopp and Woods [13], or Kyte and Doolittle [14] have been used to predict secondary structure, which might indicate regions of the protein located on the surface of the molecule and hence available to antibody. However, a simple scan of the amino acid sequence for clusters of charged or polar residues seems to be effective for locating surface struc- tures for most studies. In this regard, it is interesting to note that peptides containing proline residues in addition to the charged or polar residues appear to be particularly effective in eliciting antibodies that react with the intact protein. Perhaps this is related to the observation that proline residues are typically found at exposed corners or bends in the polypeptide chain and such bends are often accessible to antibodies. Fourth, as is true for small molecules in general, most peptides must be coupled to a carrier molecule to elicit a strong immune response. Hence, the peptide sequence must contain a functional group through which it can be coupled to a carrier, though that group need not be part of the protein sequence. The choice of carrier or mode of coupling does not appear to be critical. We routinely couple through cysteine groups or through free amino groups using gluteraldehyde as the coupling agent. One caution that must be observed when using gluteraldehyde is that it modifies the free amino groups of lysine residues and may thereby modify a key antigenic residue. As far as carriers are concerned, keyhole limpet hemocyanin, bovine serum albumin, edestin, and synthetic carriers have all been used successfully. Of course, the ultimate use of the antibodies does place certain restrictions on the choice of carrier. (For example, one would not couple a peptide to bovine serum albumin and then use the resulting antibodies to assay cells grown in tissue culture media containing fetal calf serum.) Finally, several regions of proteins appear to be particularly amenable to the synthetic peptide immunogen approach. That is, peptides corresponding to the amino or carboxy terminus of a protein generally elicit protein-reactive antibodies. Perhaps the conformation of a peptide coupled to a carrier protein closely resembles the conformations seen in the terminal regions of the intact protein. Also, Pfaff et al. [15] and Rowlands et al. [11] have found that peptides corresponding to an α-helical region that is exposed on the surface of the protein are quite effective in eliciting protein-reactive antibodies. Perhaps some computer modeling would be helpful to predict such α-helices.

Overall, through application of these criteria, more than 80% of the peptides we have synthesized have elicited antibodies that could react with the native protein. This suggests that one should be able to select one or two peptides from a protein sequence and be relatively confident that a protein-reactive serum will be elicited.

References

1. Sutcliffe JG, Shinnick TM, Green N, Lerner RA (1983) Antibodies that react with predetermined sites on proteins. Science 219:660–666
2. Lerner RA (1984) Antibodies of predetermined specificity in biology and medicine. Adv Immunol 36:1–44
3. Walter G, Hutchinson MA, Hunter T, Eckhardt W (1982) Purification of polyoma virus medium-size tumor antigen by immunoaffinity chromatography. Proc Natl Acad Sci USA 79:4025–4029
4. Adams JM, Kemp DJ, Bernard O, Gough N, Webb E, Tyler B, Gerondakis S, Cory S (1981) Organization and expression of murine immunoglobulin genes. Immunol Rev 59:5–32
5. Cheng HL, Blattner FR, Fitzmaurice L, Mushinski JF, Tucker PW (1982) Structure of genes for membrane and secreted murine IgD heavy chains. Nature 296:410–415

6. Feldman LT, Nevins JR (1983) Localization of the adenovirus E1A$_2$ protein, a positive-acting transcriptional factor in infected cells. Mol Cell Biol 3:829–838
7. Alexander H, Johnson DA, Rosen J, Jerabek L, Green N, Weissman IL, Lerner RA (1983) Mimicking the alloantigenicity of proteins with chemically synthesized peptides differing in single amino acids. Nature 306:697–699
8. Rief, AF, Allen JM (1964) The AKR thymic antigen and its distribution in leukemias and nervous tissues. J Exp Med 120:413–433
9. Schneider W, Slaughter CJ, Goldstein JL, Anderson RGW, Capra JD, Brown MS (1984) Use of antipeptide antibodies to demonstrate external orientation of the NH$_2$-terminus of the low density lipoprotein receptor in the plasma membrane of fibroblasts. J Cell Biol 97:1635–1640
10. Gerin JL, Alexander H, Shih JW-K, Purcell RH, Dapolito G, Engle R, Green N, Sutcliffe JG, Shinnick TM, Lerner RA (1983) Chemically synthesized peptides of hepatitis B surface antigen duplicate the d/y specificities and induce subtype-specific antibodies in chimpanzees. Proc Natl Acad Sci USA 80:2365–2369
11. Rowlands DJ, Clarke BE, Carroll AR, Brown F, Nicholson BH, Bittle JL, Houghten RA, Lerner RA (1983) Chemical basis of antigenic variation in foot and mouth disease virus. Nature 306:694–697
12. Chou PY, Fasman GD (1978) Prediction of the secondary structure of proteins from their amino acid sequence. Adv Enzymol 47:45–148
13. Hopp TP, Woods KR (1981) Prediction of protein antigenic determinants from amino acid sequences. Proc Natl Acad Sci USA 78:3824–3828
14. Kyte J, Doolittle RF (1982) A simple method for displaying the hydropathic character of a protein. J Molec Biol 157:105–109
15. Pfaff E, Mussgay M, Bohn HO, Schulz GE, Schaller H (1982) Antibodies against a preselected peptide recognize and neutralize foot and mouth disease virus. Eur Mol Biol J 1:869–874

Vaccines and Antiviral Therapy

CHAPTER 40
Eradication and Possible Reintroduction of Smallpox

FRANK FENNER

Origins and Global Spread of Smallpox

Smallpox was recognized in India and China from the early years of the Christian era [1]. By the tenth century it had spread to Korea, Japan, and probably Indochina to the east of the Eurasian land mass, and to Asia Minor, North Africa, and the Iberian Peninsula to the west. Repeated introductions into Europe during the Crusades led to its establishment there as an endemic disease, and by the fifteenth century it was a common disease throughout Europe, associated with an overall case-fatality rate of about 25%.

The European colonists of the Americas and Africa took smallpox with them, with disastrous effects on the Amerindians throughout North and South and Central America and on the Hottentots in southern Africa. It is not unreasonable to suggest that without smallpox as an ally, neither Cortes nor Pizarro would have conquered the indigenous people of Mexico and Peru, respectively. Three outbreaks occurred in Australia between 1789 and 1870 [2], and occasional epidemics occurred when the disease was introduced into island populations in the Indian and Pacific Oceans. By the end of the seventeenth century, it was a cosmopolitan disease, absent only from small islands and some countries with a small and scattered population.

Prevention by Inoculation and Vaccination

The first attempt to mitigate the severity of smallpox arose in about the tenth century, probably independently in India and China. This procedure consisted of inoculating pus from a smallpox pustule into the skin, or, in China, by nasal insufflation. The result was an attack of smallpox that was less severe than "natural smallpox" (but that nevertheless had a case-fatality rate of 1–

2%), and gave lifelong protection against the naturally spreading disease. In 1796, Edward Jenner [3,4] showed that it was possible to protect against smallpox by inoculating pus from a cowpox pustule instead of a smallpox pustule, with no mortality and a trivial lesion. The new practice, which was called *vaccination* (from *vacca*, "cow"), rapidly became popular, and where it was assiduously practiced, as in Sweden, it immediately reduced the incidence of smallpox.

Because cowpox was a rare, sporadic disease of cows (since its natural reservoir hosts are probably wild rodents [5]), Jenner developed arm-to-arm vaccination as a method of maintaining a supply of the virus. This was supplanted by scarification of the cow in the 1860s; later sheep and water buffalos were also used. During the nineteenth century, compulsory childhood vaccination was introduced in many countries and greatly reduced the incidence of the disease. By 1900, endemic smallpox had been eliminated in some small countries in Europe, but, although its fatal impact was reduced elsewhere, it remained a common disease in most countries of the world.

Reporting of cases was never good; however, by 1920 it was possible to gauge the countries in which endemic smallpox still occurred. The gradual reduction in the numbers of such countries is illustrated in Figure 40.1. From the time of its formation in 1948, the World Health Organization regularly encouraged all member states to control smallpox within their borders by instituting mass vaccination. Freeze-dried vaccine, introduced in the French African colonies in the 1920s and available generally by 1955, greatly improved the efficacy of vaccination in tropical countries.

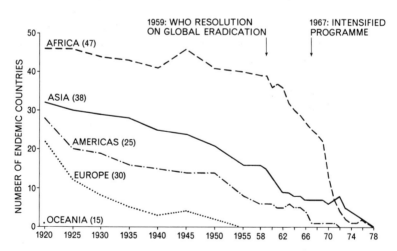

Figure 40.1. The numbers of countries in each continent in which smallpox was endemic between 1920 and 1977. The figures in brackets indicate the number of countries in each continent from which data were obtained. (From Fenner et al., *Smallpox and Its Eradication*, World Health Organization, in press, with permission.)

Achievement of Global Eradication

In 1958, encouraged by the eradication of smallpox from Europe and North America, the USSR delegation introduced a resolution to the World Health Assembly proposing global eradication of smallpox; this resolution was passed in 1959. Many countries in Asia and South America then established national smallpox eradication programs, and some success was achieved in eliminating smallpox from some smaller countries. However, by 1966 it was clear that international encouragement and some donated vaccine were not enough to eliminate smallpox from the developing countries of Asia and Africa, and the Intensified Smallpox Eradication Programme was established by WHO in 1967, with the goal of eradicating smallpox worldwide within a decade. After an enormous effort that directly involved 79 countries in which smallpox was endemic or which were at special risk [6], the last case of endemic smallpox in the world was recorded in Somalia in October of 1977. The factors that rendered this achievement possible were both biological and sociopolitical, and are summarized in Table 40.1.

The major changes instituted by the Intensified Smallpox Eradication Programme which made this achievement possible were: (a) achievement of a

Table 40.1. Biological and sociopolitical factors contributing to the eradication of smallpox.

Biological factors

There was no animal reservoir of variola virus.

Recurrence of infectivity never occurred.

There were very few subclinical cases, and those that occurred were of low infectivity.

Cases did not become infectious until the onset of rash, and were not infectious during the prodromal period.

Only one serotype of variola virus existed.

An effective and very stable vaccine was available.

Sociopolitical factors

Smallpox was a severe disease, with high mortality and serious sequelae.

Eradiction from the countries of Europe and North America showed that global eradication was an attainable objective.

The occurrence of imported cases and the costs of quarantine and vaccination for travelers provided a strong financial incentive for developed countries to achieve global eradication.

There were no social or religious barriers to the recognition of cases.

The Intensified Smallpox Eradication Unit of WHO had inspiring leaders and enlisted devoted health workers.

collaborative effort with all countries in which smallpox was endemic or were at risk; (b) provision of adequate amounts of freeze-dried vaccine, the quality of which was assured by regular testing by WHO; (c) development of the bifurcated needle as an improved method of administering vaccine in developing countries; and (d) replacement of the aim of "80% vaccination" by surveillance and containment of outbreaks [7].

Certification of Eradication

Apart from the incalculable benefits of eliminating a disease that had once caused 10% of all deaths each year in a city like London, as well as elsewhere in the world, the direct financial benefits of eradicating smallpox worldwide lay in the suspension of routine vaccination and elimination of the necessity for the provision of international vaccination certificates and accompanying airport inspections. However, national health authorities could be persuaded to discard such time-honored procedures only if they were convinced that smallpox had been eradicated. To accomplish this, a system of "certifying" the eradication of smallpox from countries, regions, and continents was established by WHO. Certification depended upon the opinion expressed by a group of international experts in public health, epidemiology, and virology, after inspection of records of prescribed precertification activities lasting for at least two years after the last known case of smallpox, that the surveillance systems in operation in each country were such that if there had been a case of smallpox within the previous two years, it would have been recognized. From 1977 to 1979, these certification activities were supervised by the Global Commission for the Certification of Smallpox Eradication, whose report [6] was accepted by the World Health Assembly in May 1980. Following this, vaccination against smallpox ceased except for investigators at special risk, as in laboratories working with orthopoxviruses that could infect humans.

Possible Reintroduction of Smallpox

The two most obvious factors that can lead to the reintroduction of a viral disease after its apparent elimination—recurrent infectivity and an animal reservoir—do not occur with variola virus. However, there are two other hypothetical sources for a return of smallpox, namely persistence of viable virus, either in the environment or in laboratory stocks.

Absence of Recurrent Infectivity
Abundant clinical experience, including immunosuppressive treatment and immunosuppressive diseases among persons who had recovered from smallpox, attested to the absence of latency and recurrent infectivity.

Absence of an Animal Reservoir

The absence of any indication of an animal reservoir in countries that had eliminated smallpox, often many years before, and careful search for an animal reservoir by epidemiologists engaged in the global smallpox eradication campaign, encouraged the belief that there was no animal reservoir. But this possibility had not been conclusively excluded in tropical Africa, a part of the world notable for its rich fauna and the large number of unique endemic diseases. Deep concern was therefore aroused when cases of smallpoxlike disease were recognized in West and Central Africa in 1970, in the absence of any indication of persisting smallpox [8,9]. Virological investigation proved that this disease was in fact caused by a distinct species of *Orthopoxvirus*, monkeypox virus. However, during investigations of the ecology of monkeypox in Zaire, four isolations of variola virus were made, apparently derived from the tissues of a chimpanzee, a monkey, a squirrel, and a rat, among some 61 wild animals tested. The investigators that were involved subsequently recovered variola virus from stocks of monkeypox virus held in their laboratory and hypothesized that variola virus was a "white variant" of monkeypox virus [10,11]. However, tests by others of tissues from over 1000 wild animals captured in Zaire were all negative [7], and other virologists were unable to confirm the results reported with monkeypox virus [12,13]. After careful appraisal of all the evidence, the conclusion was reached that these findings had resulted from laboratory contamination [7] and that there was no evidence that there was an animal reservoir of variola virus.

Persistence of Viable Variola Virus

In about the tenth century, the practice developed of inoculating pus from smallpox cases as a protection against the natural disease *(variolation)*. It was widely practiced in England in the eighteenth century and was used in China in the 1950s and 1960s, and in Afghanistan, Ethiopia, and Pakistan in the 1970s [7]. Laboratory tests with scabs had shown that in temperate climates viable virus could persist for at least 13 years [14].

However, extensive interrogation of those practicing variolation in Afghanistan, China, and Pakistan (where stored material was used, rather than recent smallpox pustules as in Ethiopia) revealed that fresh pus or scab material was added to the store, or fresh preparations were made, at least annually. Tests on some 20 samples of variolators' material failed to reveal viable virus in samples more than nine months old. The continued absence of outbreaks of smallpox in countries where variolators practiced, at intervals now of 21 years (China), 12 years (Afghanistan and Pakistan), and 10 years (Ethiopia) after the last case, renders the likelihood of reintroduction of smallpox from this source vanishingly remote. The same considerations apply to the hypothesized persistence of viable virus in coffins, burial shrouds, etc., of patients who died of smallpox.

Laboratory Stocks of Variola Virus

Stored in a deep-freeze cabinet, variola virus will remain viable indefinitely. Although laboratory workers are protected by vaccination, four outbreaks of smallpox have been ascribed to laboratory sources, the most recent in Birmingham, England, in August–September 1978, ten months after the last case of endemic smallpox in the world. Conscious of this risk, WHO embarked on a campaign to reduce the number of laboratories holding variola virus. In response to an inquiry in 1976, no fewer than 76 laboratories reported that they held such stocks. By persuasion, this number has now been reduced to two, the WHO collaborating centers for smallpox diagnosis in Atlanta, USA, and Moscow, USSR [7]. These are microbiologically and "militarily" highly secure laboratories, from which the risk of escape of the virus is minimal. Of course, there may be secret stores of the virus maintained somewhere in the world for use in biological warfare; there is no way of knowing of such a store, if it exists.

Conclusion

Following Jenner's discovery of protection against smallpox by the inoculation of a related orthopoxvirus that produced only a trivial skin lesion, the incidence of smallpox was gradually reduced wherever vaccination was practiced. This practice, combined with surveillance and containment of outbreaks, led to the global eradication of endemic smallpox in 1977.

Several hypothetical sources for a possible reintroduction of smallpox— recurrence of infectivity in humans, an animal reservoir, persistence in the environment, and escape from one of the two known laboratory repositories— have been examined and dismissed. The possibility remains of retention of viable variola virus at a biological warfare depot somewhere in the world, but this problem is not amenable to investigation.

References

1. Hopkins DR (1983) Princes and Peasants: Smallpox in History. University of Chicago Press, Chicago
2. Fenner F (1984) Smallpox, "the most dreadful scourge of the human species." Its global spread and recent eradication. Med J Aust 2:728–735; 841–846
3. Jenner E (1798) An inquiry into the causes and effects of the variolae vaccinae, a disease discovered in some of the western counties of England, particularly Gloucestershire, and known by the name of the cow pox. Reprinted in Camac CNB (ed.) Classics of Medicine and Surgery. Dover, New York, 1959, pp 213– 240
4. Baxby D (1981) Jenner's Smallpox Vaccine. The Riddle of the Origin of Vaccinia Virus. Heinemann, London
5. Baxby D (1982) The natural history of cowpox. Bristol Medico-chirurg J 97:12– 16

6. World Health Organization (1980) The Global Eradication of Smallpox. Final Report of the Global Commission for the Certification of Smallpox Eradication. History of International Public Health, No 4. World Health Organization, Geneva

7. Fenner F, Henderson DA, Arita I, Jezek Z, Ladnyi ID, (1987) Smallpox and Its Eradication. World Health Organization, Geneva (in press)

8. Ladnyi ID, Ziegler P, Kima E (1972) A human infection caused by monkeypox virus in Basankusu Territory, Democratic Republic of the Congo. Bull WHO 46:593–597

9. Breman JG, Kalisa-Ruti, Steniowski MV, Zanotto E, Gromyko AI, Arita I (1980) Human monkeypox, 1970–79. Bull WHO 58:165–182

10. Marennikova SS, Shelukhina EM (1978) Whitepox virus isolated from hamsters inoculated with monkeypox virus. Nature 276:291–292

11. Marennikova SS, Shelukhina EM, Maltseva NN, Matsevich GR (1979) Monkeypox virus as a source of whitepox viruses. Intervirology 11:333–340

12. Dumbell KR, Archard LC (1980) Comparison of white pock (h) mutants of monkeypox virus with parental monkeypox and with variola-like viruses isolated from animals. Nature 286:29–32

13. Esposito JJ, Nakano, JH, Obijeski J (1985) Can variola-like viruses be derived from monkeypox virus? An investigation based on DNA mapping. Bull WHO 63:695–703

14. Wolff HL, Croon JJAB (1968) The survival of smallpox virus (variola minor) in natural circumstances. Bull WHO 38:492–493

CHAPTER 41
Immunogenicity of Vaccine Products and Neutralizing Antibodies

ERLING NORRBY

Virion immunogens induce a spectrum of antibodies of different specificities. Some of these antibodies can inhibit the cellular expression of the viral genome, i.e., neutralize virus infectivity. Inhibition of virus infectivity is a highly specific reaction, and serologic typing of virus is therefore generally performed by neutralization tests. Neutralizing antibodies play an obvious role in restriction of virus replication, not only in cell cultures but also in vivo. For this reason, potential vaccine products are evaluated for their capacity to induce neutralizing antibodies. Although this approach provides the best guideline at the current stage, it has some inherent caveats. This review provides examples of some of the limitations in the employment of neutralization tests in vaccine studies.

The Concept of Virus Neutralization

Considerable efforts have been devoted to explaining the in vitro phenomenon of virus neutralization. In spite of this, many questions remain to be clarified [1]. The concept of neutralization is many-faceted. Three categories of neutralization phenomena can be distinguished:

1. *Intrinsic neutralization.* This means that the antibody induces some changes in the virion, which prevent expression of the genetic potential. These changes can be direct (immediate) or indirect. In the latter case, the antibody anchored on the virion surface mediates the attachment of other factors, such as complement or anti-immunoglobulin, including rheumatoid factor, which are required for neutralization. The effect is referred to as *neutralization enhancement* or *indirect neutralization* of sensitized virus.
2. *Extrinsic neutralization.* Two examples can be given. Antibodies may react

with sites that are nearby viral attachment proteins. As a consequence, virion adsorption to cells may be blocked by steric hindrance. In the case of enveloped viruses, antibodies attached to peplomers can mediate attachment of complement. Topical lysis of the membrane may ensue with destruction of virus infectivity as a consequence.

3. *Pseudoneutralization*. Because virions are multivalent and the antibodies are divalent or multivalent, their interaction in certain relative proportions may lead to aggregation by lattice formation. This will reduce the number of infectious units. It should be noted that the possibility of observing pseudoneutralization is dependent on the kind of test employed. The plaque neutralization test readily identifies aggregation phenomena, but this is not effectively scored when infectivity of antigen–antibody aggregates is evaluated by reading cytopathic effects in monolayer cell cultures.

As mentioned the role of different neutralization phenomena in vitro has not been clarified. Regarding knowledge about phenomena of importance for restricting the spread of virus in vivo, we are even more ignorant.

One Antigenic Site Involved in Neutralization

Viruses of a comparatively simple surface immunogen construction such as nonenveloped viruses like picornaviruses or parvoviruses and enveloped viruses like togaviruses and rhabdoviruses may carry on their surface, repeated units of a single antigenic site involved in neutralization. In the case of foot-and-mouth disease (FMD) virus, the protein VP1, located at the icosahedral vertices, is responsible for inducing neutralizing antibodies. Further dissection of antigenic sites in the VP1 protein (which is 213 amino acids long) by use of cyanogen bromide fragments and synthetic peptides has revealed that the regions 146–154 and to some extent 201–213 carry the antigenic sites [2]. Similar studies have been performed on the VP1 protein of poliovirus type 3 by use of synthetic peptides and also by nucleotide sequencing of non-neutralizable variants selected by propagation of virus in the presence of monoclonal antibodies [3]. It was found that amino acids 89–100 of VP1 represented the major antigenic site involved in neutralization. Corresponding studies of poliovirus type 1 have shown that neutralization may involve multiple independent sites [4]. Thus even relatively simple viruses may display complex relationships in neutralization. As a corollary, more complex viruses would be expected to show even more complicated relationships.

Togavirus envelopes contain two distinct glycoproteins, E1 and E2. Antibodies to E1 may inhibit hemagglutination and show cross-reactivity with members of the same genus, e.g., alphaviruses, whereas antibodies to E2 are type specific and can neutralize virus in vitro. Passive protection experiments utilizing monoclonal antibodies showed that both antibodies directed against E1 and those directed against E2 can prevent lethal viral encephalitis [5]. Antibodies to E2 neither neutralized nor sensitized virions, but could mediate complement-dependent cytolysis. Another example of protection against a lethal virus infection by nonneutralizable monoclonal antibodies was shown with vesicular stomatitis virus [6]. This rhabdovirus member has only one surface glycoprotein. The antibodies that reacted with this protein but lacked capacity to neutralize virus, also could mediate complement-dependent cytolysis.

More than One Antigenic Site Involved in Neutralization

Of the more complex viruses, one nonenveloped representative—adenovirus—and one enveloped representative—paramyxovirus—will be discussed. There are three main surface components in adenovirions: (a) nonvertex capsomers (hexons), (b) vertex capsomers (penton bases), and (c) vertex projections (fibers) [7]. The hexons carry a type-specific antigen exposed at the surface which plays a dominant role in virus neutralization. Antibodies against the penton base or fibers do not neutralize the virus, but when antibodies of *both* specificities are present simultaneously, a slight neutralization is observed. Both antibodies effectively sensitized virions, and addition of anti-immunoglobulin caused a pronounced neutralization [8]. It is not known to what extent antibodies against more than one adenovirus structural component are needed to provide protection in infected hosts.

Paramyxoviruses have two major surface glycoproteins: (a) the hemagglutinin–neuraminidase (HN) [or hemagglutinin (H) in morbilliviruses], and (b) fusion (F) components. The role of these components in the induction of immunity has been extensively studied [9]. Prior to these studies, important observations were made in the development and use of inactivated vaccines against measles virus in the 1960s [10]. These inactivated vaccines, prepared by formalin treatment and also by splitting the virus with Tween 80 and ether, induced readily measurable titers of neutralizing antibodies. It was therefore expected that protection against disease should be achieved because passive immunization is a highly effective measure. However, this was not so. Instead, conditions for development of immune pathological complications were established. Failure of protection was due to the destruction of crucial immunogenic parts of the F components by the inactivating procedures employed. Although antibodies to the H component suffice to neutralize virus effectively in vitro, antibodies to the F component are needed to prevent development of disease. However, the disease observed was different from regular measles and is referred to as *atypical* measles. The sensitization with H antigen is the likely cause of the immune pathological events. Dog immunization experiments with purified H and F components of the closely related morbillivirus, distemper virus, have been performed [11]. Exposure of immunized dogs to virulent distemper virus generally allowed a replication of virus in the animals, but only dogs with anti-H immunity displayed symptoms, albeit mild. We may therefore infer that the F antigen of paramyxoviruses is a more important subvirion immunogen than the H antigen.

Formalin-inactivated mumps virus vaccine also lacks an effectively immunogenic F component [10], but hitherto no cases of atypical mumps have been identified in individuals who have received this kind of vaccine. Immunopathological phenomena may possibly also be of relevance for the accentuated disease seen after immunization with formalin-inactivated respiratory syncytial (RS) virus vaccine. However, it has recently become apparent that RS virus shows some very distinctive features as compared to other paramyxoviruses. Thus in the case of this virus, it is antibodies against F and not against G components (the assumed H equivalent in RS) that play a major role in in vitro neutralization. Furthermore, subtypes of RS virus have been identified, whose possible role in repeated infections is further discussed below.

More than One Population of Infectious Particles

In general, viruses occur only in one infectious form: the *virion*. However, subvirion structures occasionally show infectivity; examples are isolated nucleocapsids and, in particular, isolated nucleic acid from virions of the kind not associated with polymerase enzymes. A special situation of two-particle infectivity is found in vaccinia virus [12]. The two different infectious forms are (a) intracellular naked particles (which so far have been the object of most poxvirus studies), and (b) extracellular enveloped particles. As has been recently shown, the latter particle represents the final product in viral morphogenesis also in vivo [13]. The envelope of the virus, which includes the hemagglutinin, appears to carry all antigens of importance for immune protection. Against this background, it is now possible to explain decades of failures to produce an inactivated poxvirus vaccine. In all these studies, the infectious intracellular naked particles were used, under the assumption that these represented the intact virions. High titers of antibodies neutralizing the naked particles could be induced, but no protection against disease was achieved. It is possible that in other virus systems as well, subvirion infectious particles relevant to the problem of vaccine development may occur. A case in point is iridoviruses, e.g., African swine fever virus.

Subtypes of Viruses in Repeated Infections

From a practical standpoint, the most important aspect of immunological cross-reactions between two viruses is whether infection with one virus provides protective immunity against the second virus. Thus, although minor variations can be distinguished between strains of antigenically stable monotypic viruses such as measles, the different strains appear categorically to induce protective homotypic immunity. In other cases, viruses have diverged to show varying degrees of cross-neutralization. It is a pragmatic problem to decide at which degree of difference strains become subtypes and subtypes become distinct types. This problem is not uniformly approached. Three examples will be discussed.

1. As concerns RS virus, strain variation in cross-neutralization tests was observed already in the 1960s, but the virus still is considered to be monotypic. Recent studies [14] show that strains can be divided into two groups, which differ in epitope characteristics of four different structural components determined by reactivity with monoclonal antibodies. The terminology *subtype A* and *B* RS viruses was therefore introduced. However, it remains to be clarified whether the occurrence of subtypes plays a decisive role in the pathogenesis of RS virus infections and in the occurrence of repeated infections. Early attempts to develop an RS vaccine did not meet with success [15]. Children who received a formalin-inactivated product developed a more severe form of disease when they were exposed to wild virus. The occurrence of subtypes, which differ extensively in the G protein and also to some extent in the F protein, makes the interpretation of sensitization events in vaccines more complex. The selection of subvirion antigens to be included in a subunit or synthetic vaccine needs to be carefully evaluated.

2. Dengue viruses have been separated into four different types [16], which show varying degree of cross-reaction in neutralization tests. The issue as to whether the dengue hemorrhagic fever and shock syndrome may reflect immune pathological events resulting from secondary infection with a new type of dengue fever virus, has been extensively discussed but remains controversial.

3. Influenza virus, in particular type A, is notoriously antigenically unstable, and both antigenic drift and shift occur. For historic reasons, all influenza A virus strains are said to belong to the same type, although by present-day terminology the immunogenic distinctiveness of their surface components, the H and N (neuraminidase) antigens, would motivate segregation into separate types. The importance of immune experience of prior influenza A virus infections on pathogenetic events in connection with new infections is a complex issue [17]. The H antigen plays a dominating role in immune protection. However, antibodies to the N component, although nonneutralizable (but sensitizing), can confer in vivo protection against disease in animals. It has been proposed that N antigen immunization might be used to induce an infection-permissive immunity (cf. the analogy with the paramyxovirus F component immunization, discussed above). An argument against the use of this kind of influenza vaccine might be the observation that reinfection with an influenza A virus by antigenic shift has a new H antigen, but a conserved N antigen appears be effective in causing disease.

Subtypes of Viruses in Persistent Infections

Antigenically unstable viruses can be of particular importance in persistent infections. By consecutive changes of surface antigen properties, the virus infection can expand in the presence of a previously mounted immune response. This is well documented by infections with certain retroviruses. The best example is the repeated episodes of disease caused by equine infectious anemia (EIA) virus [18]. Obviously, development of a vaccine against this type of antigenically labile persistent virus is exceedingly difficult. Tests for neutralizing antibodies will not provide effective guidance. Instead, a comprehensive picture of the extent of antigen variations needs to be accumulated, with the hope that certain more stable antigenic determinants may be identified. The fact that after several bouts of activated endogenous infections with EIA virus, eventually an effective immunity accumulates, motivates a certain optimism in the tackling of this kind of problem. Similar consideration may pertain to endeavors to immunize against visna and, of considerable current interest, human T-cell leukemia/lymphoma (HTLV-III) retroviruses. An additional problem with the antigenically unstable HTLV-III virus is the poor development of neutralizing antibodies after infection with this agent [19]. Apparently, cases of HTLV-III infection that develop into a full-blown terminal AIDS do not manage to control the infection. On the other hand, there is a fraction of persistently HTLV-III infected individuals who do not develop disease. Thus there may be posibilities for establishing endogenously evolving, and consequently also exogenously induceable defense mechanisms for management of the infection.

Viruses with a Poor Capacity to Induce Neutralizing Antibodies

The poor capacity of HTLV-III virus to induce neutralizing antibodies was already mentioned. This poor capacity to induce neutralizing antibodies is seen also in certain other replication-competent nontransforming retroviruses. One example is caprine arthritis–encephalitis virus. In this case, it was shown recently that immunization together with inactivated *Mycobacterium tuberculosis* could overcome the lack of capacity of goats to mount a neutralizing antibody response [20]. Use of these neutralizing antibodies demonstrated also that the caprine arthritis–encephalitis virus is antigenically unstable. It should be mentioned that all the four retroviruses discussed in this chapter belong in the *Lentivirus* genus and have exceptionally large surface glycoproteins.

Another chronic infection with a nonneutralizable virus that was extensively studied previously is Aleutian disease in minks. The disease is caused by a parvovirus [21], and infected animals develop plasmocytosis and hypergammaglobulinemia. The gammopathy may be of restricted heterogeneity or be monoclonal, and presumably to a considerable extent represents virus-specific antibodies. These antibodies do not neutralize the virus, but sensitize virions. Addition of anti-immunoglobulin to virion–antibody mixtures in vitro can bring about neutralization. Attempts were made to prepare a formalin-inactivated antigen materials from selected organs of infected minks. When it was used for immunization, essentially no antibody response was detectable. Challenge of immunized animals with live virus showed that they were more susceptible to infection and developed more severe symptoms. The basis for this accentuation of immune pathological events could be a selective destruction of crucial immunogenic sites, as in the case of paramyxoviruses. A logical approach to immune intervention in the case of Aleutian disease is, however, difficult to formulate.

Epilogue

In the above discourse, only the humoral aspect of immune protection against virus infections was considered. Obviously the cellular arm of the immune defense system may also play a role in conjunction with antibodies or on its own. Further, the discussions have departed from the assumption of a relatively complete understanding of the dominating surface antigens of viruses. Recent evidence for possible additional transmembranous structural proteins in ortho- and paramyxoviruses emphasizes that still other antigens of protective immunobiological importance may need to be disclosed. As we move into an era of synthetic vaccines, one particular requirement will be the availability of detailed knowledge on the relative qualitative and quantitative importance of virion surface antigens in immune protection. The terms *neutralizing epitopes* and *antigenic sites* are frequently used in molecular characterization of protective regions of structural proteins. As discussed in this multitargeted review this concept should be used with particular caution. In some cases, components that are poor inducers of neutralizing antibodies may be the subvirion component of choice for immunization. Although a considerable amount of important information remains to be delineated (including optimal conditions for immunization with low-molecular-weight antigen determinants), the reader should remember that many of the effective vaccines used today were developed under conditions of considerable ignorance with respect to the immunobiology of virion antigens.

References

1. Mandel B (1978) Neutralization of animal viruses. *In* Lauffer MA, Bang FB, Maramorosch K, Smith KM (eds) Advances Viral Research, vol 23. Academic Press, New York, pp 205–268
2. Bittle JL, Houghten RA, Alexander H, Shinnick TM, Sutcliffe JG, Lerner RA, Rowlands DJ, Brown F (1982) Protection against foot and mouth disease by immunization with a chemically synthesized peptide predicted from the viral nucleotide sequence. Nature 298:30–33
3. Ferguson M, Evans DMA, Magrath DJ, Minor PD, Almond JW, Schild GC (1985) Induction by synthetic peptides of broadly reactive, type-specific neutralizing antibody to poliovirus type 3. Virology 143:505–515

4. Emini EA, Kao SY, Lewis AJ, Crainic R, Wimmer E (1983) Functional basis of poliovirus neutralization determined with monospecific neutralizing antibodies. J Virol 46:466–474

5. Schmaljohn AL, Johnson ED, Dalrymple JM, Cole GA (1982) Non-neutralizing monoclonal antibodies can prevent lethal alphavirus encephalitis. Nature 297:70–72

6. Lefrancois L (1984) Protection against lethal virus infection by neutralizing and nonneutralizing monoclonal antibodies. Distinct mechanisms of action in vivo. J Virol 51:208–214

7. Pettersson U (1984) Structural and nonstructural adenovirus proteins. *In* Ginsberg HS (ed) The Adenoviruses. Plenum Press, New York, pp 205–270

8. Norrby E, Wadell G (1972) The relationship between the soluble antigens and the virion of adenovirus type 3. VI. Further characterization of antigenic sites available at the surface of virions. Virology 48:757–765

9. Örvell C, Norrby E (1985) Antigenic structure of paramyxoviruses. *In* van Regenmortel MHV, Neurath AR (eds) Immunochemistry of Viruses: The Basis for Serodiagnosis and Vaccines. Elsevier Science, Amsterdam, The Netherlands, pp 241–264

10. Norrby E (1983) Viral vaccines: The use of currently available products and future developments. Arch Virol 76:163–177

11. Norrby E, Utter G, Örvell C, Appel JG (1986) Protection against canine distemper virus in dogs after immunization with isolated fusion protein. J Virol 58:536–541

12. Boulter EA, Appleyard G (1973) Differences between extracellular and intracellular forms of poxvirus and their implications. *In* Melnick JC (ed) Prog Med Virol, vol 16. S Karger, Basel, pp 86–108

13. Payne LG, Kristensson K (1985) Extracellular release of enveloped vaccinia virus from mouse nasal epithelial cell in vivo. J Gen Virol 66:643–646

14. Mufson MA, Örvell C, Rafnar B, Norrby E (1985) Two distinct subtypes of human respiratory syncytial (RS) virus. J Gen Virol. 66:2111–2124

15. Kim HW, Canchola JG, Brandt CD, Pyles G, Chanock RM, Jensen K, Parrott RH (1969) Respiratory syncytial virus disease in infants despite prior administration of antigenic inactivated vaccine. Am J Epidemiol 89:422–434

16. Schlesinger RW (1977) Dengue viruses. *In* Gard S, Hallaner C (eds) Virology Monographs, vol 16. Springer-Verlag, Wien, pp 1–132

17. Kilbourne ED (1978) Influenza as a problem in immunology. J Immunol 120:1447–1452

18. Montelaro RC, Parekh B, Orrego A, Issel CJ (1984) Antigenic variation during persistent infection by equine infectious anemia virus, a retrovirus. J Biol Chem 259:10539–10544

19. Robert-Guroff M, Brown M, Gallo RC (1985) HTLV-III-neutralizing antibodies in patients with AIDS and AIDS-related complex. Nature 316:72–74

20. Narayan O, Sheffer P, Griffin DE, Clements J, Hess J (1984) Lack of neutralizing antibodies to caprine arthritis–encephalitis lentivirus in persistently infected goats can be overcome by immunization with inactivated *Mycobacterium tuberculosis*. J Virol 49:349–355

21. Porter DD, Larsen AE, Porter HG (1972) The pathogenesis of Aleutian disease of mink II. Enhancement of tissue lesions following the administration of a killed virus vaccine or passive antibody. J Immunol 109:1–7

CHAPTER 42

Prospects for Development of a Vaccine Against HTLV-III Disease

DANI P. BOLOGNESI, PETER J. FISCHINGER, AND
ROBERT C. GALLO

Introduction

The primary etiologic role of human T-cell leukemia/lymphoma virus (HTLV-III) in acquired immunodeficiency syndrome (AIDS) is now indisputable. From its apparent origin in Equatorial Africa, the virus presumably found its way to Haiti and Belgium in the late 1970s (Figure 42.1) from where it has spread with lightning speed throughout North America and Europe. In the United States alone, estimates on the number of individuals infected with the virus range from 500,000 to 2,000,000, and it is likely to be twice this number worldwide. The difficulties in obtaining more accurate figures are due in large measure to the existence of a clinically healthy carrier state, coupled with the variable incubation periods of several months up to as long as seven years, before signs of disease occur. We also know very little about prevalence and distribution of the virus in the African continent.

Based on the accrual of cases of AIDS, it would appear that the doubling time of the epidemic continues to be about twelve months. This translates into more than 10,000 new cases next year with a mortality greater than 85%, when residual survivors beyond two years are considered. The question of spread into the general population from the current restricted risk groups remains significant, particularly if one considers the pattern of infection in Central Africa, which does not appear to be restricted to the currently accepted high-risk groups in the United States and Europe.

The combination of these factors emphasizes the imminent danger of a rapidly expanding virus infectious pool and dictates the necessity for effective countermeasures against the virus. Clearly, prevention via vaccine production is a major consideration and is the focus of this discussion. We will review here what preventive measures have been effective against retroviruses in

Figure 42.1. HTLV-III probably originated in Central Africa and spread to the Western World in the manner suggested by the diagram.

animal models and discuss the prospects for vaccine development against HTLV-III.

Retrovirus Antigens That Elicit Protective Antibody

The virion component that is responsible for host range, infectivity, interference subgroups, and neutralization patterns is its major envelope glycoprotein (gp) [1]. Because of its strategic location on the virion surface, the gp represents the principal target for immune attack against the virus. Immunization with the gp alone elicits strong neutralizing antibody, which can be of type, group, or interspecies specificity, depending on the host and the regimen of inoculation [1]. The potential for eliciting broad reactivity is of particular importance, given the extensive variation in virus isolates of the HTLV-III family, most of which occurs within the envelope *(env)* gene [2]. Of particular interest is that in the absence of immunization, most naturally occurring immune reactivities display only a type-specific neutralization pattern and cannot protect against challenge with a suitably distant variant. In several retrovirus models, where non-cross-neutralizing virus types occur, immunization with sufficient quantities of homogeneous gp of a single virus type can elicit immune responses reactive to widely different retrovirus groups [1].

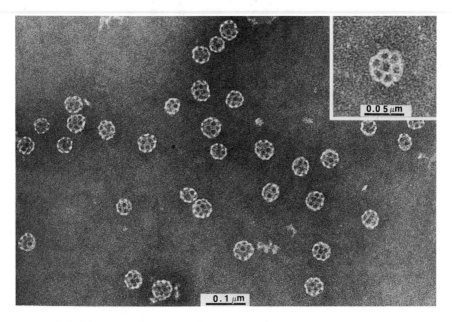

Figure 42.2. Electron micrograph of an HTLV-III ISCOM preparation. (Courtesy of Dr. M. Gonda, Frederick Cancer Research Facility.)

Is the Manner of Antigen Presentation Important for Protective Immunity?

Although monomeric gp is capable of inducing protective immunity, relatively large quantities of antigen were required to accomplish this reproducibly [3]. In some systems, however, purified gp alone was not protective and only occasionally was able to induce neutralizing antibodies (unpublished observation). On the other hand, antigen mixtures containing gp obtained from cell culture supernatants following serum starvation have been licensed for use in cats on the basis of veterinary trials where they proved to be better than 50% effective against live virus challenge [4,5]. Both in the mouse and the cat leukemia virus systems, a linkage of the major gp to its naturally contiguous transmembrane protein (tmp) resulted in an antigen complex that was significantly superior to gp alone in eliciting protection from primary infection [6].

Recently, as reported by Morein and colleagues [7], the capture of virus envelope components (both gp and tmp) by glycoside lattices through hydrophobic interactions with tmp generates a multimeric matrix structure (ISCOM; immune stimulating complex) which is an unusually powerful immunogen. An HTLV-III ISCOM is shown in Figure 42.2. As compared to monomeric gp, an ISCOM preparation containing an equal amount of gp elicits at least a tenfold increase in the protective end point [8]. It may be that these more complex structures are more easily recognized, presented, taken up, and processed by macrophages for antigen presentation to the immune system.

Significance of the Immune Response to HTLV-III in Humans

It is now well recognized that individuals infected with HTLV-III respond to the virus with a vigorous humoral immunity [9]. This forms the basis for the current screening tests used by blood banks and infectious disease clinics. A number of questions have been raised about the significance of this response, particularly because the ability to isolate virus from a patient is independent of the currently measurable antibody detected in enzyme-linked immunosorbent assay (ELISA), and Western blot tests. Even more perplexing is the recent finding that virus-neutralizing antibody exists in high titer even in late stages of disease [10]. The inability to correlate the immune response with the presence or absence of disease raises doubts about the existence of natural protective immunity. On the other hand, the presence of such immunity is suggested by the disproportionately high number of individuals who have been exposed to HTLV-III yet have not succumbed to disease. Will this healthy virus carrier state eventually give way to disease

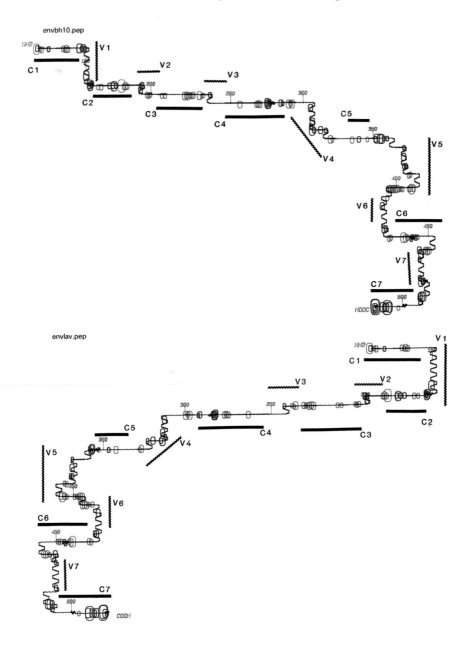

Figure 42.3. Predicted secondary structure of the extracellular envelope proteins of four AIDS viruses. Extracellular envelope glycoproteins of BH-10, LAV-1a (LAV = lymphadenopathy associated virus), ARV-2 (ARV = AIDS related virus), and HAT-3 were compared by computer analysis. Blue circles denote hydrophilic regions, and red circles hydrophobic regions. Positions of β turns are indicated by turns of the chain (⇌), α helices as bold coils (∿∿), β sheets as close zigzags

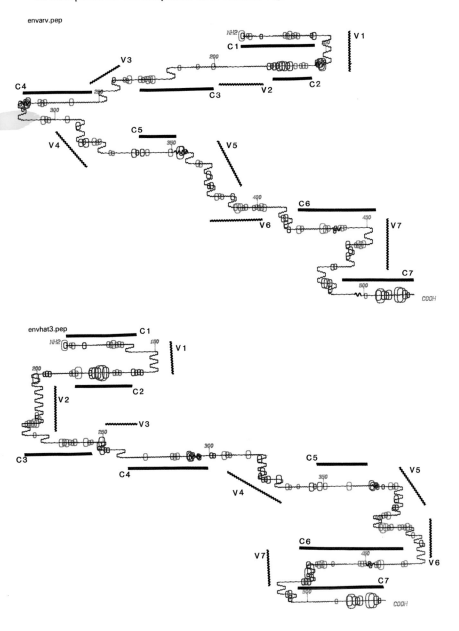

(〰〰), and random coils as wide zigzags (〜〜). The positions of the constant (C) and variable (V) regions are derived from the primary amino acid sequences. β-Turns adjacent to β sheets or α-helical regions in a hydrophilic environment are likely candidates for antigenic sites. C7 is a good example of a conserved region showing such antigenicity.

progression? Current projections from several studies indicate that of persons who seroconverted two to three years ago, about 10% have developed clinical AIDS and that another 10–20% have symptoms and signs of the AIDS related complex (ARC).

There are several possible explanations for this phenomenon. Two will be discussed here. The first is that the immune response that is detected may have arisen too late in relation to the infectious process. That is, once the virus has taken hold, the presence of a neutralizing environment is not sufficient to control the spread of the virus within the patient. This is especially expected to be the case if the mode of virus transmission is by cell-to-cell contact as well as by free infectious virus. In fact, cell-to-cell transfer is a plausible mode during the most common infectious process of sexual transmission from individual to individual. If this indeed occurs, protective immunity would have to include antibodies that were cytotoxic or a cell-mediated immune response directed against infected target cells. Neither of these has thus far been detected in affected individuals.

Special Considerations for Vaccine Development Against HTLV-III

HTLV-III virus glycoprotein has several unique features that need to be considered when one is devising strategies for vaccine development [11]. First, it is a very large molecule, of 120,000 daltons (gp120), in comparison to most retroviruses, with the exception of the gp of lentiviruses. The HTLV-III gp is also covered by an unusually extensive amount of carbohydrate, i.e., consisting of 32 potential glycosylation sites on the *env* gene, such that about 50% of its apparent molecular weight on sodium dodecyl sulfate (SDS) gels consists of polysaccharide side chains [12]. Finally, there is extensive variability among HTLV-III substrains within the envelope region, suggesting that the virus may have developed a protective mechanism to escape the host immune response, again reminiscent of members of the lentivirus subfamily of retroviruses [13].

With these points in mind, two major questions present themselves: Is the sugar umbrella a detriment to successful vaccine development because it represents a protective layer for epitopes located on the amino acid backbone that are targets for neutralization? A corollary to this would be, are deglycosylated forms of the gp better immunogens, or at least equally effective ones, for induction of neutralizing antibodies? This possibility, recently suggested by the studies of Elder et al. [14], is important to consider if recombinant products are to be eventually used. Alternatively, some glycosylation may be critical for the appropriate folding of the protein chain, especially if the epitopes represent noncontiguous regions. This latter possibility has been implicated in the elicitation of the G IX murine leukemia virus gp70 type-specific antibody [15].

The second issue, namely the envelope variability, also requires particular attention. The major question, in this regard, is, do common conserved regions exist among the various substrains, which themselves represent target epitopes for neutralization? A recent comparative analysis [16] of the envelope regions among widely different isolates indeed suggests that such regions may exist even in fully glycosylated products (Figure 42.3). Should deglycosylated forms prove to be more effective, the number and quality of such regions as immunogens may increase.

Approaches for a Vaccine Against HTLV-III Disease

Other than the question of what constitutes an effective human vaccine, there are the issues of safety, homogeneity, quantity, and economy. Such considerations lead one immediately to favor molecular engineering approaches for one's primary emphasis. This is not unlike the current strategies for developing vaccines against hepatitis B or herpesviruses [17]. To be decided initially is whether producing recombinant products in bacteria, yeast, or mammalian cells is the method of choice for the HTLV-III vaccine preparation. This will depend on the role that carbohydrate plays in the immunogenicity trials that use native or modified virion-derived antigens discussed above. Possibly related to this is the form of the molecule displaying the optimal neutralization epitopes shared among the substrains. If a sufficient quantity or quality of common epitopes proves to be a problem so that the neutralizing antibody is not widely protective, one will be faced with the difficult task of a cocktail of immunogens originating from representative strains.

If linear epitopes that are effective targets for neutralization can be demonstrated, the possibility of synthetic peptides comes immediately to mind. However, on the basis of current experiences with synthetic peptides as immunogens, it is likely that special modes of presentation are necessary [18] in order to generate protective immune responses against live virus challenges. As indicated above for the native gp, one may wish to consider the ISCOM approach for both the peptide and recombinant materials.

Another attractive method would be to use infectious recombinant viruses. Typically, a nonessential region of vaccinia virus, a classical vaccine virus with a known safety record, would be used as the locus for the desired gene, which would then be expressed under the control of vaccinia promoters. The resulting construct would consist of autonomously replicating vaccinia virus, which would also express large amounts of properly processed inserted antigen [19]. An attractive feature would be that a number of related gp regions could be introduced in sequence so that antibodies could be made to each of these, as has been recently demonstrated [20]. Because there are other considerations that might mitigate against using vaccinia as a vector (i.e., prior immunization of the population, possible pathogenicity in certain individuals), nonpathogenic strains of herpes- or adenoviruses are also being considered.

A final possibility is the use of anti-idiotypes as immunogens. An *anti-idiotype* represents the internal image of the relevant epitome within the antigen-combining regions of the antibody. In several experimental systems, a proper combination of native antigen and an anti-idiotype to the neutralizing antibody induces a powerful and long-lasting protective response within several compartments of the immune system [21].

Conclusions

All of the feasible approaches are predicated on a limited number of antigenic variants of HTLV-III as well as on the assumption that genetic drift and shift are not insurmountable. However, a significant degree of genetic heterogenicity exists among various HTLV-III isolates, especially in the *env* region of the genome, so that essentially no two virus isolates are identical [2]. As is the case with some lentiviruses, it is possible that significant new genetic variants can occur during the course of disease, even in a single host. However, studies have so far indicated that this phenomenon may be unusual if it occurs at all [22]. Should this occur, however, it would suggest that vaccine preparations would have to be complex to encompass many antigenic variants. Counter to that pessimism is that fact that in both the mouse and the cat retroviral systems, the exposure of the animal to an isolated gp can lead to broadly reactive antibodies, which neutralize distantly related virus types through conserved antigenic sites [1,6]. The possibility of enhancing this type immunogenicity of gp by its inclusion in ISCOM matrices provides an avenue for development of a safe and effective vaccine against human retroviruses.

It is therefore reasonable that as we gain knowledge of the structure of HTLV-III envelope components along with access to modern biotechnology, we can be optimistic about development of effective vaccines against HTLV-III disease. Primate animal models for HTLV-III infection [23] and at least one disease model with a related primate virus (STLV-III) [24] are now available for critical testing of vaccine strategies.

References

1. Schafer W, Bolognesi DP (1977) Mammalian C-type oncornaviruses: Relationships between viral structural and cell-surface antigens and their possible significance in immunological defense mechanisms. Contemp Top Immunobiol 6:127–167
2. Hahn BH, Gonda MA, Shaw GM, Popovic M, Hoxie JA, Gallo RC, Wong-Staal F (1985) Genomic diversity of the acquired immune deficiency syndrome virus

HTLV-III: Different virus exhibit greatest divergence in their envelope genes. Proc Natl Acad Sci USA 82:4813–4817

3. Hunsmann G, Moenning V, Schafer W (1975) Properties of mouse leukemia virus. IX. Active and passive immunization of mice against Friend leukemia with isolated viral GP$_{71}$ glycoprotein and its corresponding antiserum. Virology 66:327–329
4. Lewis MG, Mathes LE, Olson RG (1981) Protection against feline leukemia by vaccination with a subunit vaccine. Infect Immun 34:888–894
5. Olson RG (1982) United States Patent No. 4,332,793
6. Hunsmann G, Schneider J, Schulz A (1981) Immunoprevention of Friend virus-induced erythroleukemia by vaccination with viral envelope glycoprotein complexes. Virology 113:602–612
7. Morein B, Sundquist B, Hoglund S, Dalsgaard K, Osterhaus S (1984) Iscom, a novel structure for antigenic presentation of membrane proteins from enveloped viruses. Nature 308:457–462
8. Osterhaus A, Weijer K, Uytdehaag F, Jarrett O, Sundquist B, Morein B (1985) Induction of protective immune response in cats by vaccination with feline leukemia virus iscom. J Immunol 135:591–596
9. Sarngadharan MG, Popovic M, Bruch L, Schupbach J, Gallo RC (1984) Antibodies reactive with human T-lymphotropic retroviruses (HTLV-III) in serum of patients with AIDS. Science 224:506–508
10. Robert-Guroff M, Brown M, Gallo RC (1985) HTLV-III neutralizing antibodies in patients with AIDS and AIDS-related complex. Nature 316:72–74
11. Kitchen LW, Barin F, Sullivan JL, McLane MF, Brettler DB, Levine PH, Essex M (1984) Aetiology of AIDS-antibodies to human T-cell leukemia virus (type III) in haemophiliacs. Nature 312:367–369
12. Ratner L, Haseltine W, Patarca R, Livak KJ, Starcich B, Josephs SF, Doran ER, Rafalski JA, Whitehorn EA, Baumeister K, Ivanoff L, Petteway SR Jr, Pearson ML, Lautenberger JA, Papas TS, Ghrayeb J, Chang NT, Gallo RC, Wong-Staal F (1985) Complete nucleotide sequence of the AIDS virus, HTLV-III. Nature 313:277–284
13. Scott JV, Stowring L, Haase AT (1979) Antigenic variation in visna virus. Cell 18:321–327
14. Elder JH, McGee JS, Alexander S (1986) J Virol, in press
15. Pierotti M, DeLeo AB, Pinter A et al. (1981) The G$_{IX}$ antigen of murine leukemia virus: An analysis with monoclonal antibodies Virology 112:450–455
16. Starcich BR, Hahn BH, Shaw GM et al. (1986) Cell, in press
17. Blumber BS, London WT (1985) J Natl Cancer Inst 71:267–273
18. Tainer JA, Getzoff ED, Alexander H, Houghten RA, Olson AJ, Lerner RA (1984) The reactivity of anti-peptide antibodies is a function of the atomic mobility of sites in a protein. Nature 312:127–134
19. Paoletti E, Lipinskas BR, Samsonoff C, Mercer S, Panicali D (1984) Construction of live vaccines using genetically engineered poxviruses: Biological activity of vaccinia virus recombinants expressing the hepatitis B virus surface antigen and the herpes simplex virus glycoprotein D. Proc Natl Acad Sci USA 81:193–197
20. Perkus ME, Piccini A, Lipinskas BR, Paoletti E (1985) Recombinant vaccinia virus: Immunization against multiple antigens. Science 229:981–984
21. McNamara M, Kohler H (1984) Immunol Rev 79:87–91
22. Wong-Staal F (in preparation)

23. Alter HJ, Eichberg JW, Masur H, Saxinger WC, Gallo R, Macher AM, Lane HC, Fauci AS (1984) Transmission of HTLV-III infection from human plasma to chimpanzees: An animal model for AIDS. Science 226:549–552
24. Daniel MD, Letvin NL, King NW, Kannagi M, Sehgal PK, Hunt RD, Kanki PJ, Essex M, Desrosiers RC (1985) Isolation of T-cell tropic HTLV-III-like retrovirus from macaques. Science 228:1201–1204

CHAPTER 43

Prospects for a Varicella–Zoster Vaccine

MICHIAKI TAKAHASHI

Until recently there has been little effort to develop an attenuated live varicella vaccine, mainly because varicella is usually a mild disease in normal children and because varicella–zoster (VZ) virus causes latent infection which might lead to zoster in later life. In addition, technical difficulties in managing VZ virus, the poor virus yield in tissue culture, and the extremely labile infectivity of the virus under the usual conditions have been obstacles preventing development of a vaccine.

In 1974, we reported development of a live varicella vaccine [1,2]. The Oka strain of VZ virus, isolated from vesicles of a 3-year-old boy who developed typical varicella symptoms, was serially passaged 11 times in human embryonic lung cells at 34°C and 12 times in guinea pig embryo cells at 37°C. The virus was then passaged in human diploid cells (WI-38 cells or MRC-5 cells) several times in order to prepare the experimental vaccine. The vaccine virus was found to have the following properties [3]:

1. Temperature sensitivity. The vaccine virus was thermosensitive at 39°C.
2. Characteristic host range. The vaccine strain showed a higher ratio of infectivity in guinea pig embryo fibroblast (GPEF) to infectivity in human embryo fibroblast (HuEF), than any wild-type strains.

These results suggest that the vaccine virus is a variant of VZ virus, at least with respect to themosensitivity and host range [3].

Differences in the migration patterns of DNA of vaccine virus and other wild strain have been found after cleavage with restriction endonucleases. Using endonucleases Hpa I, Bam HI, and Bgl I, Martin et al. [4] showed significant differences between the DNA cleavage patterns of the vaccine strain (Oka) and wild strain. Using Hpa I, we also observed differences between DNA cleavage patterns of the vaccine strain (Oka), and wild strains [3].

Tests of the relative infectivities on GPEF and HuEF and of the DNA cleavage pattern obtained with Hpa I on nine clinical isolates from vaccinees with acute leukemic who developed symptoms of varicella or zoster proved useful in differentiating the vaccine strain from wild strains [3].

Early Appearance of Cell-mediated Immunity (CMI) After Vaccination and Its Apparent Correlation with the Protective Effect of Prompt Vaccination Shortly After Exposure

A VZ virus skin-test antigen was first developed by Kamiya et al. [5] to assess CMI. Later, an improved soluble skin-test antigen was prepared from culture fluid [6].

The times of conversion in the VZ virus skin test in children with natural varicella infection versus vaccinated children were compared. In natural infection, skin test reactions were negative before the appearance of a rash, but consistently positive after appearance of a rash. On the other hand, in about half the vaccinated children, the skin reaction was positive as early as four days following vaccination [7]. Furthermore, a positive skin reaction appeared 4–7 days earlier than that of neutralizing antibody after vaccination [8]. This early appearance of cellular immunity after vaccination seems to be closely related to the protective effect of prompt vaccination of susceptible children shortly after their contact with varicella patients. Indeed, household contacts or children in the ward have been protected from clinical varicella by vaccination within about 72 hours after exposure to varicella patients [9,10].

Role of Cell-mediated Immunity in Prevention of Clinical Varicella Infection

We have followed up VZ virus infection in a population of infants who were under one year of age during three outbreaks of varicella in a semiclosed domiciliary institution for infants. Preexisting antibody did not prevent development of illness, although the severity of disease was somewhat less in infants with a higher antibody titer. We also found that all the infants who developed clinical symptoms gave a negative skin test reaction before infection, and that the level of antibody or the cellular immune response to subsequent infection was not altered by preexisting antibody, though it was somewhat lower in the group with mild symptoms [11].

After administration of the live varicella vaccine, the VZ skin response of the infants became positive, and all of them were resistant to subsequent clinical varicella epidemic in the same institution [7]. Thus it may be concluded that CMI to VZ virus plays a major role in preventing clinical varicella infection.

Use of a Live Varicella Vaccine for Diseased Children Including High-risk Children

The salient feature of the live varicella vaccine (Oka strain) is that it can be used for high-risk children—i.e., those with acute leukemia, those who have been receiving immunosuppressive therapy, and those with other diseases—as well as for normal children.

In general, in Japan, children with acute leukemia had to meet the following requirements in order to receive vaccination: (a) They had to be in remission. (b) Their CMI, assessed with phytohemagglutinin (PHA) or other reagents, had to be normal. (c) Administration of all anticancer drugs except 6-mercaptopurine (6MP) was suspended from one week before to one week after vaccination.

Of 254 patients who were vaccinated under the above stipulations, 47 (18.5%) showed mild to moderate (usually mild) clinical reactions. Most of the vaccinees gave immune responses by the immune adherence hemagglutinin (IAHA) test or the fluorescent antibody to membrane antigen (FAMA) test. The protective effect in cases of confirmed household exposure was 81.8% (9/11), and the symptoms of most of the children who did develop varicella were mild [12]. Children with diseases other than malignancies, and normal children who were vaccinated, showed slight, if any, clinical reactions and good antibody responses. The protective effects in these children were over 90%, and most cases of clinical illness were mild [13,14].

Results of Clinical Trials of Varicella Vaccine in the United States and Europe

In the first reported study in the United States, 99 healthy children were immunized with the Oka strain of varicella vaccine. Their seroconversion rate was over 94%, and they showed slight, if any, clinical reaction [15]. Then a double-blind, placebo-controlled efficacy trial of the live attenuated Oka/Merck varicella vaccine was conducted on 956 children [16]. The vaccine caused only slight clinical reaction and was well tolerated. There was no clinical evidence of viral spread from vaccinated children to sibling controls. During the nine-month surveillance period, the vaccine was 100% efficacious in preventing varicella [16].

In a clinical trial of Oka/Merck varicella vaccine in children with acute leukemia, conducted by the NIH Study Group, there was serologic evidence of an immune response in approximately 80% after one dose, and in more than 90% after two doses. Mild-to-moderate clinical reactions were observed in 2 of 53 (4%) vaccinees who had completed chemotherapy, and in 49 of 138 (36%) vaccinees who had been receiving maintenance chemotherapy. The attack rate of clinical varicella in household exposure of vaccinees was

18% (4/22), all cases of clinical illness being extremely mild. Thus it was concluded that the varicella vaccine was approximately 80% effective in preventing clinical varicella in children with leukemia, and completely effective in preventing severe varicella in the high-risk group [17]. A similar result was reported by Brunell et al. [18].

In Europe, clinical trials with the varicella vaccine (Oka strain) were conducted in both high-risk and normal children, and results similar to those found in Japan and the United States were reported [19]. At the end of 1984, license was given for use of varicella vaccine of the Oka strain in Belgium, West Germany, Switzerland, Luxemburg, Portugal, and Austria.

Attempts to Isolate VZ Virus from Blood Cells of Naturally Infected and Vaccinated Children

Ozaki et al. [20] demonstrated that VZ virus could be recovered from mononuclear cells of varicella patients with no underlying diseases just after the onset of rash. Later, Asano et al. [21] showed that virus could be isolated for five days before onset of rash. In contrast, no VZ virus could be recovered from a total of 27 children 4–14 days after vaccination, even with a large dose of 5000 plaque-forming units (PFU) [22]. These results suggest that the vaccine virus replicates far less than wild-type VZ virus in the susceptible organs and blood cells, but still induces cellular and humoral anti-VZ immunities. These characteristics seem be related to the attenuated nature of the vaccine virus.

Incidence of Zoster After Vaccination

A major question has been whether the vaccine virus becomes latent, resulting in later development of zoster. A long-term follow-up of vaccinated normal children will be required to answer this question definitively. Children with acute leukemia tend to develop zoster early after natural infection. Careful comparative studies on groups of vaccinated and naturally infected children with acute leukemia showed that the incidence of zoster in the vaccinated group was the same or less than that in the naturally infected group and that the clinical symptoms of zoster in the vaccinated group were usually mild and not clinically troublesome, whereas in some patients in the naturally infected group, symptoms were moderate or severe [12,13]. Brunell et al. also observed that zoster developed in 19 of 76 children with acute leukemia who *had* varicella, but did not develop *after vaccination* in any of 48 children with acute leukemia ([23]; Brunell, personal communication). Thus there is no reason to suspect that zoster is more likely to develop after vaccination than after natural infection.

Immunities Inducible by Inactivated VZ Vaccine

Guinea pigs immunized with live varicella vaccine virus showed positive delayed-type hypersensitivity and a lymphocyte transformation response to viral antigen as well as a neutralizing (NT) antibody response, whereas those immunized with heat-inactivated vaccine virus or soluble VZ antigen showed only a NT antibody response of the same degree as that with live vaccine virus [24]. CMI to VZ virus could be induced by inactivated VZ virus only in combination with Freund's adjuvant. As CMI plays a major role in preventing clinical varicella infection, development of a safe and potent immunopotentiator is a prerequisite for clinical application of possible subunit VZ vaccines.

Prospects for Active Immunization with Live Varicella Vaccine

Many views have been expressed on active immunization with the live varicella vaccine [25–29]. The chief concerns about the use of this vaccine have been directed to the possible development of zoster resulting from latent infection with the vaccine virus and to its safety in high-risk children.

However, the latency of the vaccine virus leading to development of zoster in later life poses no severe threat, as discussed above, and the safety of vaccine use in high-risk children has been demonstrated in many clinical trials.

As pointed out by Kempe and Gershon [27], the risk/benefit ratio is of vital importance in any experimental endeavor involving human beings. Currently available data indicate that the potential benefits of live VZ vaccine outweigh the potential hazards, particularly in high-risk children. The live varicella vaccine is useful and effective for preventing varicella in immunocompromised children as well as in normal children.

References

1. Takahashi M, Otsuka T, Okuno Y, Asano Y, Yazaki T, Isomura S (1974) Live vaccine used to prevent the spread of varicella in children in hospital. Lancet ii:1288–1290
2. Takahashi M, Okuno Y, Otsuka T, Osame J, Takamizawa A, Sasada T, Kubo T (1975) Development of a live attenuated varicella vaccine. Biken J 18:25–33
3. Hayakawa Y, Torigoe S, Shiraki K, Yamanishi K, Takahashi M (1984) Biological and biophysical markers of a varicella vaccine strain (oka strain): Identification of clinical isolates from vaccine recipients. J Infect Dis 149:956–963
4. Martin JH, Dohner DE, Wellinghoff WJ, Gelb LD (1982) Restriction endonuclease analysis of varicella zoster vaccine virus and wild type DNAs. J Med Virol 9:69–76

5. Kamiya H, Ihara T, Hattori A, Iwasa T, Sakurai M, Izawa T, Yamada A, Takahashi M (1977) Diagnostic skin test reactions with varicella virus antigen and clinical application of the test. J Infect Dis 136:784–788

6. Asano Y, Shiraki K, Takahashi M, Nagai H, Ozaki T, Yazaki T (1981) Soluble skin test antigen of varicella–zoster virus prepared from the fluid of infected cultures. J Infect Dis 143:684–692

7. Takahasahi M, Baba K (1984) A live varicella vaccine: Its protective effect and immunological aspects of varicella–zoster virus infection (1984). *In* de la Maza LM, Peterson EM (eds) Medical Virology vol 3. Elsevier Scientific Publishing, pp 255–278

8. Baba K, Yabuuchi H, Okuni H, Takahashi M (1978) Studies with live varicella vaccine and inactivated skin test antigen: Protective effect of the vaccine and clinical application of the skin test. Pediatrics 61:550–555

9. Asano Y, Nakayama H, Yazaki T, Kato R, Hirose S, Tsuzuki K, Ito S, Isomura S, Takahashi M (1977) Protection against varicella in family contacts by immediate inoculation with live varicella vaccine. Pediatrics 59:3–7

10. Asano Y, Nakayama H, Yazaki T, Ito S, Isomura S, Takahashi M (1977) Protective efficacy of vaccination in children in four episodes of natural varicella zoster in the ward. Pediatrics 59:8–12

11. Baba K, Yabuuchi H, Takahashi M, Ogra P (1982) Immunologic and epidemiologic aspects of varicella infection acquired during infancy and early childhood. J Pediatr 100:881–885

12. Kamiya H, Kato T, Isaji M, Oitani K, Ito M, Ihara T, Sakurai M, Takahashi M (1984) Immunization of acute leukemic children with a live varicella vaccine (Oka strain). Biken J 27:99–102

13. Yabuuchi H, Baba K, Tsuda N, Okada S, Nose O, Seino Y, Tomita K, Ka K, Mimaki T, Ogawa M, Kenesaki T, Yoshida A, Takahashi M (1984) A live varicella vaccine in a pediatric community. Biken J 27:43–49

14. Horiuchi K (1984) Chickenpox vaccination of healthy children: Immunological and clinical responses and protective effect in 1978–1982. Biken J 27:37–38

15. Arbeter AM, Starr SE, Weibel RE, Plotkin SA (1982) Live attenuated varicella vaccine: Immunization studies with the Oka strain in healthy children. J Pediatr 100:886–893

16. Weibel RE, Neff BJ, Kuter BJ, Buess HA, Rothenberger CA, Fitzgerald AJ, Connor KA, McLean AA, Hilleman MR, Buynak EB, Scolnick EM (1984) Live attenuated varicella virus vaccine: Efficacy trial in healthy children. N Engl J Med 310:1409–1415

17. Gershon AA, Steinberg SP, Gelb L, Galasso G, Borkowsky W, LaRussa P, Ferrara A (The National Institute of Allergy and Infectious Diseases Varicella Vaccine Collaborative Study Group) (1984) Live varicella vaccine: Efficacy for children with leukemia in remission. JAMA 252:355–362

18. Brunell PA. Shehab Z, Geiser C, Waugh JE (1982) Administration of live varicella vaccine to children with leukemia. Lancet ii:1069–1072

19. Andre FE (1984) Summary of clinical studies with the Oka live varicella vaccine produced by Smith Kline RIT. Biken J 27:89–98

20. Ozaki T, Ichikawa T, Masui Y, Nagai T, Asano Y, Yamanishi K, Takahashi M (1984) Viremic phase in nonimmunocompromised children with varicella. Pediatrics 104:85–87

21. Asano Y, Itakura N, Hiroishi Y, Hirose S, Nagai T, Ozaki T, Yazaki T, Yamanishi

K, Takahashi M (1985) Viremia is present in incubation period in nonimmuno-compromised children with varicella. J Pediatr 106:69–71

22. Asano Y, Itakura N, Hitoishi Y, Hirose S, Ozaki T, Kuno K, Nagai T Yazaki T, Yamanishi K, Takahashi M (1985) Virus replication and immunologic responses in naturally infected children with varicella–zoster virus and in varicella vaccine recipients. J Infect Dis 152:863–868
23. Brunell PA, Taylor-Wiedman J, Geiser CF, Friedman L, Lydick E (1986) The risk of zoster in children with leukemia who received varicella vaccine as compared to those who had chicken pox. Pediatrics 77:53–56
24. Shiraki K, Yamanishi K, Takahashi M (1984) Delayed-type hypersensitivity and in vitro lymphocyte response in guinea pigs immunized with a live varicella vaccine. Biken J 27:19–22
25. Brunell PA (1977) Protection against varicella. Pediatrics 59:1–2
26. Sabin SA (1977) Varicella–zoster virus vaccine. JAMA 238:1731–1733
27. Kempe CH, Gershon AA (1977) Commentary. Pediatrics 60:803–804
28. Plotkin SA (1977) Varicella vaccine: Plotkin's plug. Pediatrics 59:953–954
29. Gershon AA (1978) Varicella–zoster virus: Prospects for active immunization. Am J Clin Pathol 70:170–174

Chapter 44
Prospects for Treatment and Prevention of Latent Herpes Simplex Virus Infection

JAMES F. ROONEY AND ABNER LOUIS NOTKINS

Typically, a primary infection with herpes simplex virus (HSV) is followed by the establishment of latent infection in the sensory ganglia of the affected dermatome. At the site of inoculation, the virus replicates in epithelial cells and then travels by retrograde axonal transport to the local sensory ganglia. There the virus causes a productive infection, which is followed in several days by the establishment of a latent infection. The virus persists asymptomatically in this latent state for the life of the host, but may periodically reactivate and travel down the neuron back to the epithelial surface to cause the typical recurrent herpetic lesion [1,2].

The molecular state of the HSV genome during latency is still unknown. Some investigators believe the virus exists in an episomal form, whereas others believe that the viral DNA is integrated into the host cell genome [3]. It is also uncertain whether limited transcription of the viral genome occurs during latency [3], and the precise mechanism of reactivation of latent virus has also not been established [4]. Substantial information, however, has accumulated about the immune response to HSV and its antigenic components. HSV codes for at least 50 proteins, including several glycoproteins which are expressed in the viral envelope and on the surface of infected cells. HSV-1 and HSV-2 have five glycoproteins designated gB, gC, gD, gE, and gG. While each HSV-1 glycoprotein (except possibly gG) has immunologic cross-reactivity with the respective glycoprotein of HSV-2, gD and gB appear to share the most type-common antigenic determinants. In addition, gB and gD appear to be major targets for neutralizing antibodies, and antibodies to these glycoproteins appear during the course of natural infection [5]. Humoral and cellular immunity play critical, but distinct, roles in controlling HSV infections. Humoral immunity is important in protection against the acquisition of primary disease, but reactivation occurs even in the presence of high antibody titers [6]. This may be due to the fact that HSV spreads directly from cell-to-cell, thereby avoiding neutralization by antibody in the extracellular space. Cellular immunity and interferon are probably the most important factors in limiting the cell-to-cell spread of the infection [7]. Patients with cellular immune dysfunction suffer more severe and frequent recurrences than immunocompetent individuals. No specific cellular, humoral, or nonspecific host defense defect, however, has been identified in immunocompetent patients with frequently recurrent disease [8].

Information on the pathogenesis of infection with HSV is making it possible to formulate rational intervention directed at the different stages of the infectious cycle. This includes (a) prevention of latent infection by vaccination, (b) treatment of established symptomatic infection by chemotherapy, and (c) limiting reactivation of the virus by identification and abrogation of precipitating factors.

Prevention of Latent Infection

To be effective, a herpes vaccine not only must limit the symptoms of primary infection, but must prevent the establishment of a latent infection. This means that the virus must be neutralized prior to entry into the nerve ending. Once inside the nerve, the virus is protected from the host immune response and may establish a latent infection which has the potential to periodically reactivate.

Our understanding of the pathogenesis of latency and reactivation in humans is based in large part on research carried out in experimental animals. A mouse model of HSV infection has been adapted to study the efficacy of various HSV vaccines in protecting against the development of latent infection with wild-type virus [9]. In this model, vaccinated mice are challenged with HSV by a variety of routes, including lip, eye, ear, or footpad. After several weeks, the local sensory ganglia are explanted onto indicator cells and observed for reactivation of HSV. Studies with several HSV vaccines have demonstrated substantial but not complete protection against the development of latent ganglionic infection. In some cases, latent infection developed despite high levels of neutralizing antibody [9,10]. This suggests that the immunogenicity of a vaccine may not be the only factor that determines the degree of protection afforded by vaccination. For example, there is some evidence that the degree of protection correlates with the amount of serum exudate at the site of inoculation of the challenge virus [9]. Vaccinated mice challenged by the relatively avascular footpad route, which induced little bleeding or serum exudate, showed much less protection against the development of latent infection than identically vaccinated mice challenged by the more vascular lip route. This may be due to the fact that the serum exudate contains antibody, and the more antibody at the site of viral challenge, the more likely it is that the virus will be neutralized prior to entry into the nerve ending. Thus the state of the epithelium at the site of viral exposure during the course of a natural infection may influence the outcome of the infection. Because there is still little precise information as to what happens at the challenge site in a natural infection in humans, the true efficacy of an HSV vaccine can be determined only by clinical trials in humans.

Although an HSV vaccine may prove useful in preventing the establishment of a latent infection, it will probably not be effective in reducing the frequency of recurrence in patients already latently infected. This assumption is based on the fact that immunity induced by natural infection is not effective in preventing subsequent recurrences. For a vaccine to be effective in decreasing recurrences, it would have to stimulate the immune system to a greater degree or in a different manner than that induced by natural infection. Because no immune deficit has been documented in patients with frequent recurrences, no immunological marker exists for the development of a vaccine to prevent recurrences of HSV [11].

If an HSV vaccine is to be safe and effective in preventing latent infection, it must be immunogenic, be of low virulence, not induce latency (or at least not reactivate), and not cause transformation of cells. As the oncogenic potential of HSV in humans is not known, many investigators have suggested that a safe vaccine for HSV should be free of viral DNA or at least free of transforming sequences [12] (which have not been conclusively identified).

Several new approaches have been taken toward developing such a vaccine. The first involves purification of subunit preparations by affinity chromatography, using monoclonal antibodies to select for HSV glycoproteins gD or gB. Administered without adjuvant, these purified glycoproteins are immunogenic in mice. HSV-1 gD provides greater protection than HSV-1 gB against lethal challenge with either HSV-1 or HSV-2 [13,14]. Protection against the development of latency and duration of immunity studies have not been reported.

A second new approach toward development of a vaccine for HSV involves the use of synthetic peptides. The antigenic domains of HSV-1 gD and HSV-2 gD have been determined, and synthetic peptides of varying lengths have been modeled after the N-terminal amino acids 1–23, which encompass one of the type-common antigenic domains [15]. These peptides are not immunogenic alone; however, when coupled to a carrier protein (keyhole limpet hemocyanin) and injected with adjuvant, they induce an antibody response which in most cases is capable of neutralizing both HSV-1 and HSV-2 in vitro. The highest antibody titers and greatest protection against HSV challenge are noted with the longest peptides, but none of the peptides provides the degree of protection offered by the intact glycoproteins. Studies on protection against the development of latency have not yet been reported.

A third approach involves the cloning of glycoprotein genes into expression vectors for production of large amounts of purified protein. HSV gD has been cloned into a variety of expression vectors, including *Escherichia coli*, yeast, and certain mammalian cell lines [16]. HSV-1 gD and HSV-2 gD produced in *E. coli* are nonglycosylated, are flanked by bacterial plasmid sequences, and are truncated to increase the level of expression and decrease the toxicity to the bacterial vector. The antigenic domains of the glycoproteins are preserved. These recombinant fusion proteins (containing both bacterial and glycoprotein sequences) are immunogenic in mice and rabbits. Antibodies from mice vaccinated with either the HSV-1 or HSV-2 recombinant proteins neutralize both HSV-1 and HSV-2 in vitro. Preliminary studies show that the HSV-1 gD recombinant protein is protective against lethal challenge in mice with HSV-2. Protection against latency and duration of immunity studies have not yet been reported. The safety of any bacterial sequences in the recombinant fusion protein will have to be determined prior to use of this product as a vaccine in humans. HSV-1 gD has also been expressed in Chinese hamster ovary cells. The protein product is glycosylated and is probably very similar to native HSV-1 gD. By deletion of the anchor sequence of the glycoprotein gene, the product is secreted from the mammalian cell vector into the surrounding media. This glycoprotein is immunogenic in mice and protects against lethal challenge with HSV-1 [17]. An advantage of both synthetic peptide and recombinant protein vaccines is their presumed lack of oncogenicity due to the absence of viral DNA.

A shortcoming of many subunit and peptide vaccines, thus far, has been the mandatory use of adjuvants and the need for repeated injections to achieve adequate protection. Live vaccines, which replicate in the host, induce stronger and longer-lasting humoral and cellular immunity than inactivated preparations. Several new approaches have been taken toward the development of a live HSV vaccine. One approach involves the selective deletion of genes coding for undesirable properties of the virus. If the genes coding for virulence, latency, and transformation could be identified, it might be possible to alter, by genetic engineering, the HSV genome to create a stable strain that did not contain these genes [18]. In fact, several HSV-1 deletion mutants have been created that are attenuated in virulence [11,19]. These strains are genetically stable, induce a good immune response, and are protective in animals against lethal challenge with HSV-1 and HSV-2. However, these mutants show variable protection against the development of latency with challenge virus and are only partially attenuated in ability to induce a latent infection themselves.

Another approach for a live vaccine involves the cloning of HSV glycoprotein genes into an unrelated expression vector such as vaccinia. A vaccinia virus recombinant possessing the HSV-1 gD has been constructed and tested in mice, with almost 100% protection noted against lethal challenge with HSV-1 and HSV-2, and substantial protection noted against development of latency with homologous virus [10]. Experiments examining the protection afforded by vaccination with a vaccinia HSV-1 gB recombinant are underway. The genome of vaccinia is large and will accommodate the insertion of several foreign genes allowing for the potential creation of a recombinant that could protect against several infectious agents (e.g., HSV-1, HSV-2, hepatitis) with a single vaccination. Perhaps more important, the vaccinia recombinant does not induce latency and only a small portion of the HSV genome is present, lessening the potential risk of transformation.

Some serious questions, however, have been raised with regard to the potential use of vaccinia recombinants as vaccines. The ability of patients to respond to a booster dose of the same or another recombinant would be an important factor in determining the overall utilization of recombinant vaccinia vaccines. A significant percentage of the world's population has already been vaccinated with vaccinia as a smallpox vaccine, raising the question of efficacy of a vaccinia recombinant in this population. The risk of transmission of a vaccina recombinant from vaccinees to nonvaccinated immune-impaired contacts must also be considered. In addition, severe postvaccination complications occur on rare occasions with vaccinia, especially in primary vaccinees [20]. Recent studies have shown that vaccinia thymidine kinase-deficient recombinants are less virulent in mice than the standard vaccinia strains [21], raising the hope that there might be a similar reduction of complications with use of these recombinants in humans.

It seems likely that, within the next several years, one or more of the potential HSV vaccines will undergo clinical trials in humans to establish their true efficacy.

Treatment of Established Latent Infection

Antiviral chemotherapy has provided a significant reduction in the severity of symptoms in patients already infected with HSV. Acyclovir, a nucleoside analogue of guanosine, is the first antiviral agent effective in the treatment of both systemic and mucocutaneous infection with HSV [22]. This drug is taken up preferentially by HSV-infected cells, is activated primarily by virally encoded thymidine kinase, and exhibits selective inhibition of virally encoded DNA polymerase [23]. Because of its specificity for virally encoded enzymes, acyclovir is relatively nontoxic to the host. The drug has proven effective in the treatment of primary mucocutaneous HSV infections, allowing for a significant reduction in pain, duration of viral shedding, and time to healing [22].

Acyclovir has also been effective, when given in a daily prophylactic dosage, in reducing the frequency and severity of HSV recurrences in both immunocompetent and immunocompromised patients. The drug is thought to act by blocking viral replication in epithelial tissues once reactivation has occurred. The drug does not appear to act on the latent form of the virus, and in fact, once the drug is stopped, the frequency of recurrences is similar to that prior to drug therapy. Some investigators postulate that in the absence of antiviral agents, reseeding of the ganglia may occur with each reactivation, and that the frequency of reactivation may depend, in part, on the number of neurons in the ganglia that are infected [24,25]. If this is so, perhaps very long-term prophylactic treatment with acyclovir will decrease the total number of neurons seeded and thereby reduce the frequency of recurrences once therapy is stopped. Objective evidence to support this hypothesis has not yet been obtained.

Much concern has been expressed about the possible emergence of strains of HSV resistant to acyclovir. In the laboratory, several types of drug-resistant mutants of HSV have been isolated by passage of virus in low concentrations of drug. Thus far, only thymidine kinase mutants have been isolated from clinical specimens. Whether these mutants or others may emerge as a significant clinical problem is not yet known [26]. Many other antiviral agents with activity against HSV are currently being evaluated, but thus far none has been shown to act on the latent form of the virus.

Identification of Stimuli That Trigger Reactivation of HSV

A third approach being taken to achieve control of latent infections with HSV involves the identification of factors that "trigger" reactivation of the latent virus. It has long been postulated that factors such as fever, sunlight, stress, menses, or trauma can act as reactivating stimuli. Recent studies have documented that, in patients with recurrent disease, exposure to ultraviolet B light in an area of previous recurrence can cause reactivation 60% to 70% of the time [27,28]. This provides evidence that ultraviolet light, and possibly sunlight, may act as a trigger for reactivation of latent HSV. A similar approach may be used in evaluating other factors, such as local tissue trauma, for their ability to cause reactivation. If the stimuli that trigger reactivation can be objectively identified, it may be possible to formulate interventions for avoidance of these stimuli or selective drug suppression to prevent recurrences. In addition, the UV light model provides an opportunity to study the early phases of reactivation prior to the onset of symptoms. It is possible that investigation of the changes in skin produced by ultraviolet light or other stimuli could lead to an understanding of the mechanisms and mediators involved in reactivation. This human model also provides a system for directly testing the ability of a variety of drugs to block reactivation of HSV. For example, it is known that prostaglandins play a role in the inflammatory response in skin following UV light exposure, and it has been postulated that prostaglandins may be involved in the reactivation of HSV. Perhaps antiprostaglandins, sunblockers, or other agents that block local inflammatory changes will prove to be therapeutically useful in preventing UV-induced reactivation of HSV.

Concluding Comments

In summary, the prospect of preventing and treating latent HSV infections by vaccines, chemotherapy, and identification and abrogation of reactivating stimuli is now becoming a reality. Inactivation or elimination of the viral genome from the latently infected host, however, remains a more difficult task. Perhaps as more is known about the molecular biology of the viral

DNA during latency, new interventions may be formulated to effect this goal. For example, if gene therapy becomes practical, it may someday be possible to insert into latently infected neurons plasmid constructs that express mRNA antisense to selected HSV transcripts [29]. In this way, one could inhibit reactivation of HSV by blocking transcripts required for viral reactivation.

References

1. Baringer JR (1981) Latency of herpes simplex and varicella zoster viruses in the nervous system. *In* Nahmias AJ, Dowdle WR, Schinazi RF (eds) The Human Herpesviruses: An Interdisciplinary Perspective. Elsevier, New York, Amsterdam, pp 201–205
2. Openshaw H, Sekizawa T, Wohlenberg C, Notkins AL (1981) The role of immunity in latency and reactivation of herpes simplex virus. *In* Nahmias AJ, Dowdle WR, Schinazi RF (eds) The Human Herpesviruses: An Interdisciplinary Perspective. Elsevier, New York, pp 289–296
3. Cantin EM, Puga A, Notkins AL (1984) Molecular biology of herpes simplex virus latency. *In* Notkins AL, Oldstone MBA (eds) Concepts in Viral Pathogenesis, Vol. 1. Springer Verlag, New York, pp 172–177
4. Blyth WA, Hill TJ (1984) Establishment, maintenance, and control of herpes simplex virus (HSV) latency. *In* Rouse BT, Lopez C (eds) Immunobiology of Herpes Simplex Virus Infection. CRC Press, Boca Raton, FL, pp 9–32
5. Norrild B (1985) Humoral response to herpes simplex virus infections. *In* Roizman B, Lopez C (eds) The Herpesviruses vol 4. Plenum Press, New York, pp 69–86
6. Reeves WC, Corey L, Adams HG, Vontver LA, Holmes KK (1981) Risk of recurrence after first episodes of genital herpes: Relation to HSV type and antibody response. N Engl J Med 305:315–319
7. Notkins Al (1974) Commentary: Immune mechanisms by which the spread of viral infections is stopped. Cell Immunol 11:478–483
8. Straus SE (1985) Biology and immunology of herpes simplex. *In* Straus SE (moderator) Herpes simplex virus infection: Biology, treatment, and prevention. Ann Intern Med 103:404–419
9. Price RW, Walz MA, Wohlenberg C, Notkins AL (1975) Latent infection of sensory ganglia with herpes simplex virus: Efficacy of immunization. Science 188:938–940
10. Cremer KJ, Mackett M, Wohlenberg C, Notkins AL, Moss B (1985) Vaccinia virus recombinant expressing herpes simplex virus type 1 glycoprotein D prevents latent herpes in mice. Science 228:737–740
11. Meigner B (1985) Vaccination against herpes simplex virus infections. *In* Roizman B, Lopez C (eds) The Herpesviruses vol 4. Plenum Press, New York, pp 265–296
12. Nahmias AJ, Roizman B (1973) Infection with herpes-simplex viruses 1 and 2 (third of three parts). N Engl J Med 289:781–789
13. Long D, Madara TJ, Ponce de Leon M, Cohen GH, Montgomery PC, Eisenberg RJ (1985) Glycoprotein D protects mice against lethal challenge with herpes simplex virus types 1 and 2. Infect Immun 43:761–764
14. Chan WL (1983) Protective immunization of mice with specific HSV_1 glycoproteins. Immunology 49:343–352

15. Eisenberg JR, Cerini PC, Heilman CJ, Joseph AD, Dietzschold B, Golub E, Long D, Ponce de Leon M, Cohen GH (1985) Synthetic glycoprotein D-related peptides protect mice against herpes simplex virus challenge. J Virol 56:1014–1017

16. Watson RJ, Enquist L (1985) Genetically engineered herpes simplex virus vaccines. *In* Melnick JL (ed) Progress in Medical Virology vol 31. Karger, Basel, pp 84–108

17. Lasky LA, Dowbenko D, Simonson C, Berman PW (1984) Production of an HSV subunit vaccine by genetically engineered mammalian cell lines. *In* Chanock RM, Lerner RA (eds) Modern Approaches to Vaccines: Molecular and Chemical Basis of Virus Virulence and Immunogenicity. Cold Spring Harbor Laboratory, Cold Spring Harbor, New York pp 189–194

18. Roizman B, Warren J, Thuning CA, Fanshaw MS, Norrild B, Meignier B (1982) Application of molecular genetics to the design of live herpes simplex virus vaccines. Dev Biol Stand 52:287–304

19. Roizman B, Arsenakis M (1985) Genetic engineering of herpes simplex virus genomes for attenuation and expression of foreign genes. *In* Quinnan GV Jr (ed) Vaccinia Viruses as Vectors for Vaccine Antigens. Elsevier, New York, pp 211–223

20. Arita I, Fenner F (1985) Complications of smallpox vaccination. *In* Quinnan GV Jr (ed) Vaccinia Viruses as Vectors for Vaccine Antigens. Elsevier, New York, pp 48–60

21. Buller RML, Smith GL, Cremer K, Notkins AL, Moss B (1985) Decreased virulence of recombinant vaccinia virus expression vectors is associated with a thymidine kinase-negative phenotype. Nature 317:813–815

22. Seidlin M (1985) Natural history and treatment of mucocutaneous infections. *In* Straus SE (moderator) Herpes simplex virus infection: Biology, treatment, and prevention. Ann Intern Med 103:404–419

23. Whitley R (1985) A perspective on the therapy of human herpesvirus infections. *In* Roizman B, Lopez C (eds) The Herpesviruses, vol 4. Plenum Press, New York, pp 339–369

24. Klein RJ (1985) Initiation and maintenance of latent herpes simplex virus infections: The paradox of perpetual immobility and continuous movement. Rev Infect Dis 7:21–30

25. Walz MA, Yamamoto H, Notkins AL (1976) Immunological response restricts number of cells in sensory ganglia infected with herpes simplex virus. Nature 264:554–556

26. Nusinoff-Lehrman S (1985) Antiviral resistance. *In* Straus SE (moderator) Herpes simplex virus infection: Biology, treatment, and prevention. Ann Intern Med 103:404–419

27. Spruance S (1985) Pathogenesis of herpes simplex labialis: Experimental induction of lesions with UV light. J Clin Microbiol 22(3):366–368

28. Perna J, Mannix M, Notkins AL, Straus SE (1986) Reactivation of latent herpes simplex virus infection by ultraviolet light: A human model. (Submitted for publication)

29. Izant J, Weintraub H (1985) Constitutive and conditional suppression of exogenous and endogenous genes by anti-sense RNA. Science 229:345–352

CHAPTER 45
Prospects for Improved Immunization Against Influenza

EDWIN D. KILBOURNE

Soon after the first isolation of influenza A virus from humans in 1933, effective artificial immunization against the disease was accomplished empirically with inactivated and attenuated viral vaccines. It is now known that immunity to influenza is mediated principally by antibodies to epitopes of the hemagglutinin (HA) and neuraminidase (NA) external glycoproteins of the influenza virion. Either antigen alone is effective, but only antibody to the HA is neutralizing and prevents infection. At its simplest, immunity to influenza can be achieved by parenteral injection of the HA or its components, or even by the administration of HA-specific antibody. Not inappropriately, therefore, the HA of influenza virus is the best studied of any viral protein. As a consequence, its three-dimensional structure has been determined, its antigenic sites have been localized with monoclonal antibody, and its biologic functions in cellular attachment and entry have been defined [1] (see Wilson, Chapter 2, this volume). Knowledge of its nucleotide-deduced amino acid sequence has encouraged the search for immunogenic oligopeptides of the HA protein, as well as their synthesis, with the expectation that such components can be used as highly specific vaccines. Alternatively and ideally, oligopeptides from conserved regions of the molecule might induce broad immunity to a wide range of virus variants.

The approach to influenza vaccination outlined above is attractive in its simplicity and reductionism. However, except as it addresses the goal of broadened immunogenicity, this approach is essentially only a further and potentially less immunogenic refinement of the unsatisfactory inactivated whole virus or subvirion component vaccines of the present. Improved immunization against influenza may require more than the identification and utilization of the major immunogenic virion protein and its principal epitopes for control of an epidemiologically complex disease caused by a virus with extraordinary evolutionary adaptability. Only in small part is the problem of influenza vaccination a problem in immunogenicity.

This review will (a) emphasize the unique problems presented by influenza and its adaptable virus, (b) identify the multiple components that comprise immunity in influenza, and (c) summarize current and new approaches to artificial induction of this immunity.

The Virus

Influenza is caused by any of three antigenically discrete viral types A, B, and C, whose alphabetical rank order corresponds to their epidemiologic and clinical importance. Of these, only influenza A viruses cause natural infection in animals and pandemics in humans—two circumstances that are probably related [2]. These extraordinarily well-characterized viruses contain 8 RNA segments of nonmessage sense that code for 7 structural and at least 3 nonstructural proteins. Of these, the hemagglutinin (HA) and neuraminidase (NA) external glycoproteins are inserted as spikelike structures into the virion lipid envelope and comprise the principal immunizing antigens of the virus. The viral core is delimited by the M protein, within which lie the RNA segments complexed in discrete RNPs with the nucleoprotein (NP). Both M and NP are type-specific antigens (i.e., common to all type A or type B variants) that participate in immunologic reactions but do not confer antibody-mediated immunity. Little is known of the immunogenic properties of the three proteins of the RNA polymerase complex (PB1, PB2, and PA) or of the nonstructural proteins M2, NS1, and NS2. The recent demonstration of M2 in the cell membrane [3] suggests its possible participation in immunospecific cytolytic reactions.

Influenza A viruses of humans include three HA subtypes, H1, H2, and H3, and two NA subtypes, N1 and N2, that have circulated cyclically since 1890. Presently, H1N1 and H3N2 subtypes are cocirculating. Although antigenically variable, influenza B and C viruses do not have distinguishable subtypes. Present vaccines contain H1N1, H3N2, and influenza B viruses.

Unique Problems in Vaccination Presented by Influenza

Although influenza viruses are not uniquely mutable, but share with other RNA viruses a high intrinsic mutation rate [4], they are unique in their apparent requirement for continual antigenic change to effect their survival in human populations (Table 45.1). The result of this sequential antigenic drift is recurrent regional epidemics, as virus strain or variant-specific immunity is bypassed by the changing virus. Because immunity induced by present inactivated vaccines is short-lived, new vaccines incorporating new antigens must be continually fabricated.

Major antigenic change in human influenza viruses results from the in-

Table 45.1. A unique virus presents unique problems in vaccination.

Viral attribute	Result	Epidemiologic event	Vaccination problem(s)
Mutable RNA genome and antigenic selectability[a]	sequential antigenic variation as reciprocal of rising host-population antibody response	interpandemic (epidemic) influenza	need for continual vaccine revision or induction of polyvalent or broadened *intrasubtypic* immunity
Segmented genome	capability for genetic reassortment and introduction of "new" antigens into host population	pandemic influenza	need for totally new vaccine or broadened *intratypic* immunity

[a]Antigenic mutability is inherent in all RNA viruses; influenza viruses seem unique in the natural emergence and selection of such variants.

corporation, through genetic reassortment, of HA genes from animal viruses [5]. This antigenic shift confronts the human population with an antigenically novel virus with which it has had no previous experience. The result, pandemic disease, is a threat that occurs irregularly at 10 to 40-year intervals and requires antigenically new vaccines or new approaches to the induction of broadened immunity.

Components of Immunity in Influenza

Immunity in influenza is mediated principally by humoral (serum) antibody to the HA [6]. HA antibodies neutralize the virus and prevent infection. They may be virus strain (variant)-specific, reactive with many variants within a subtype (homosubtypic), or, infrequently, with HAs of other subtypes (heterosubtypic) (Table 45.2).

Neuraminidase antibodies are also protective, but do not directly neutralize virus. Rather they restrict its replication in multicycle infection. Heterosubtypic cross-reactivity of NA antigens has not been demonstrated, perhaps because of limited study. Antibodies to other viral proteins do not measurably influence viral replication nor are they correlated with immunity.

Table 45.2. Potential components of influenza immunity.

Viral protein		Homologous (variant-specific)	Homosubtypic (heterovariant)	Homotypic (heterosubtypic)	Heterotypic
		Antigenic cross-reactivity			
HA	antibody	+	+	±[c]	+[d]
	CMI	+	+	+	
NA	antibody	+	+		
	CMI	+[a]			
NP	antibody			+	+[d]
	CMI		+	+	
M	antibody			+	+
	CMI	+[a]			
PB2	antibody				
	CMI	+[a,b]			
unknown	CMI				+[e]

Key: CMI, immunity mediated through T-helper or cytotoxic T lymphocytes or both.
[a]Only variant-specific reactivity was examined.
[b]From ref. 10. [c]Cross-reactions occur with low frequency. [d]A–B cross-reactions are mediated through HA$_2$ and NP [8]. [e]A–B cross-reactivity is probably T-helper mediated in a secondary in vivo reaction [9].

The potentially important role of cell-mediated immunity (CMI) by helper (Th) and cytotoxic (Tc) lymphocytes has received increasing attention (Table 45.2). HA-specific reactions involve sites other than those recognized by antibody and often are heterosubtypic (Table 45.2). Most cytotoxic effects, however, appear to be mediated through the NP antigen [7] (see Braciale and Braciale, Chapter 20, this volume). Intertypic cross-reactions between influenza A and B viruses are rare and probably have no practical significance.

Doubts about the feasibility of inducing lasting immunity to influenza should be dispelled by the experience of 1977 when an H1N1 virus returned virtually unaltered after an absence of 27 years. A pandemic of the young ensued, but those more than 25 years of age who had had earlier infection with the virus in the 1946–57 period were solidly immune.

The desirability of live virus immunization, with its capacity to induce local as well as serum antibody and better CMI, has long been appreciated. Categorically, live virus vaccines can be defined as viruses bearing temperature-sensitive (ts), cold-adapted (ca), host-range (hr), or deletion mutations that restrict replication to an immunogenic, but non-disease-producing level (Table 45.3). These attenuating mutations can be transferred between viruses of the same or different subtype as new variants arise (mutant reassortments) [11], or mutants of viruses already reassorted for other characteristics can be employed (reassortment mutants) [12]. In experimental use, these genetically defined and engineered viruses have been promising. However, vaccine viruses, no less than wild-type viruses, are plagued with intrinsic

Table 45.3. Antigenic presentation in contemporary and proposed influenza vaccines.

Nonreplicating antigens	*Replicating antigens*
Inactivated virus[a] (high yield reassortants)	Empirically attenuated virus
Subvirion components[a] (disrupted virus or HA–NA)	Genetically engineered virus ts, ca, hr, or deletion mutants reassortants
Individual antigens (purified or cloned)	reassortant-mutants
Individual epitopes (oligopeptides)	Cloned genes replicated in vaccinia
"Internal image" anti-idiotype antibody	

Combined (diphasic) approach	
Inject:	Infect with:
NA (purified or cloned)	Wild or attenuated virus

[a]Vaccines licensed in the United States.

genetic instability, so that vaccines with the proper balance of immunogenicity and nontoxicity have not yet been licensed in the United States.

The introduction of the influenza virus HA gene into vaccinia virus represents the ingenious attainment of monoantigenic immunization against a single influenza virus protein by a replicating vehicle that also stimulates HA-specific cytotoxic T cells [13]. But attaining this objective through infection with the most complex of animal viruses hardly simplifies the problem of influenza vaccination.

Nonreplicating Antigens

Advocates of nonreplicating antigens can point to the safety of inactivated virus and subunit vaccines during the past 40 years, but must recognize the narrow specificity and lesser duration of immunity induced by them. Subunit virus preparations, although less toxic, are also less immunogenic. Not only antibody response, but also CMI is less effectively stimulated by HA–NA component vaccines [14].

The identification of four principal antigenic sites on the HA molecule [1] has led to the synthesis of oligopeptides representing these and other regions of the protein, with the expectation that such peptides might function effectively as immunogens. Furthermore, a new approach to widening immunologic response is suggested by synthesis of a peptide able to generate antibodies that interact with regions of the HA which are not immunogenic or accessible when the protein is the immunogen [15]. Some have reported that conjugated synthetic peptides induce the formation of antibody reactions with HA [16] or can even effect marginal reduction of infectivity in mice [17]. Others have failed to demonstrate reactivity of antibodies against most peptides with virus, even those containing amino acid sequences from key antigenic sites [18]. On present evidence, synthetic oligopeptides must be viewed as interesting probes for dissection of the immune response, but not as immediate candidates for improved influenza virus vaccines.

It is likely that the newly proposed anti-idiotype or internal image vaccines [19] would share the potential of oligopeptides in their specificity and low toxicity, but similarly will require adjuvanted immunogenicity. Unless either approach identifies a significant antigenic site common to all influenza A viruses, the major need in influenza vaccination will not be met.

Recognizing that the best immunization for influenza remains natural infection, modulation rather than prevention of natural infection (*infection-permissive* vaccination) with influenza virus NA, is an attractive concept. Antibody to NA limits but does not prevent infection. Similar effects might be attained with submolecular antigens of the HA or other viral proteins incapable of evoking neutralizing antibody response [20].

There is no categorical reason that enduring broad immunity cannot be induced by hyperimmunization with one or several purified antigens introduced by either natural or parenteral routes. The apparent broader cross-reactivity of epitopes recognized by T-helper and cytotoxic T cells deserves further exploration. Candidate live virus vaccines are now attenuated on the basis of more rational principles, but as mutable agents always will require close monitoring of their immunogenicity and potential for virulence and transmissibility.

As the complexities of immunity in influenza are further explored, new approaches to its artificial induction will be found. But strategies at present must be pluralistic and should address the unique epidemiology of a protean virus.

References

1. Wiley DC, Wilson IA, Skehel JJ (1981) Structural identification of the antibody binding sites of Hong Kong influenza hemagglutinin and their involvement in antigenic variation. Nature 289:373–378
2. Kilbourne ED (1975) Epidemiology of influenza. In Kilbourne ED (ed) The Influenza Viruses and Influenza. Academic Press, New York, p 483
3. Lamb RA, Zebedee SL, Richardson CD (1985) Influenza virus M_2 protein is an integral membrane protein expressed on the infected-cell surface. Cell 40:627–633
4. Holland J, Spindler K, Horodyski F, Grabau E, Nichol S, VandePol S (1982) Rapid evolution of RNA genomes. Science 215:1577–1585
5. Scholtissek C, Rohde W, von Hoyningen V, Rott R (1978) On the origin of the human influenza virus subtype H2N2 and H3N2. Virology 87:13–14
6. Couch RB, Kasel JA, Six HR, Cate TR, Zahradnik JM (1984) Immunological reactions and resistance to infection with influenza virus. In Stuart-Harris C, Potter CW (eds) The Molecular Virology and Epidemiology of Influenza. Academic Press, New York p 119
7. Townsend ARM, Skehel JJ (1984) The influenza A virus nucleoprotein gene controls the induction of both subtype specific and cross-reactive cytotoxic T cells. J Exp Med 160:552–563
8. Graves PN, Schulman JL, Young JF, Palese P (1983) Preparation of influenza virus subviral particles lacking the HA1 subunit of hemagglutinin: Unmasking of cross-reactive HA2 determinants. Virology 126:106–116
9. Sitbon M, Gomard E, Hannoun C, Levy J-P (1983) Anti-influenza human T killer cells present an intertypic activity anti-A and -B type viruses in a secondary reaction in vitro. Clin Exp Immunol 54:49–58
10. Bennink JR, Yewdell JW, Gerhard W (1982) A viral polymerase involved in recognition of influenza virus-infected cells by a cytotoxic T-cell clone. Nature 296:75–76
11. Chanock RM, Murphy BR (1979) Genetic approaches to control of influenza. Perspect Biol Med 22:S37–S48
12. Schiff GM, Linnemann CC, Shea L, Lange B, Rottee T (1975) Evaluation of a live, attenuated recombinant influenza vaccine in high school children. Infect Immun 11:754–757

13. Bennink JR, Yewdell JW, Smith GL, Moller C, Moss B (1984) Recombinant vaccinia virus primes and stimulates influenza hemagglutinin-specific cytotoxic T cells. Nature 311:578–579
14. Webster RG, Askonas BA (1980) Cross-protection and cross-reactive cytotoxic T cells induced by influenza virus vaccines in mice. Eur J Immunol 10:396–401
15. Wilson IA, Niman HL, Houghten RA, Cherenson AR, Connolly ML, Lerner RA (1984) The structure of an antigenic determinant in a protein. Cell 37:767–778
16. Green N, Alexander H, Olson A, Alexander S, Shinnick TM, Sutcliffe JG, Lerner RA (1982) Immunogenic structure of the influenza virus hemagglutinin. Cell 28:477–487
17. Shapira M, Jibson M, Muller G, Arnon R (1984) Immunity and protection against influenza virus by synthetic peptide corresponding to antigenic sites of hemagglutinin. Proc Natl Acad Sci USA 81:2461–2465
18. Nestorowicz A, Tregear GW, Southwell CN, Martyn J, Murray JM, White DO, Jackson DC (1985) Antibodies elicited by influenza virus hemagglutinin fail to bind to synthetic peptides representing putative antigenic sites. Mol Immunol 22:145–154
19. Dreesman GR, Kennedy RC (1985) Anti-idiotypic antibodies: Implications of internal image-based vaccines for infectious diseases. J Infect Dis 151:761–765
20. Kilbourne ED (1984) Immunization strategy: Infection-permissive vaccines for the modulation of infection. In Chanock RM, Lerner RA (eds) Modern Approaches to Vaccines. Cold Spring Harbor Laboratory, New York, p 269

CHAPTER 46

Cleavage Site Mutant As a Potential Vaccine

Morio Homma

Introduction

Cleavage of the envelope glycoprotein of paramyxoviruses [1–3] and influenza viruses [4–7] is accomplished by a host protease. By this means, virus becomes activated, causes hemolysis, cell fusion, and infectivity. The presence of the protease allows the virus to replicate in multiple cycles. In contrast, in the absence of protease, replication is limited to a single cycle.

A protease activation mutant of Sendai virus has been obtained. The respiratory route is the natural pathway of infection and the glycoprotein of the mutant virus is not cleaved by the host protease of the lung of mice but can be cleaved in vitro by protease from other sources. When the mutant is administered into mice after its initial activation in vitro, its replication becomes restricted in the lung to a single cycle, and no appreciable disease results. Nevertheless, the mutant induces immunity, as such inoculated animals successfully handle a specific challenge with the wild-type virulent virus. Although various kinds of live attenuated vaccines have been prepared, this observation with a *protease mutant virus* represents the first attempt to use the principle of protease activation for vaccination. This chapter reviews experimental results in this area and discusses the availability of such mutants for vaccines, along with their strengths and limitations.

Activation of Paramyxoviruses by Protease

Members of the paramyxovirus group have two glycoproteins on the surface of their envelopes, HANA and F. The HANA glycoprotein has both hem-agglutinating and neuraminidase activities and is responsible for adsorption to the host cell receptors [8–10]. The F glycoprotein mediates virus entry into the cell by fusing the virus envelope with the plasma membrane of host cells. The biochemical reaction is the proteolytic cleavage of the F glycoprotein into disulfide-linked glycoprotein subunits, F_1 and F_2 [1–3]. The proteolytic cleavage exposes a new N terminus on F_1 glycoprotein. The amino acid sequence at this site is conserved among different paramyxoviruses [11]. The structure of the new N terminus is characterized by its high degree of hydrophobicity, and this structure is indispensable for the interaction of the virus envelope with the cytoplasmic membrane.

Sendai viruses grown in variety of tissue culture cells, possess an uncleaved precursor of F glycoprotein and hence, are biologically inactive; they lack the ability of envelope fusion and show neither infectivity, nor hemolysing nor cell-fusing activities. They do retain full hemagglutinating and neuraminidase activity. These viruses can be activated by a treatment with trypsin, which cleaves the F glycoprotein into F_1 and F_2 subunits [1,2]. The embryonated chicken egg has been chosen initially through empirical observations as the preferred host for Sendai virus and has been used for passage of seed virus. Retrospectively, this is because the yield of biologically active virus results from cleavage of F glycoprotein by protease present in the chorioallantoic fluids [12]. Similar results occur with certain primary tissue culture cells [13]. These observations collectively indicate that the presence of the activating enzymes determines the host range and organ tropism of Sendai virus [12,14]. Similar results have been obtained with Newcastle disease virus [15].

Protease-mediated Pneumotropism in Mice

Mice infected with the active form of wild-type Sendai virus develop signs of respiratory infection, consisting of nasal discharge, difficulty in breathing, and cyanosis, which are associated with lowered body weight. Most infected mice die in one week. The lung is swollen and congestive, and focal consolidation becomes visible. Histologic changes show peribronchial infiltration with large mononuclear cells, swelling of the bronchial epithelium, and bleeding into the interstitium. Immunoperoxidase staining reveals viral antigens in the epithelial cells of the trachea, bronchus, and bronchiolus, but not in cells of the alveolus. These findings indicate that the target cells of Sendai virus in the lungs of mice are confined to bronchial epithelium. The receptors for Sendai virus are distributed rather widely in various organs of mice, e.g., gastrointestinal tract, kidney, spleen, heart, brain, and lymph nodes [16]. Neither replication of the virus nor pathologic change in organs other than lung is, however, detectable in natural or experimental infection of mice, i.e., intranasally or intraperitoneally [17]. These observations also indicate that the pneumotropism of Sendai virus is not determined merely by the presence and distribution of the receptors for this virus. Development of the histologic changes and increase of the antigen-producing bronchial cells daily after infection indicate that replication of Sendai virus in the lung occurs in multiple steps. This is confirmed by measuring the time course of infectivity in the lungs of mice. These findings, together with the fact that the progeny virus recovered from the lung is of the active form, suggest that there is trypsin-like protease in the lung of mice activating the progeny virus through the cleavage of F glycoprotein. This was further shown by observations obtained using minced lung blocks prepared from mice infected with active form of Sendai virus intranasally for two hours, and cultured in a CO_2 incubator [18]. With this culture system, the in vivo situation of the mouse lung in terms of the development and distribution of the viral antigens was mimicked. Progeny virus recovered from this culture system was always in the active form and was accompanied by the cleavage of F glycoprotein into F_1 and F_2. Hence, the activating mechanism found in vivo was preserved in the lung block culture. Incorporation of various kinds of protease inhibitors—tosyllysylchloromethylketone, leupeptin, soybean trypsin inhibitor, and antipapain—in the culture inhibited viral activation. Inactivated Sendai virus never grew in the lung of mice, indicating that the activating enzyme in the lungs is not accessible to the free inactive virus. The cleavage in the lung occurs intracellularly in the bronchial cells or at the plasma membrane.

Cleavage Site Mutants

Scheid and Choppin described Sendai virus mutants that exhibit altered specificities to protease activation and altered host range in vitro. Included were mutants activated by either chymotrypsin or elastase or no longer activated by trypsin [19]. Similarly, we obtained a trypsin-resistant mutant, designated TR-2 [20], after infecting LLC-MK2 cells with wild-type Sendai virus in the presence of chymotrypsin and subsequent fivefold passaging. TR-2 is resistant to trypsin but is activated by chymotrypsin, which cleaves glycoprotein F into F_1 and F_2 subunits. Otherwise, no differences are found between the wild-type virus and TR-2 mutant by sodium dodecyl sulfate (SDS) gel electrophoresis patterns of polypeptides, buoyant density, or one-step growth profile in LLC-MK2 cells. The TR-2 mutant grows in multiple steps in the chorioallantoic cavity of the embryonated chicken egg, but only when chymotrypsin is included. As in the case of inactive wild-type Sendai virus, inactive TR-2 neither grew in nor caused disease in the lungs of mice. TR-2 initiates replication in the bronchial epithelium if activated by chymotrypsin before inoculation into mice. Because the progeny virus is not activated by proteolytic enzymes present in the lungs of mice, the replication of TR-2 in the lung is limited to a single cycle of replication. Following TR-2 inoculation, dose-dependent pathologic changes are found in peribronchial areas. Viral antigens are detectable in a dose-dependent manner in epithelial cells of the respiratory tract after immunoperoxidase staining of lung tissue.

Protection of Mice by TR-2

Although TR-2 does not bring about clinical illness or lung lesions in inoculated mice, nevertheless high titers of hemagglutination-inhibiting antibody (IgG) activity against wild-type Sendai virus are formed. Thereafter, when these mice are challenged with wild-type Sendai virus after three or more weeks, they are resistant to disease initiated by the wild-type progeny virus. Lung consolidation does not occur, and histologic changes of injury are not found. Level of the serum antibody increases further after the second challenge. Secretory immunoglobulin A antibody on the surface of the respiratory tracts is believed to be important for protection against respiratory viral infections [21]. Bronchoalveolar lavages obtained from the mice three weeks after the infection with TR-2 mutant and prior to challenge with wild-type virus contain neutralizing antibody, with the majority of activity in the IgA class. Protection against wild-type virus challenge is dependent upon the inoculum dose because a 50% protective dose is reached with 4.5×10^6 plaque-forming units (PFU) against the challenge virus of 20 median lethal doses (LD_{50}). At present, there is no evidence that protease activation mutants acquire altered organ tropism.

Conclusion

The use of cleavage site mutant, as described here for TR-2, as a live vaccine candidate has been demonstrated in an experimental in vivo system. What are the advantages and disadvantages of such an approach? Limited replication of TR-2 to a single cycle in the host animal reduces the frequency of the generation of revertants. Further, if shedding of virus used for vaccination occurs, it will not cause the disease because the progeny vaccine virus is produced only in inactive form. In addition, the resistance to a specific protease provides a marker of the administered virus, and, any reversion, if it occurs, can easily be detectable by in vitro testing. A strong protective immune response is induced, despite the single cycle of replication, and the response is equivalent to that observed with ordinary attenuated live vaccine. However, compared to attenuated live vaccines, a relatively higher dose of TR-2 may be required. An additional concern is that TR-2 is potentially virulent and hence may cause some illness when applied in large amounts, although this has not yet been seen. Nevertheless, the principle of the protease activation mutant like TR-2 may be applicable to other paramyxoviruses and orthomyxoviruses that are activated by trypsin or plasmin through proteolytic cleavage of hemagglutinin glycoprotein [5,6], and represents a novel and potentially efficient method for vaccination.

Acknowledgments

This research was supported by a Grant-in-Aid for Scientific Research from the Ministry of Education, Science, and Culture, Japan. Much of this work was done in collaboration with M. Tashiro.

References

1. Homma M, Ohuchi M (1973) Trypsin action on the growth of Sendai virus in tissue culture cells. III. Structural difference of Sendai virus grown in eggs and in tissue culture cells. J Virol 12:1457–1465
2. Scheid A, Choppin PW (1974) Identification of biological activities of paramyxovirus glycoproteins. Activation of cell fusion, hemolysis and infectivity by proteolytic cleavage of an inactive precursor protein of Sendai virus. Virology 57:475–490
3. Nagai Y, Klenk H-D, Rott R (1976) Proteolytic cleavage of viral glycoproteins and its significance for the virulence of Newcastle disease virus. Virology 72:494–508
4. Lazarowitz SG, Compans RW, Choppin PW (1971) Influenza virus structural and non-structural proteins in infected cells and their plasma membranes. Virology 46:830–843
5. Klenk H-D, Rott R, Orlich M, Bloedorn J (1975) Activation of influenza A viruses by trypsin treatment. Virology 68:426–439
6. Lazarowitz SG, Choppin PW (1975) Enhancement of the infectivity of influenza A and B viruses by proteolytic cleavage of the hemagglutinin polypeptide. Virology 68:440–454
7. Garten W, Bosch FX, Linder D, Rott R, Klenk H-D (1981) Proteolytic activation of the influenza virus hemagglutinin: The structure of the cleavage site and the enzyme involved in cleavage. Virology 115:361–374
8. Scheid A, Caliguiri LA, Compans RW, Choppin PW (1972) Isolation of myxovirus glycoproteins. Association of both hemagglutinating and neuraminidase activities with the larger SV5 glycoprotein. Virology 50:640–652
9. Tozawa H, Watanabe M, Ishida N (1973) Structural components of Sendai virus. Serological and physicochemical characterization of hemagglutinin subunit associated with neuraminidase activity. Virology 55:242–253
10. Scheid A, Choppin PW (1974) The hemagglutinating and neuraminidase (HN) protein of a paramyxovirus: Interaction of neuraminic acid in affinity chromotography. Virology 62:125–133
11. Scheid A, Graves MC, Silver SM, Choppin PW (1978) Studies on the structure and functions of paramyxovirus glycoproteins. *In* Mahy BWJ, Barry RD (eds) Negative Strand Viruses and the Host Cell. Academic Press, London, p 181
12. Muramatsu M, Homma M (1980) Trypsin action on the growth of Sendai virus in tissue culture cells. V. An activating enzyme for Sendai virus in the chorioallantoic fluid of the embryonated chicken eggs. Microbiol Immunol 24:113–122

13. Silver SM, Scheid A, Choppin PW (1978) Loss on serial passage of rhesus monkey kidney cells of proteolytic activity required for Sendai virus activation. Infect Immun 20:235–241
14. Ishida N, Homma M (1978) Sendai virus. Adv Virus Res 23:349–383
15. Garten W, Berk W, Nagai Y, Rott R, Klenk H-D (1980) Mutational changes of the protease susceptibility of glycoprotein F of Newcastle disease virus: Effect on pathogenicity. J Gen Virol 50:135–147
16. Takano M, Itoh Y, Maeno K, Shimokata K, Yamamoto T, Hara K, Iijma S (1980) In vivo distribution of cells with virus receptor. 2. Distribution of cells with HVJ-receptor. (In Japanese). Abstracts of the General Meeting of the Society of Japanese Virologists p. 2027
17. Tashiro M, Homma M (1983) Pneumotropism of Sendai virus in relation to protease-mediated activation in mouse lungs. Infect Immun 39:879–888
18. Tashiro M, Homma M (1983) Evidence of proteolytic activation of Sendai virus in mouse lung. Arch Virol 77:127–137
19. Scheid A, Choppin PW (1976) Protease activation mutants of Sendai virus: Activation of biological properties by specific proteases. Virology 69:265–277
20. Tashiro M, Homma M (1985) Protection of mice from wild-type Sendai virus infection by a trypsin-resistant mutant, TR-2. J Virol 53:228–234
21. Green GM, Jakab GJ, Low RB, Davis GS (1977) Defense mechanism of the respiratory membrane. Am Rev Respir Dis 115:479–514

CHAPTER 47
Prospects for a Synthetic Foot-and-Mouth Disease Vaccine

FRED BROWN*

Foot-and-mouth disease is one of the most devastating diseases of farm animals. Cattle, sheep, pigs, and goats are all affected, and productivity losses are usually estimated to be as high as 25%. In addition, indirect losses due to embargoes on trading can be crippling to a country whose economy depends on the export of farm products. The disease occurs in most parts of the world, and only North America, Australasia, Britain, and Japan can be considered disease-free. Control is by slaughter in those countries where the disease does not normally occur, but in endemic areas control is by vaccination. This is a large undertaking, and more than 1.5 billion doses of vaccine are used each year. Consequently, the methods used for preparing vaccines and their quality control are of considerable importance to the livestock industry. In this article, methods for preparing current vaccines are described briefly, but attention is focused on the prospects for engineering a synthetic vaccine based on the knowledge gained from studies on the molecular biology of the virus.

Current Vaccines

Present-day vaccines are prepared by inactivation of virus grown in the baby hamster kidney cell line, BHK 21 [1] or in surviving bovine tongue epithelium fragments [2]. About 10 ug of virus particles are sufficient to protect a cow against subsequent challenge with 100,000 times the median infective dose (ID_{50}) of the homologous virus. The vaccines are usually applied adsorbed

on to aluminum hydroxide gel, in the presence of low concentrations of sa-
ponin, or emulsified with incomplete Freund's adjuvant.

In general, conventional vaccines of this type have proved efficient in
controlling the disease. Since the application of a comprehensive vaccination
policy in many of the countries of Western Europe in the mid 1950s, the
incidence of the disease there has fallen dramatically (Figure 47.1), and only
a small number of outbreaks now occur each year. In other parts of the
world, the situation is not so encouraging because of the greater difficulties
in administering the vaccine.

Vaccines are usually effective for about 6–12 months. This has led to the
adoption of vaccination campaigns, which are conducted at regular intervals
within this time span. The intervals vary from country to country. In Ar-
gentina, for example, compulsory vaccination is conducted every four to six
months, whereas in Europe it is usually on an annual basis.

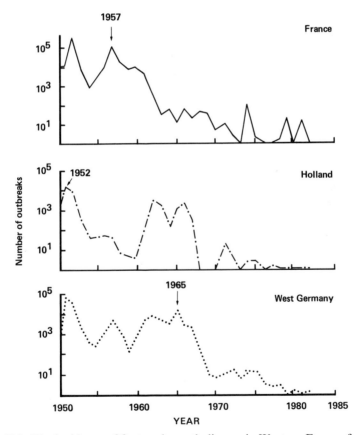

Figure 47.1. The incidence of foot-and-mouth disease in Western Europe from 1950
to 1982.

Problems with Current Vaccines

Despite the undoubted success of vaccination in controlling the disease when the quality of the vaccines is good and when they are applied in well-conducted campaigns, some problems remain. In the first place, the vaccines are relatively unstable, and care must be taken to preserve their immunizing activity. This requires a cold chain, which is difficult to ensure, particularly under field conditions in large tropical countries. Secondly, the virus exhibits considerable antigenic variability. The virus occurs as seven major serotypes: 0, A, C, SAT1, SAT2, SAT3, and Asia 1. There is no cross-protection against viruses of the other serotypes, even when an animal is solidly immune against viruses of the homologous serotype. Moreover, there is sufficient antigenic variation between isolates within a serotype to make it necessary to monitor outbreak strains for their serologic relatedness to the available vaccines. This is usually done by comparing the neutralizing activity of sera from vaccinated animals against the outbreak virus and the homologous vaccine strains. In general, from the manufacturers' point of view, changing the virus for vaccine production is to be avoided if possible because it may mean abandoning a vaccine strain whose characteristics in terms of growth, stability, and immunizing activity are well established, and replacing it with a strain whose properties may be much less satisfactory. Nevertheless, it sometimes becomes imperative to do this because of the wide serologic difference between the new outbreak strain and the available vaccine strains.

Vaccines with a wide antigenic spectrum are clearly desirable so that the difficulties associated with intratypic variation can be overcome. There are examples of viruses with these properties, and a comparison of their molecular structure with that of viruses having a narrow antigenic spectrum could conceivably enable us to design appropriate immunogens, either as synthetic peptides or as new viruses obtained by site-specific mutagenesis.

Molecular Biology of the Virus

Virus-specified Particles

Any rational design of improved vaccines depends on an understanding of the molecular structure of the immunogens that elicit the protective immune response. The structure of the virus and of the virus-specified antigens that are found in virus harvests has been elucidated over the last twenty or so years, and a large amount of information is now available. Virus harvests contain four virus-specified particles: (a) the infective 146S particle, consisting of one molecule of single-stranded RNA (molecular weight, about 2.6×10^6) and 60 copies of each of four proteins, VP1–VP4; (b) a so-called "empty particle" sedimenting at 75S, with the same size and shape as the infective particle, but consisting only of the four proteins VP1–VP4, in this case with VP2 and VP4 linked covalently; (c) a 12S protein subunit consisting of five copies of VP1, VP2, VP3; and (d) the RNA polymerase. The major immunogenic component is the intact virus particle [3]. However, the empty particle possesses some immunogenic activity, provided it is "stabilized" by fixation with low concentrations of formaldehyde [4]. The 12S particle has very low activity, of the order of 1% or less of that of the virus particle, and no immunogenic activity has been found for the RNA polymerase.

Location of Immunogenic Site(s)

Several observations have allowed the major immunogenic site on the virus to be located:

1. Treatment of the virus particle with trypsin dramatically reduces its ability to attach to host cells and hence its infectivity; in addition, with some strains the immunogenic activity is also considerably lower after treatment with the enzyme [5]. Examination of the proteins of the treated particles shows that VP1 is cleaved into two fragments whose total molecular weight is only slightly smaller that that of the intact molecule, whereas the other three proteins and the RNA are unaffected. This result implicates VP1 as the major immunogen of the virus particle.

2. Disruption of the virus into its constituent proteins and RNA, followed by testing of the individual proteins shows that only VP1 possesses any immunogenic activity [6]. The level of activity is very low compared with that of the intact virus particle. Nevertheless, this appears to confirm the observations made in the trypsin cleavage experiments.

3. Fractionation of the products obtained by cleaving VP1 in situ with different proteolytic enzymes or, after isolation, with cyanogen bromide has revealed that possible immunogenic sites are located at amino acid residues 146–154 and 200–213 [7].

4. Comparison of the amino acid sequences of VP1 of viruses from different serotypes and subtypes reveals that, although most of the amino acids are conserved, there are three regions of high variability, two of which include the residues 146–154 and 200–213 referred to in the preceding paragraph [8]. The third hypervariable region at residues 41–60 is slightly hydrophobic and therefore is unlikely to be on the surface of the molecule, whereas the 140–160 and 193–213 residues are hydrophilic. On the assumption that the serologic variability can be correlated with sequence variability, it seems that sequences 140–160 and 193–213 would be likely immunogenic sites.

This information has led to two approaches for a "synthetic" immunogen. The first is the biosynthesis of VP1 in *Escherichia coli*. This approach depended on the knowledge, gained from biochemical mapping studies, of the region of the RNA genome coding for the protein. When complementary DNA corresponding to this segment of the RNA genome is expressed in *E. coli*, very high yields of VP1 are obtained, often being as much as 20% of the total bacterial protein [9]. Protection of cattle and pigs against challenge infection was obtained by injecting two doses each of 250 μg of the synthetic VP1. Nevertheless, the synthetic VP1 has very low immunogenic activity as compared with the virus particle, and it seems unlikely that this approach to a synthetic vaccine will be feasible unless the expressed protein can be made to take up the conformation that VP1 assumes on the virus particle.

The second approach has depended on the location of the immunogenic sites on VP1. Peptides corresponding to 20-mers of the entire protein, i.e., 1–20, 21–40 . . . 181–200, and 200–213 have been tested for their ability to elicit neutralizing antibody and, in some instances, their ability to protect animals against experimental infection. The hypervariable region 141–160 elicits high levels of neutralizing antibody and protects guinea pigs against experimental infection. The same region has also been shown to afford protection in a small number of cattle. The second hydrophilic and hypervariable region, 200–213, also elicits neutralizing antibody in guinea pigs but at a much lower level than the 141–160 region. Nevertheless, this region may be crucial in its interaction with 141–160 as part of a conformational immunogenic site (see below). A third region, 161–180, also elicited low levels of neutralizing antibody.

Most of the experiments have been done with the 141–160 peptide coupled to a carrier protein such as keyhole limpet hemocyanin, tetanus toxoid, or bovine serum albumin. The method of coupling has also been varied [10]. From these experiments, no clearly "best method" has emerged either for the coupling method or for the carrier to be used. Indeed, we have obtained convincing evidence that good neutralizing antibody responses can be obtained without the use of carriers [10]. This finding could be of crucial importance in the eventual application of peptides, particularly in humans.

In addition to the organic synthetic approach, sequences containing the 141–160 peptide have been biosynthesized in *E. coli*. The Genentech–Plum Island group in the United States have expressed a cDNA clone corresponding to residues 130–157 of VP1 isolated from the DNA obtained by reverse transcription of the virus RNA, after linking it to the *trp*LE gene. This product will protect cattle against challenge infection after a single injection of 250 μg fusion protein, a weight corresponding to 25 μg peptide (J.J. Callis, personal communication).

In our own experiments, a chemically synthesized DNA corresponding to the amino acid sequence 142–160 was inserted into a vector designed to express fusion proteins with the carboxy terminal 1015 amino acids of β-galactosidase. The hybrid protein, which is soluble, has full β-galactosidase enzyme activity and can be purified by affinity chromatography by use of antibody against β-galactosidase. The purified product elicits neutralizing antibody in mice and guinea pigs at a level that would be expected to afford protection against challenge [11].

Priming with Peptides

One of the problems with which manufacturers of foot-and-mouth disease vaccines are confronted is the need to meet the requirement that a steer that has been given *one* injection of the vaccine should be protected at 21 days against challenge with $100,000 \times ID_{50}$ of the homologous virus. This requirement is far more stringent than is demanded of any other inactivated vaccine. For most inactivated vaccines, more than one dose is given, with the consequent boosting effect that is obtained with the second and subsequent injections. However, the daunting task of organizing and carrying out more than one injection of foot-and-mouth disease vaccine resulted in the acceptance of the present-day requirements. Consequently, the possibility of giving more than one dose by making use of triggered release technology is attractive. With foot-and-mouth disease virus, however, the instability of the particles, particularly on drying, does not allow the application of this technology.

With the observation that peptides will elicit a protective immune response, the difficulties associated with the instability of the virus particles for this type of experiment are essentially overcome. We have found in experiments with guinea pigs that killed virus particles or the 141–160 peptide will prime for a neutralizing antibody response on a subsequent injection of peptide [12]. Even when the amounts of virus particles or peptide used are insufficient to elicit a neutralizing antibody response in a single injection, the priming is sufficient to stimulate a good neutralizing response on the second injection.

These observations have allowed us to investigate the possibility of using triggered release as a second method for obtaining high levels of neutralizing antibody. In preliminary experiments we have found that peptide, suitably encapsulated, can be released several days after injection to give a booster effect in animals primed with the peptide. Development of this approach could lead to high, protective levels of antibody with very small quantities of peptide.

Future Trends

The current foot-and-mouth disease vaccines are effective in the control of the disease, as amply demonstrated by the dramatic fall in number of outbreaks in Western Europe since comprehensive vaccination programs were introduced in the 1950s. However, the position in the rest of the world, particularly in tropical countries, is not so encouraging. Part of this relative lack of success can be attributed to the instability of the vaccine under the field conditions prevailing in those countries. In addition, because current vaccines depend for their safety on the consistent and reliable inactivation of infective virus, there is always the danger of introducing the disease into the nearby environment. This means that modern, high-security buildings are required for the production of the vaccines. For these reasons alone, it is worthwhile to explore the possibility of furnishing a vaccine that is indefinitely stable and completely innocuous. The peptide approach affords the opportunity to develop such a vaccine, and the priming experiments in conjunction with improved triggered release mechanisms point the way to a completely new approach to vaccination against foot-and-mouth disease.

References

1. Mowat GN, Chapman WG (1962) Growth of foot-and-mouth disease virus in a fibroblast cell line derived from hamster kidneys. Nature 194:253–255
2. Frenkel HS (1947) La culture du virus de la fièvre aphteuse sur l'épithelium de la langue des bovidés. Bull Off Int Epizoot 28:155–162
3. Brown F, Crick J (1959) Application of agar gel diffusion analysis to a study of the antigenic structure of inactivated vaccines prepared from the virus of foot-and-mouth disease. J Immunol 82:444–447
4. Rowlands DJ, Sangar DV, Brown F (1975) A comparative chemical and serological study of the full and empty particles of foot-and-mouth disease virus. J Gen Virol 26:227–238
5. Wild TF, Burroughs JN, Brown F (1969) Surface structure of foot-and-mouth disease virus. J Gen Virol 4:313–320
6. Laporte J, Grosclaude J, Wantyghem J, Bernard S, Rouze P (1973) Neutralization en culture cellulaire du pouvoir infectieux du virus de la fièvre aphteuse par des sérums provenant de porcs immunisés à l'aide d'une proteine virale purifiée. C R Acad Sci 276:3399–3401
7. Strohmaier K, Franze R, Adam K-H (1982) Localization and characterization of the antigenic portion of the foot-and-mouth disease virus protein. J Gen Virol 59:295–306
8. Bittle JL, Houghten RA, Alexander H, Shinnick TM, Sutcliffe JG, Lerner RA, Rowlands DJ, Brown F (1982) Protection against foot-and-mouth disease by immunization with a chemically synthesized peptide predicted from the viral nucleotide sequence. Nature 298:30–31
9. Kleid DG, Yansura D, Small B et al (1981) Cloned viral protein vaccine for foot-and-mouth disease: Responses in cattle and swine. Science 214:1125–1129
10. Bittle JL, Worrell P, Houghten RA, Lerner RA, Rowlands DJ, Brown F (1984) Immunization against foot-and-mouth disease with a chemically synthesized peptide In Chanock RM, Lerner RA (eds) Modern Approaches to Vaccines: Molecular and Chemical Basis of Virus Virulence and Immunogenicity. Cold Spring Harbor Laboratory, Cold Spring Harbor, New York, 1984 p 103–107
11. Winther MD, Allen G, Bomford RH, Brown F (1986) Neutralizing antibodies elicited by a biosynthetic peptide expressed from the chemically synthesized DNA corresponding to an antigenic region of foot-and-mouth disease virus. J Immunol 136:1835–1840
12. Francis MJ, Fry CM, Rowlands DJ, Brown F, Bittle JL, Houghten RA, Lerner RA (1985) Immunological priming with synthetic peptides of foot-and-mouth disease virus. J Gen Virol 66:2347–2354

CHAPTER 48
Anti-idiotype Vaccines Against Viral Infections*

HILARY KOPROWSKI AND DOROTHEE HERLYN

The so-called third generation of viral vaccines usually comprises immunochemically or biochemically purified components of the virus, synthetic peptides, or products obtained by cloning genes that encode viral antigens. Although viral-component vaccines offer clear advantages over live or inactivated vaccines in terms of safety, purity, and specificity, the immunochemical or biochemical isolation of viral antigens in large quantities is often difficult, and products obtained by peptide synthesis or gene cloning may lack the tertiary structure of the original antigen and may thus be unable to elicit a specific immune response.

A more recent novel approach to the development of antiviral vaccines involves the use of anti-idiotype (anti-Id)† antibodies. Anti-Id antibodies are specific for determinants, called *idiotypes*, on variable regions of immunoglobulin heavy and/or light chains. Urbain et al. [1] defined the sequential sets of antibodies interconnected in a functional cascade as follows: (a) Ab_1, the Id on an immunoglobulin V region in response to and specific for a given antigen; (b) Ab_2, the anti-Id antibody population raised against V region of Ab_1; (c) Ab_3, a population of anti-anti-Id antibodies raised against V regions of Ab_2.

Implicit in the network theory of Jerne [2] is the postulate that V regions of Ab_2 can mimic antigenic structures recognized by V regions of Ab_1. This

---

*A more detailed presentation of studies related to anti-idiotype immunization may be found in ref. 33.

†Abbreviations: Ab, antibody; Ab_1, antibody against epitope; Ab_2, anti-idiotype (Id) antibody; Ab_3, anti-anti-idiotype (Id) antibody; Ag, antigen; CRI, cross-reactive idiotype; CTL, cytotoxic T lymphocyte; DTH, delayed-type hypersensitivity; HBVs, hepatitis B virus surface antigen; Id, idiotype.

observation suggested the use of these anti-Id antibodies in lieu of or in addition to antigen in priming for immunity. The original observation that priming of an animal host with anti-Id rather than with antigen leads to an immune response was made by Eichmann and Rajewsky [3]. The observation that anti-Id can mimic the biological functions of an antigen was made by Sege and Peterson [4] who found that anti-Id induced by immunization with anti-insulin Ab_1 mimics the function of insulin. Later on, Strosberg [5] published a similar observation in the adrenergic hormone system.

Among the Ab_2 molecules, two populations can be distinguished: one that bears an *internal image of the antigen* and one that binds to *recurrent or cross-reactive idiotypes* (CRI) [6,7]. The ratio of the Ab_2 bearing the internal image of the antigen to all other Ab_2 *not* bearing internal images is, in fact, decisive for the practical application of Ab_2 in immunization without antigen. The higher the ratio of anti-CRI Ab_2 to internal image Ab_2, the lesser the chance of obtaining antigen-specific Ab_3. CRI are expressed in antibody populations derived from different individuals of the same species, or even from individuals belonging to different species in response to the same antigen. Thus, expression of a particular CRI by antibodies against a given antigen is often closely associated with the binding specificity of these antibodies. Both internal-image Ab_2 and anti-CRI Ab_2, thus, often (a) react with Ab_1 derived from various species and binding to the same epitope; detect antigen-inhibitable Ids on Ab_1; and (c) elicit Ab_3 across species barriers. However, immunization with anti-CRI Ab_2^+ stimulates the production of a population of Ab_3 in which Ab_3 binding to Ag (Ab_3 Ag^+) and Ab_3 not binding to Ag (Ab_3 Ag^-) exist, but the Ab_3 Ag^+ fraction may be very small as compared to Ab_3 Ag^-. Immunization with internal-image Ab_2, on the other hand, is likely to change the ratio in favor of the Ab_3 Ab^+ population. Successful immunization of an animal host with Ab_2 leads to the development of antibodies that bind to Ab_2. A fraction of the population of these antibodies that does not share idiotypes with Ab_1 and thus does not inhibit binding of Ab_2 to Ab_1 will unlikely bind the antigen. Another fraction of Ab_3 usually less frequently elicited represents antibodies that are expected to be exactly complementary to the antigen binding site of Ab_2, which in turn may be complementary to the binding site of Ab_1. Of these Ab_3 induced by internal image of the antigen, some are Ag^+, but only few are also Id^+; most of these Ab_3 generated express new idiotypic determinants [8].

In summary, both internal-image vaccines and anti-CRI Ab_2 may stimulate an antigen-specific immune response, but internal-image Ab_2 are expected to be superior to anti-CRI because stimulation of an immune response with Ab_2 that mimics antigen should be very specific; furthermore, these complementary structures could, in principle, be made in connection with *any* frameworks, i.e. with any V-region genes, and are thus not even restricted to a given species. The probability of stimulating antigen-specific B cell clones is far greater when internal-image Ab_2 are used as opposed to anti-CRI Ab_2.

Stimulation of Antiviral Immune Responses by Anti-Id

We now will provide examples of the use of Ab_2 for immunization against viral agents. Table 48.1 summarizes the results of several studies in which antiviral immunity was induced by anti-Id antibodies.

Hepatitis B Virus

Kennedy et al. [9] raised polyclonal Ab_2 in rabbits against human Ab_1 that bind to hepatitis B virus surface antigen (HBVs). The Ab_2 cross-reacted with various anti-HBVs Ab_1 derived from most individuals of six different species. Following immunization with Ab_2, BALB/c mice produced low levels of Ab_3 with binding specificity identical to the original Ab_1. TheAb_3 response was considerably enhanced when the animals were boosted with viral antigen following priming with Ab_2 [10,11]. The Ab_3 inhibited the binding of Ab_2 to its homologous Ab_1, indicating that Ab_3 shared idiotypes with Ab_1 [12]. It is unclear whether the polyclonal Ab_2 preparation used in these studies is directed against CRI and/or consists of internal-image Ab_2. Immunization of animals of various other species with Ab_2 might clarify this point. However, the failure of Ab_2 to bind to all anti-HBVs sera derived from different species suggests the absence of internal-image Ab_2 in the described preparation. Unfortunately, the possible protective effects of Ab_2 immunization could not be tested in these studies because HBV is not pathogenic for mice. Since the submission of this manuscript the authors (R. Kennedy et al. Science 232: 220, 1986) have demonstrated protective effects of Ab_2 immunizations in chimpanzees.

Rabies Virus

Reagan et al. [13,14] have produced Ab_2 in rabbits against five monoclonal Ab_1, each of which binds to a different epitope on the rabies virus glycoprotein [15,16]. Four of the Ab_1 had strong virus neutralizing activity, and one neutralized infectivity poorly. Of the five different Ab_2 preparations raised against the respective Ab_1, three contained antibodies that bound within or close to the antigen-combining site of Ab_1; whereas in the two other preparations, combining site-specific Ab_2 could not be detected [13]. Each Ab_2 bound only to its homologous Ab_1. The absence of cross-reactions among the idiotypic regions of the Ab_1 was not surprising because each Ab_1 bound to a different epitope on the viral glycoprotein. Two of the five Ab_2 induced Ab_3 having virus-neutralizing activity in ICR mice. This activity could be removed by

Table 48.1 Induction of immunity against viruses by anti-Id antibodies.[a]

Virus	Ab₁			Ab₂							Immunity induced by Ab₂	
	Species of origin	Clonality	Binding specificity	Species of origin	Clonality	Binding specificity	Mouse strain	Humoral/cell-mediated	Induction by Ab₂ alone or by Ab₂ + antigen	Specificity	Effect on challenge with virus	Reference
Hepatitis B	human	polyclonal	hepatitis B surface antigen group a determinant	rabbit	polyclonal	Cross-reactive Id on human, chimpanzee, swine, goat, rabbit, guinea pig, mouse, but not chicken anti-hepatitis virus Ab	BALB/c	polyclonal Ab₃	Ab₂ alone; but enhancement by subsequent antigen exposure	Hepatitis B surface antigen group a only	nd[b]	10,11,12
Rabies	BALB/c mouse	monoclonal	various nucleo-capsid glyco-proteins	rabbit	polyclonal	bind to homologous Ab₁ only	ICR	neutralizing polyclonal Ab₃	Ab₂ alone; but enhancement by subsequent antigen exposure	identical or different from Ab₁, depending on source of Ab₂	protection	13,14
Sendai	B10.D2 mouse	monoclonal[c]	Sendai virus (not further specified)	B10.D2	monoclonal	bind to 37% of the various Sendal	B10.D2 and other strains	cytotocix T cells	Ab₂ alone	syngeneic and allogeneic Sendal	protection	17

Virus	Ab₁		Haplotype	Ab₂		virus-specific T-helper cell clones	of different haplotype		Ab₂ (DTH) or Ab₂ hybridomas (CTL)	virus infected, but not uninfected targets		Ref.
Reo	monoclonal	reovirus type 3 hemagglutinin	BALB/c	monoclonal	bind to homologous Ab₁ only	BALB/c and mice of various other Igh allotypes. CTL: BALB/c	DTH, CTL	Ab₂ (DTH) or Ab₂ hybridomas (CTL)	DTH: reovirus type 3 HA only. CTL: reovirus type 3 HA (variant viruses not tested)	nd[d]	18
	BALB/c											
Polio	monoclonal	poliovirus type II	BALB/c	monoclonal	Cross-reactive Id on anti-poliovirus Ab from mice, rabbits, guinea pigs, and humans	BALB/c	neutralizing polyclonal Ab₃	Ab₂ alone	poliovirus type II (other types not tested)	none	21
	BALB/c											

[a] Abbreviations: CTL, cytotoxic T lymphocytes; DTH, delayed-type hypersensitivity; HA, hemagglutinin; nd, no data.

[b] Since the submission of this manuscript the authors (R. Kennedy et al. Science 232:220,1986) have demonstrated protective effects of Ab₂ immunizations in chimpanzees.

[c] Instead of Ab₁ a Sendai virus-specific T-helper cell clone was used for generation of Ab₂.

[d] Since the submission of this manuscript the authors (G. Gaulton and M. Greene Ann Rev Immunol 4:253,1986) have demonstrated protective effects of Ab₂ immunizations in mice.

absorption of the Ab_3 with the homologous Ab_2. In one case, the Ab_3 showed neutralizing activity despite the absence of this activity in the corresponding Ab_1 at the beginning of the idiotypic cascade [13]. These surprising findings cannot be easily explained because of our incomplete knowledge of the mechanism of virus neutralization. Indeed, further studies along these lines may provide a better clue to understanding the steps leading to virus neutralization.

Whereas immunization with Ab_2 alone was not sufficient to protect mice against a lethal challenge with rabies virus, priming with Ab_2 followed by a booster immunization with rabies virus vaccine conferred protection at a level comparable to that obtained after conventional rabies virus vaccination [14], probably by priming of a subcompartment of the immune system.

Sendai Virus

In a different approach, Ertl and Finberg [17] used a Sendai virus-specific helper T-cell clone to induce syngeneic monoclonal Ab_2. These Ab_2 were able to elicit an antigen-specific cytotoxic T lymphocyte (CTL) response in various mouse strains, indicating that the expression of this particular idiotype was neither H-2 nor Igh-1 (Ig heavy chain) restricted. Fewer Ab_2-induced CTL clones than virus-induced CTL clones were H-2 restricted, i.e., 21% of the virus-induced CTL clones lysed allogeneic target cells as compared to 61% of the CTL clones induced by Ab_2. Mice immunized with Ab_2 were as efficiently protected as Sendai virus-immunized mice against a lethal challenge with this virus.

These results demonstrate that induction of CTL by Ab_2, unlike the stimulation by viral antigen, is possible without processing and presentation in context of H-2 antigen. Furthermore, Ab_2-induced CTL effector functions are less H-2-restricted as compared to antigen-stimulated CTL functions. Whether these differences reflect different mechanisms of CTL induction by Ab_2 and by viral antigen and/or differences in the cell populations stimulated remains to be determined.

Ertl and Finberg [17] have suggested that Ab_2 stimulates CTL either directly, because of idiotypic structures shared between CTL and helper T cells, and/or indirectly via the induction of idiotype-bearing antigen-specific helper T cells. Most importantly, this study suggests the potential value of Ab_2 as an antiviral immunity against a subsequent lethal infection.

Reovirus

Analogous to the Sendai virus system, reovirus-induced Ab_2 elicited specific CTL responses in mice [18]. Syngeneic monoclonal Ab_2 was raised in BALB/

c mice against a neutralizing monoclonal anti-reovirus hemagglutinin Ab_1. The Ab_2, like the viral hemagglutinin, was shown to block both viral binding to neuronal cells and CTL-mediated lysis of virally infected target cells [19,20]. Unlike the Sendai virus system, Ab_2 in the reovirus system alone led to only minimal stimulation of CTL activity, and presentation of Ab_2-inducing hybridoma cells was necessary to obtain a significant response [18]. A delayed-type hypersensitivity (DTH) response, on the other hand, could be induced by Ab_2 alone. Whereas the DTH response was shown to be serotype specific, the specificity of the CTL response and the question of whether target cell recognition by these CTL was H-2 restricted have not been investigated. Although the CTL response induced by Ab_2 was comparable to that induced by immunization with virus only, the potential usefulness of this Ab_2 as an antiviral vaccine awaits evaluation in protection experiments. Since the submission of this manuscript the authors (G. Gaulton and M. Greene, Ann Rev Immunol) have demonstrated protective effects of Ab_2 immunizations in mice.

Poliovirus

Syngeneic monoclonal Ab_2 was raised against an idiotype on a neutralizing protective monoclonal Ab_1 with specificity for poliovirus type II [21]. The Ab_2 detects a combining site-related, interspecies-cross-reactive idiotype present on anti-poliovirus type II antibodies from rats, mice, guinea pigs, and humans. Therefore, the authors suggested that the Ab_2 bears an internal image of the viral epitope. The Ab_2 induced virus-neutralizing Ab_3 in syngeneic mice; the specificity of this response, however, has not been investigated, nor has the idiotypic specificity of the Ab_3 been compared with that of the Ab_1 at the beginning of the cascade. Further, the poliovirus-neutralizing Ab_3 failed to protect the mice against a lethal challenge with poliovirus, perhaps, as the authors suggest, because only low levels of virus-neutralizing Ab_3 were induced.

Discussion

In the selection of the viral determinant to which one would like to produce Ab_2, the choice should fall on a determinant that binds antibodies involved in the protection of the host against infections. Furthermore, this determinant should be immunodominant for all individuals. In practice, we would recommend for the production of Ab_2, to select those monoclonal Ab_1 that share idiotypes with a large variety of polyclonal antiviral antibodies derived from individuals who recovered from a viral infection or were immunized successfully against the infection. Ab_2 produced against such Ab_1 are most likely to protect the immunized host against exposure to a variety of different strains of the virus. For practical purposes, Ab_2 reacting with idiotypes on antibodies derived from different species and exhibiting identical binding specificity may be the best choice for vaccination protocols because such Ab_2 most likely bear an internal image of the antigen.

Because the frequency of inducing Ab_2 that bear an internal antigen image [22] is rather low, it is certainly advantageous to again produce monoclonal Ab_2. Ab_2 carrying the internal image of the antigen can be distinguished from anti-CRI Ab_2 by analysis of the Ab_3 population produced in response to Ab_2 immunization. This analysis would involve, first, the demonstration of Ids shared by Ab_3 and Ab_1 in assays where the binding of Ab_2 to Ab_1 is inhibited by Ab_3. This is followed by absorption of Ab_3 to antigen and subsequent testing of the non-absorbed Ab_3 molecules for the expression of Ids shared by Ab_1 in the inhibition assays. If the Ab_2 used to generate Ab_3 bears an internal image of the antigen, then all monoclonal Ab_3 molecules analyzed as described above should bind to antigen, and absorption of polyclonal Ab_3 to antigen should remove all Id^+ molecules. If, on the other hand, only a fraction of polyclonal Ab_3 binding to the combining site of Ab_2 also bind to the antigen, then one has to assume that Ab_2 consist of anti-Id reacting with CRI and possibly of some Ab_2 molecules bearing an internal image of the antigen.

It has been reported [3] that repeated immunizations with anti-Id may result in expansion of the B cell compartment and thus its memory, without leading to secretion of antibodies. The priming effect of anti-Id could then be detected by memory stimulation, i.e., protection against challenge with a lethal dose of the infectious agent.

After the Ab_2 is selected, the optimal immunization protocols to induce protective immunity can be determined by varying the dose of Ab_2 [9,16,23], its state of aggregation (soluble or aggregated [9]), the type of adjuvant, and perhaps the isotype of Ab_2 [24]. Thus, administration of high doses of Ab_2 may lead to immunosuppression [23], possibly due to induction of Id-specific suppressor T cells [25–27]; Ab_2 given in soluble form produced a predominantly IgM anti-HSV response in mice, whereas Ab_2 in aggregated form elicited an IgG response [28].

Ab_2 vaccines can be easily prepared either from hybridoma culture supernatants or from sera of animals immunized with Ab_1. Because the same structure of antigenic determinant on Ab_2 is presented to the immune system in a molecular environment (i.e., on an immunoglobulin molecule) clearly different from that of conventionally prepared antigens, stimulation of Ab_2 of lymphocyte clones that are tolerant to stimulation by the original antigen may be expected. An immune response not observed after stimulation with antigen but observed after stimulation with Ab_2 was shown when anti-Id to antibodies directed against capsular antigens of pathogenic bacteria, primed neonatal mice for immunity upon subsequent immunization with bacterial antigen, leading to protection of the mice against lethal challenge with live bacteria. In contrast, priming of neonates with antigen alone did not induce protective immunity [29,30].

Although we have emphasized the use of internal-image Ab_2 for induction of antigen-specific immune responses, it has recently been demonstrated that even an Ab_2 directed against a private Id on Ab_1 was able to elicit antigen-specific immune responses across species barriers [31,32].

The use of anti-Ids as unconventional vaccines is of particular relevance in diseases where the viral agents, e.g., human T-cell leukemia/lymphoma virus (HTLV), are highly infectious and pathogenic. In this instance, fear of using inactivated virions or even viral components for immunization purposes may justify efforts to produce anti-Id that is devoid of any danger for the vaccinated individual.

Although one may hope that the internal image of an antigen will be useful as a vaccine for immunizing humans and animals, it is also evident that our present knowledge of the functions of the immune system and its reaction in the organism to antigens or their internal image before and during the process of infection is still incomplete.

References

1. Urbain J, Francotte M, Franssen JD, Hiernaux J, Leo O, Moser M, Slaoui M, Urbain-van Santen G, van Acker A, Wikler M (1983) From clonal selection to immune networks: Induction of silent idiotypes. *In* Bona CA, Köhler H (eds) Immune Network. Ann N Y Acad Sci 418:1–8
2. Jerne NK (1974) Towards a network theory of the immune system. Ann Immunol (Inst Pasteur) 125C:373–389
3. Eichmann K and Rajewsky K (1975) Induction of T and B cell Immunity by anti-idiotypic antibody Eur J Immunol 5:661–666
4. Sege K, Peterson, PA (1978) Use of anti-idiotypic antibodies as cell surface receptor probes. Proc Natl Acad Sci USA 75:2443–2447
5. Strosberg AD (1983) Anti-idiotype and anti-hormone receptor antibodies. Springer Semin Immunopathol 6:67–78
6. Lindemann J (1973) Speculation on idiotypes and homobodies. Ann immunol (Inst Pasteur) 124C:171–184
7. Bona CA, Heber-Katz E, Paul NE (1981) Idiotype-anti-idiotype regulation. Immunization with a levan-binding myeloma protein leads to the appearance of auto-anti-(anti-idiotype) antibodies and to the activation of silent clones. J Exp Med 153:951–967
8. Wikler M, Urbain J (1984) Idiotypic manipulation of the rabbit immunoresponse

against *Micrococcus luteus*. *In* Kohler H, Urbain J, Cazanave PH (eds) Idiotypy in Biology and Medicine. Academic Press, New York, pp 219–241

9. Kennedy RC, Ionescu-Matiu I, Sanchez Y, Dreesman GR (1983c) Detection of interspecies idiotypic cross-reactions associated with antibodies to hepatitis B surface antigen. Eur J Immunol 13:232–235

10. Kennedy RC, Dreesman GR (1984) Enhancement of the immune response to hepatitis B surface antigen. In vivo administration of anti-idiotype induces anti-HBs that expresses a similar idiotype. J Exp Med 159:655–665

11. Kennedy RC, Adler-Storthz K, Henkel RD, Sanchez Y, Melnick JL, Dreesman GR (1983) Immune response to hepatitis B surface antigen: Enhancement by prior injection of antibodies to the idiotype. Science 221:853–855

12. Kennedy RC, Melnick JL, Dreesman GR (1984) Antibody to hepatitis B virus induced by injecting antibodies to the idiotype. Science 223:930–931

13. Reagan KJ, Wunner WH, Wiktor TJ, Koprowski H (1983) Anti-idiotypic antibodies induce neutralizing antibodies to rabies virus glycoprotein. J Virol 48:660–666

14. Reagan KJ, Wunner WH, Koprowski H (1985) Viral antigen-independent immunization: Induction by anti-idiotypic antibodies of neutralizing antibodies to defined epitopes on rabies virus glycoprotein. *In* Dreesman GR, Bronson JG, Kennedy RC (eds) High Technology Route to Virus Vaccines. Am Soc J Microbiol Publ pp 117–124

15. Wiktor TJ, Koprowski H (1980) Antigenic variants of rabies virus. J Exp Med 152:99–112

16. Lafon M, Wiktor TJ, Macfarlan R (1983) Antigenic sites on the CVS rabies virus glycoprotein: Analysis with monoclonal antibodies. J Gen Virol 64:843–851

17. Ertl HCJ, Finberg RW (1984) Sendai virus-specific T-cell clones: Induction of cytolytic T cells by an antiidiotype antibody directed against a helper T-cell clone. Proc Natl Acad Sci USA 81:2850–2854

18. Sharpe AH, Gaulton GN, McDade KK, Fields BN, Greene MI (1984) Syngeneic monoclonal antiidiotype can induce cellular immunity to reovirus. J Exp Med 160:1195–1205

19. Noseworthy JH, Fields BN, Dichter MS, Sobotka C, Pizer E, Perry LL, Nepom JT, Greene MI (1983) Cell receptors for the mammalian reovirus. I. Syngeneic monoclonal anti-idiotypic antibody identifies the cell surface receptor for reoviruses. J Immunol 131:2533–2538

20. Kaufman RS, Noseworthy JH, Nepom JT, Finberg R, Fields BN, Greene MI (1983) Cell receptors for the mammalian reovirus. II. Monoclonal anti-idiotypic antibody blocks viral binding to cells. J Immunol 131:2539–2541

21. Uytdehaag IGCM, Osterhaus ADME (1985) Induction of neutralizing antibody in mice against poliovirus type II with monoclonal anti-idiotypic antibody. J Immunol 134:1225–1229

22. Urbain J, Slaoui M, Mariame B, Leo O (1984) Idiotypy and internal images. *In* Kohler H, Urbain J, Cazanave PH (eds) Idiotypy in Biology and Medicine. Academic Press, New York, pp 15–28

23. Rajewsky K, Takemori T (1983) Genetic expression and function of idiotypes. Ann Rev Immunol 1:569–580

24. Eichmann K (1974) Idiotype suppression. I. Influence of the dose and of the effector functions of anti-idiotypic antibody on the production of an idiotype. Eur J Immunol 4:296–302

25. Eichmann K (1975) Idiotype suppression. II. Amplification of a suppressor T cell with anti-idiotypic activity. Eur J Immunol 5:511–517

26. Kim BS (1979) Mechanism of idiotype suppression. I In vitro generation of idiotype-specific suppressor T cells by anti-idiotype antibodies and specific antigen. J Exp Med 149:1371–1378

27. Owen FL, Ju ST, Nisonoff A (1977) Presence of idiotype-specific suppressor T cells of receptors that interact with molecules bearing the idiotype. J Exp Med 145:1559–1566

28. Kennedy RC, Adler-Storthz K, Burns JW, Henkel RD, Dreesman GR (1984) Anti-idiotype modulation of herpes simplex virus infection leading to increased pathogenicity. J Virol 50:951–953

29. Sein K, Söderström, J (1984) Neonatal administration of idiotype or anti-idiotype primes for protection against *Escherichia coli* K13 infection in mice. J Exp Med 160:1001–1011

30. Rubenstein LJ, Goldberg B, Hiernaux J, Stein KE, Bona CA (1983) Idiotype-anti-idiotype regulation V. The requirement for immunization with antigen or monoclonal anti-idiotypic antibodies for the activation of $\beta2\rightarrow6$ and $\beta2\rightarrow1$ polyfructosan reactive clones in BALB/c mice treated at birth with minute amounts of anti-A48 idiotype antibodies. J Exp Med 158:1129-1144

31. Francotte M, Urbain J (1984) Induction of anti-tobacco mosaic virus antibodies in mice by rabbit anti-idiotypic antibodies. J Exp Med 160:1485–1494

32. Kennedy RC, Dreesman GR, Butel JS, Lanford RE (1985) Suppression of in vivo tumor formation induced by simian virus 40-transformed cells in mice receiving antiidiotypic antibodies. J Exp Med 161:1432–1449

33. Melchers F, Koprowski H (eds) (1985) *In* Current Topics in Microbiology and Immunology, Vol 119. Springer-Verlag, Berlin, pp 1–142

CHAPTER 49
Chemotherapy of Lassa Virus Infection

Joseph B. McCormick

Lassa fever is an acute febrile illness caused by an arenavirus (Lassa virus) first isolated in 1969 from patients in Nigeria. In 1972, the rodent reservoir of Lassa virus was identified as the multimammate rat *Mastomys natalensis*. Since then, human infection and disease have been observed in five other countries in West Africa, and related viruses have been isolated from various rodent species in four countries in Central and Southern Africa, though the roles of the latter in human infection and disease are presently unknown. Recent intensive study of Lassa fever in Sierra Leone, West Africa has led to a greater understanding of its epidemiology and clinical spectrum, and to the demonstration of effective chemotherapy with ribavirin [1]. This chapter will discuss the background and evidence of the effectiveness of ribavirin in Lassa fever.

The clinical spectrum of Lassa virus infection is widely varying in severity, from mild or asymptomatic to a severe and often fatal febrile illness [2]. Between 10% and 20% of human infections result in hospitalization; among these patients the mortality is 16%, though the mortality associated with all infections is estimated to be much lower [3]. The disease usually begins insidiously with fever, general malaise, fatigue, and headache, with myalgia and sore throat developing by the third day. In severe cases, there is prostration, with intense abdominal or retrosternal pain, vomiting, or diarrhea. The patient is usually admitted to the hospital between the fifth and sixth day of illness. Complications include hemorrhage, serous effusions, pericarditis, encephalopathy, deafness, uveitis, septic abortion, and orchitis [2]. In fatal disease, death usually follows hours after onset of irreversible shock. The vascular collapse appears to be associated with platelet and endothelial cell dysfunction characterized by loss of in vitro platelet aggregation responses and reduced prostacyclin production by vascular endothelium [4]. Disseminated intravascular coagulation does not play a role in the fatal shock associated with Lassa fever.

The disease also affects children in whom the clinical manifestations are difficult to distinguish from those of other febrile illnesses. One study in an endemic area of Sierra Leone found that 5% of febrile children who were admitted to a hospital had Lassa viremia [5]. The case fatality in children hospitalized with Lassa fever is about 12%.

Data from an extensive clinical study of Lassa fever have shown that the outcome was associated with at least two factors: (a) the levels of serum aspartate aminotransferase (AST) [6] and (b) the viremia both at the time of admission as well as throughout the course of illness [6,7]. The presence of an AST level $\geq$ 150 IU/liter on admission was associated with 58% mortality, whereas patients with an admission viremia of $\geq$ $10^{3.6}$ tissue-culture median infective dose (TCID$_{50}$/ml) had a case fatality of 73%. These observations have also been made in laboratory studies of monkeys infected with Lassa virus [8,9]. The correlation between viremia and outcome of Lassa fever suggests that any successful therapy should be associated with a reduction in viremia levels. It should be noted that the production of IgG or IgM antibody to Lassa virus in the human host is not associated with either a reduction in viremia nor recovery from Lassa fever [7].

The practical application of viremia levels for selecting Lassa fever patients with a high risk of death is limited because no rapid diagnostic test is yet available for the field. Nevertheless, viremia has been important for the analysis of treatment data. On the other hand, AST can be measured by rapid assay in Lassa fever patients in endemic areas. A therapeutic trial was therefore conducted, using AST levels on hospital admission to identify patients with a high risk of death, thus permitting an analysis of the effect of plasma and drug on outcome in a smaller group of patients than would have been possible in a trial of persons with a lower risk of death [6].

Ribavirin is a broad-spectrum antiviral drug with in vitro and in vivo activity against several different viruses [10,11]. Its mode of action has been attributed to interferences with certain virus replication steps such as capping of viral messenger RNA (togaviruses and pox viruses) and viral polymerase (influenza A virus) [12]. Ribavirin has been shown to have significant antiviral activity against Lassa virus: In vitro, levels of 32 μm inhibited Lassa virus replication in vero cells, and animals started on 30 mg/kg/day within four days of infection by Lassa virus survive [13]. The mechanism of interference with arenavirus replication has not been studied, but in the light of its effectiveness against these viruses such studies are needed.

The first randomized therapeutic trial of Lassa fever examined the effect of oral ribavirin (1 g/day for 10 days in divided doses), or 1 unit of plasma containing antibodies to Lassa virus (with an immunofluorescent antibody titer [IFA] of 1 $\geq$ 128) taken from laboratory-confirmed convalescent Lassa fever patients. Plasma containing antibody to Lassa virus had no measurable effect on mortality, whereas oral ribavirin had a measurable effect, reducing mortality four fold over untreated and plasma-treated patients [6].

In a second trial, patients with clinical signs of Lassa fever and an AST $\geq$ 150 IU/liter on admission were entered into the study and were then randomly divided into two groups. One group was treated with intravenous (IV) ribavirin given as a 2-g loading dose, followed by 1 g every 6 hours for 4 days and then ½ g every 6 hours for 6 days. The second group was treated with the same dose of ribavirin plus one unit of plasma containing antibody to Lassa virus.

Five of 30 people (17%) died in the group treated with IV ribavirin compared to seven of 32 (22%) in those treated with IV ribavirin and plasma. Both treated groups had lower fatality than the expected fatality of 55% observed in the untreated comparison group. For patients who had elevated viremias ($\geq 10^{3.6}$ TCID$_{50}$/ml), the fatality was 30% in those treated with ribavirin compared to 73% in the untreated. Fatality in untreated patients with lower viremias was 18% compared to 10% in those treated with IV ribavirin. Most importantly, for all patients, regardless of viremia, the fatality was statistically significantly reduced (eight fold) when oral or IV ribavirin therapy was initiated in the first 6 days of illness, as compared to 7 days or later in illness.

The analysis of patients treated with IV ribavirin shows suppression of viremia even in patients in whom IV ribavirin did not prevent death, and there was a reduction in viremia following therapy. The course of viremia in patients treated successfully with either oral ribavirin or plasma containing antibody to Lassa virus did not differ from that observed in untreated patients with fatal disease [6].

Evidence is therefore available that Lassa fever can be successfully treated with ribavirin, especially when therapy is started early in clinical illness. There remains, however, a group of people who either are treated too late or for whom viremia is apparently too advanced to be adequately treated by ribavirin. In these patients, we presently understand at a more fundamental level of physiology the profound shock to which they succumb [4]. There are indications that platelet and endothelial dysfunction occur, and that specific intervention with prostacyclin or one of its analogues may prevent or even reverse the endothelial failure associated with shock in Lassa fever.

Data from ribavirin studies in experimentally infected monkeys correlate well with human disease both as regards dose and outcome of ribavirin therapy [8,9,13], but data from passive antibody studies do not. Monkey or human plasma with a log neutralization index against Lassa virus of above 2.0 protects monkeys from death when administered within 4 days of infection, but in humans no significant therapeutic effect of plasma containing antibody to Lassa virus was observed in randomized trials regardless of timing of therapy.

In humans, most convalescent plasma containing Lassa virus antibodies, regardless of when collected after illness, does not have a log neutralization index equivalent to that described for plasma required for therapy or prevention of Lassa fever in experimentally infected monkeys. This may explain its failure in patients treated by plasma alone or in combination with ribavirin. It is possible that the use of concentrated human plasma with a higher neutralization index might be effective. In contrast, treatment of patients with Argentine hemorrhagic fever (caused by an arenavirus, Junin) by administration of Junin virus immune plasma within the first 8 days of illness reduced mortality from 16% to 1% [14].

On the basis of this study, it might have been anticipated that passively administered antibodies to Lassa fever would be effective. However, Junin virus is consistently neutralized in a classical constant-virus, serum-dilution, plaque-reduction test, whereas Lassa virus is not. In Argentine hemorragic fever, the appearance of immunoglobulin G antibody correlated with the disappearance of the virus, which is not the case for Lassa fever.

Although the dosage of ribavirin used in the therapy of Lassa fever is higher than previously used, the only side effect observed has been a decrease in hematocrit by an average of 10–20% over the 10-day therapeutic course [15]. Such has also been observed in studies of nonhuman primates, and in all instances the hematocrit and hemoglobin levels have returned to normal when the drug was stopped. There have been no reported occurrences of stem cell toxicity from the use of ribavirin.

The effective therapy of Lassa fever by ribavirin is the first demonstration of a lethal systemic viral disease's response to an antiviral drug. It is an illustration of the potential for antiviral therapy of many other viral diseases. As with antibacterial therapy, early treatment is a key to a successful outcome. The demonstration of the clinical efficacy of ribavirin is only a first step in the complex process required for its eventual routine use by clinicians in West Africa. Availability of the drug, education of physicians in its use, and education of the population on the need for early therapy will be required before its full effectiveness will be realized.

References

1. McCormick JB, Johnson KM (1984) Viral hemmorhagic fevers. *In* Tropical and Geographical Medicine. McGraw Hill, New York, pp 676–684
2. McCormick JB, King IJ, Webb PA et al. (1986) Lassa fever: A case control study of the clinical diagnosis and course. J Infect Dis (in press)
3. McCormick JB, Webb PA, Krebs JW, Johnson KM, Smith E (1986) Lassa fever: A prospective study of its epidemiology and ecology. J Infect Dis (in press)
4. Fisher-Hoch SP, Mitchell SW, Sasso DM, McCormick JB, Lange JV, Ramsey R (1986) Pathophysiological and immunological disturbances associated with shock in a primate model. (Submitted for publication)

5. Webb PA, McCormick JB, Bosman I et al. (1986) Lassa fever in children. R Soc Trop Med Hyg (in press)
6. McCormick JB, King IJ, Webb PA, Scribner Cl, Craven RB, Johnson KM, Elliott LH, Williams RB (1986) Lassa fever: Effective therapy with ribavirin. N Engl J Med 314:20–26
7. Johnson KM, McCormick JB, Webb PA, Smith E, Elliott LH, King IJ (1986) Lassa fever in Sierra Leone: Clinical virology in hospitalized patients. J Infect Dis (in press)
8. Jahrling PB, Peters CJ, Stephens EL (1984) Enhanced treatment of Lassa fever by immune plasma combined with ribavirin in cynomolgus monkeys. J Infect Dis 149:420–427
9. Jarhling PB, Peters CJ (1984) Passive antibody therapy of Lassa fever in cynomolgus monkeys: Importance of neutralizing antibody and Lassa virus strain. Infect Immun 44:528–533
10. Breese-Hall C, McBride JT, Walsh EE et al. (1983) Aerosolized ribavirin treatment of infants with respiratory syncytial virus. A randomized double blind study. N Engl J Med 308:1443–1447
11. Knight V, McClung HW, Wilson SZ et al (1981) Ribavirin small particle aerosol treatment of influenza. Lancet ii:945–949
12. Smith RA (1980) Mechanisms of action of ribavirin. In Ribavirin: A broad spectrum antiviral agent. Academic Press, New York, pp 99–118
13. Jarhling PB, Hesse RA, Eddy GA, Johnson KM, Callis RT, Stephen EL (1984) Lassa virus infection of rhesus monkeys: Pathogenesis and treatment with ribavirin. J Infect Dis 149:580–589
14. Maiztegui JI, Fernandez NJ, deDamilano AJ (1970) Efficacy of immune plasma in treatment of Argentine hemorrhagic fever and association between treatment and late neurological syndrome. Lancet ii:1216–1217
15. McCormick JB, Webb PA, Scribner CL, Johnson KM (1985) Chemotherapy of acute Lassa fever with ribavirin. In Proceedings of the Second Symposium on Ribavirin. Academic Press, New York (in press)

Index

Note: Page numbers with * refer to figures.